CAMBRIDGE URBAN AND ARCHITECTURAL

8   DESIGN WITH ENERGY
    THE CONSERVATION AND USE OF ENERGY

CAMBRIDGE URBAN AND ARCHITECTURAL STUDIES

General Editors

LESLIE MARTIN
Emeritus Professor of Architecture, University of Cambridge

LIONEL MARCH
Rector and Vice-Provost, Royal College of Art

VOLUMES IN THIS SERIES

1. *Urban Space and Structures*, edited by Leslie Martin and Lionel March
2. *Energy, Environment and Building*, by Philip Steadman
3. *Urban Modelling*, by Michael Batty
4. *The Architecture of Form*, edited by Lionel March
5. *The Evolution of Design*, by Philip Steadman
6. *Bruno Taut and the Architecture of Activism*, by Iain Boyd Whyte
7. *Incidence and Symmetry in Design and Architecture*, by Jenny A. Baglivo and Jack E. Graver
8. *Design with Energy: The Conservation and Use of Energy in Buildings*, by John Littler and Randall Thomas

Design with energy

The conservation and use of energy in buildings

JOHN LITTLER
and
RANDALL THOMAS

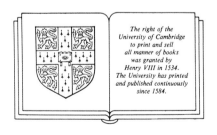

CAMBRIDGE UNIVERSITY PRESS

CAMBRIDGE
LONDON   NEW YORK   NEW ROCHELLE
MELBOURNE   SYDNEY

Published by the Press Syndicate of the University of Cambridge
The Pitt Building, Trumpington Street, Cambridge CB2 1RP
32 East 57th Street, New York, NY 10022, USA
296 Beaconsfield Parade, Middle Park, Melbourne 3206, Australia

© Cambridge University Press 1984

First published 1984

Printed in Great Britain at the University Press, Cambridge

Library of Congress catalogue card number: 83−7537

British Library cataloguing in publication data

Littler, John
Design with energy, the conservation and use of
energy in buildings. − (Cambridge urban and
architectural studies)
1. Architecture and energy conservation
I. Title II. Thomas, Randall
696   NA2542.3

ISBN 0 521 24562 1 hard covers
ISBN 0 521 28787 1 paperback

For David Bullett and Michèle Thomas

Contents

Units, symbols, abbreviations, conventions and conversion factors		*page* xi
Preface		xv

1 Energy and buildings — 1
  1.1 Introduction — 1
  1.2 Energy and the built environment – past and present — 2
  1.3 Energy and the built environment – the future — 5

2 Site planning and analysis — 14
  2.1 Introduction — 14
  2.2 Solar radiation — 16
  2.3 Wind — 24
  2.4 Soil — 33

3 Building design — 42
  3.1 Introduction — 42
  3.2 Energy demand and thermal response — 42
  3.3 The internal environment — 47
    3.3.1 Comfort — 47
    3.3.2 Temperature — 48
    3.3.3 Room air movement, ventilation and relative humidity — 52
  3.4 Size and type — 59
  3.5 Form and orientation — 60
  3.6 External and internal layout — 67
  3.7 Construction – general — 70
  3.8 Foundations and walls — 76
  3.9 Floors — 83
  3.10 Windows — 84
  3.11 Doors — 87
  3.12 Ceilings and roofs — 88

4 Passive solar design — 97
  4.1 Introduction — 97
  4.2 Elements of passive solar systems — 100
    4.2.1 Glazing — 100
    4.2.2 Single and multiple glazing — 103

	4.2.3	Insulating blinds and shutters	109
	4.2.4	Shading	116
	4.2.5	Radiation enhancement using reflectors	117
	4.2.6	Thermal storage	118
4.3	Passive solar heated buildings	123
	4.3.1	Direct gain	123
	4.3.2	Attached sun spaces	128
	4.3.3	Thermal storage walls	137
	4.3.4	Roof ponds	144
	4.3.5	Roof-space collectors	146
	4.3.6	Convective loops	147
4.4	Design methods for passive solar buildings	148
	4.4.1	Introduction	148
	4.4.2	Characteristics of a perfect design model	150
	4.4.3	Models approaching the desired degree of flexibility	150

5 Active solar heating — 159

5.1	Introduction	159
	5.1.1 Definitions	159
	5.1.2 Popularity of solar heating	159
	5.1.3 Outline of the active systems	159
	5.1.4 Approximate energy available	160
	5.1.5 Hot water compared to hot water + space heating	161
5.2	Solar air heating	161
	5.2.1 Collectors	162
	5.2.2 Air to water heat exchangers	163
	5.2.3 Dampers	163
	5.2.4 Auxiliary energy	163
	5.2.5 Pebble beds	163
	5.2.6 Controls	165
	5.2.7 Results of monitored systems	170
5.3	Solar water heating	171
	5.3.1 Outline of systems	171
	5.3.2 Collectors	173
	5.3.3 Pumps	175
	5.3.4 Heat exchangers	175
	5.3.5 Thermal storage	175
	5.3.6 Drain-down valves	176
	5.3.7 Transfer fluid	176
	5.3.8 Controllers	177
5.4	New types of system	177
	5.4.1 Evacuated tubes	177
	5.4.2 Plastic collectors	177
	5.4.3 Zeolite heating and cooling collectors	178
	5.4.4 Refrigerant-charged loop	178
5.5	Installed systems	179
5.6	Multifamily installations	182

Contents ix

| | 5.7 | Predictive methods | 182 |

6 **Space heating and ventilation** 189
  6.1 Introduction 189
  6.2 Solid-fuel heating 196
  6.3 Water-distribution systems 205
  6.4 Forced-air systems 217
  6.5 Heating with electricity 225
  6.6 Heat pumps 228
  6.7 Group schemes 234
  6.8 Conclusion 236

7 **Thermal storage** 241
  7.1 Introduction 241
  7.2 Sensible-heat storage in water 243
  7.3 Sensible-heat storage in rock 249
  7.4 Phase-change energy storage 252
  7.5 Developments in storage 255
  7.6 Conclusion 258

8 **Wind energy** 262
  8.1 Introduction 262
  8.2 Power extracted by turbines 264
    8.2.1 Power in the wind 264
    8.2.2 Variation of wind speed with location 264
    8.2.3 Variation of wind speed with height 267
  8.3 Power extracted by real turbines 268
  8.4 Types of turbine 270
  8.5 Electrical generation 278
    8.5.1 Electrical generators 278
    8.5.2 Gearboxes 280
    8.5.3 Electricity storage 280
  8.6 Use of wind-generated electricity 282
    8.6.1 AC or DC? 282
    8.6.2 Inverters 283
    8.6.3 Thermal storage 283
  8.7 Economies of scale 284
  8.8 Installed systems 284
    8.8.1 Conservation house at NCAT 284
    8.8.2 Wind-powered cottage at NCAT 287

9 **Water-supply systems** 289
  9.1 Introduction 289
  9.2 The Cambridge Autarkic House water-supply system 290

10 **Waste disposal and utilization** 293
  10.1 Introduction 293
  10.2 Aerobic systems 294

	10.3 Anaerobic systems	295
	10.4 Methane digestion	296

11 Domestic-energy saving — 304
11.1 Introduction — 304
11.2 Domestic hot water — 304
11.3 Electrical power — 306
11.4 Cooking — 307
11.5 Space heating — 308
11.6 Transportation — 308

12 Housing case studies — 310
12.1 Introduction — 310
12.2 New houses — three solar air-heated houses in Peterborough — 310
12.3 New houses — energy-efficient houses in Newnham, Cambridge — 315
12.4 Rehabilitated houses — renovation of a farmhouse to include a sun space (contributed by Peter Clegg) — 317
12.5 Rehabilitated houses — house conversion incorporating a roof-space collector (contributed by Peter Clegg) — 319
12.6 Rehabilitated houses — a nineteenth-century terraced house — 322

13 Non-domestic case studies — 326
13.1 Introduction — 326
13.2 The swimming pool, Sheiling Schools — 326
13.3 The new BRS office building — 328
13.4 Agricultural buildings — 329
13.5 School buildings (contributed by Nick Baker) — 330

Appendix 1: Weather data — 342
A 1.1 Introduction — 342
A 1.2 Solar spectrum — 342
A 1.3 Global solar radiation data for the UK — Kew and Bracknell — 343
A 1.4 Global solar radiation data on the horizontal for three UK stations — 343
A 1.5 Direct and diffuse solar radiation for Kew, UK — 343
A 1.6 Temperature data for the UK — 344
A 1.7 Wind data — 344

Appendix 2: Thermal performance — 347

Appendix 3: Interstitial condensation — 358

Index — 363

Units, symbols, abbreviations, conventions and conversion factors

(1) Principal units

(In some cases a base unit is combined with a multiplier because this is its most common form.)

°C	degree Celsius
d	day
g	gram
h	hour
ha	hectare
J	joule
K	degree kelvin
kcal	kilocalorie
l	litre
MTCE	megatonnes of coal equivalent
m	metre
mb	millibar
N	newton
p	person
Pa	pascal (N/m^2)
ppm	parts per million
s	second
t	tonne
V	volt
W	watt

The principal units are frequently used with the following multiples and submultiples:

10^{-1}	deci	d	10	deca	da
10^{-2}	centi	c	10^2	hecto	h
10^{-3}	milli	m	10^3	kilo	k
10^{-6}	micro	μ	10^6	mega	M
10^{-9}	nano	n	10^9	giga	G

(2) Symbols

(Where two meanings are given, the correct one should be evident from the context of the chapter.)

A	area
C_v	ventilation heat loss
c_p	specific heat; power coefficient of wind turbine
E	energy
G	mean rate of moisture emission; gust ratio
g_i	mean inside moisture content
g_o	mean outside moisture content
H	effective height
I	insolation
k	extinction coefficient
N	number of room air changes per hour
ΔP	indoor–outdoor vapour pressure difference
Δp	pressure drop between points (of an element of construction)
q	energy transferred to the water in a solar collector
R	thermal resistance
r_t	individual thermal resistance
r_v	vapour resistance
T	temperature
t_{ai}	inside air temperature
t_r	mean radiant temperature
t_{res}	dry resultant temperature
Δt	temperature drop
U	thermal transmittance
V	wind speed, volume
v	indoor air speed, volumetric rate of air change
Y	admittance
ρ	density
θ	tilt angle
μ	absolute viscosity

(3) Abbreviations

AIA	American Institute of Architects
BRE	Building Research Establishment
BRS	Building Research Station
BSS	British Standard Specification
CIBS	Chartered Institution of Building Services (formerly IHVE)
CSTC	Centre Scientifique et Technique de la Construction
IHVE	Institute of Heating and Ventilation Engineers
PCL	Polytechnic of Central London
RIBA	Royal Institute of British Architects

Units, symbols and abbreviations

(4) Conventions

$\sqrt{(3X)}$ denotes the square root of 3X.

(5) Conversion factors

Length
1 cm = 0.394 in.
1 m = 3.281 ft

Area
1 m² = 10.76 ft²
1 ha = 2.471 acre

Volume
1 m³ = 35.31 ft³
1 l = 0.2642 gallons (US) = 0.220 gallons (UK)

Mass
1 kg = 2.205 lb

Density
1 kg/m³ = 0.062 lb/ft³

Force
1 N = 0.2248 lb (force)

Pressure
1 Pa = 0.004 in. H_2O
1 kPa = 0.145 psi

Energy, work, heat
1 kJ   = 0.948 Btu
1 kWh   = 3414 Btu
1 GJ   = 278 kWh
1 therm = 105.5 MJ
1 tonne coal equivalent (TCE) = 26.4 GJ
1 tonne oil equivalent   = 44.7 GJ

Power
1 kW = 1.341 hp

Heat flux
1 W/m² = 0.317 Btu/(ft² h)

Thermal conductivity
1 W/(m K) = 0.578 Btu/(ft² h °F)

Heat transfer coefficient
1 W/(m² K) = 0.176 Btu/(ft² h °F)

Specific heat
1 kJ/(kg K) = 0.239 Btu/(lb °F)

Temperature
°C = (5/9) (°F − 32)

Temperature intervals
1 °C = 1.8 °F

Preface

The authors have rendered a valuable service to the building industry by the preparation of this volume with its wealth of actually built, rather than projected, examples. The book is well and soundly written with its emphasis maintained from start to finish on their basic theme, the necessity for the use in buildings of energy in many forms and the increasing desirability of learning how to minimize the use of non-renewable forms of energy in meeting those essential needs. Directed primarily towards architects, builders and owners in the United Kingdom, and consequently written within the scope of SI units, the book will be helpful to all who plan to build in latitudes north of the 50th parallel, where winter heating is more important than summer cooling.

Proper ventilation and the admission of outdoor air under suitable circumstances are not neglected, however. Building design features which are recommended by the authors are liberally illustrated by photographs and drawings of residential, institutional and commercial structures which actually exist. The soundness of their recommendations has been verified in most cases by the first-hand knowledge of the authors.

Many of the 'Energy' books which have appeared in recent years have dealt almost exclusively with non-renewable forms such as the wind and the sun. These are not neglected here but their applicability is subjected to careful and objective analysis, with the intent of giving guidance which is based upon knowledge and experience rather than upon enthusiasm alone.

The reader may well divide the book into three sections. The first, Chapters 1 through 3, deals with energy and the built environment, the building site and its energy attributes and with a general discussion of building design principles. The second section, Chapters 4 through 7, is devoted to space heating by passive and active solar systems, to conventional space heating and to thermal storage as it can be accomplished with the three currently available materials, water, rocks and phase-change substances. The third section, Chapters 8 through 13, discusses wind energy, water supply and waste disposal, energy conservation in the home and case studies for built and operating residential and institutional structures.

The appendices are filled with useful data pertaining to the atmospheric temperatures and solar intensities which prevail throughout the UK. The concluding appendix is particularly pertinent, since it deals quantitatively with the amount of moisture which is added to the indoor environment by the occupants. This is an aspect of design which is all too often overlooked until it is encountered by the individuals whose cooking, cleaning, bathing and breathing are responsible for the rise in indoor vapor pressure.

The literary styles of the two authors are so similar that the book is a first-rate example of collaboration, with each writer contributing the material with which he is most familiar. A convenient list of conversion factors is included for the assistance of readers to whom SI units are still unfamiliar. *Design with energy* is destined to become a valuable addition to the libraries of both professionals and owners who are concerned with the use and conservation of energy in buildings.

JOHN I. YELLOTT, M.B.E., P.E.
College of Architecture, Arizona State University
Tempe, AZ 85287 USA

Acknowledgements

The authors wish to thank: Paul Ruyssevelt for drawing many of the diagrams; Christopher Martin for helpful comments on the chapter on active solar heating; and Prof. John Yellott and Dr David Bullett for their helpful comments on the whole of the text.

1

Energy and buildings

1.1 Introduction

Forecasting is an art in which all are likely to be wrong but some will be more wrong than others. We expect the next few years to be a time for reflection. A variety of opinions on the conservation and use of energy in buildings now exists and a limited amount of data is available. Economic recession has increased the pressure on public and private funds; and the lack of national, let alone global, strategies for resource development has hindered development of energy conservation projects and the exploitation of alternative sources of energy. The result is an atmosphere of caution which at its worst could result in inactivity and resignation and at its best could lead to significant, if somewhat restrained, progress in the use and conservation of energy in the built environment. This may have to be enough — building, after all, is another art of the possible.

Our approach in this book has been to provide designers with a systematic framework for considering the use of energy in buildings. We have tried to cover the more important of the wide variety of topics that must be considered and to provide sufficient references to allow the reader to pursue points of particular interest in depth. A great deal of information has been produced during the last few years but virtually no overviews exist. Different research groups work independently of each other and often work with different aims in view. The overall result is a jungle of papers, data, evaluations and opinions for which we have tried to offer a guide.

Because of our past work we have a strong bias towards integration — energy use in buildings is a function of the site, form, method of construction, controls available, pattern of use by the occupants and the psychological feeling of comfort, as well as the material and energy flows through the building.

The structure of the book reflects this by starting with broader considerations of site and design, then examining specific topics such as solar and wind energy and then finishing with a fairly detailed examination of a number of applications in both domestic and non-domestic buildings.

Our preoccupations have been the areas most familiar to us and so the book is centred on northern Europe, especially the United Kingdom. However, because much of the impetus for the development of ambient energy sources comes from the United States and because of the extensive work

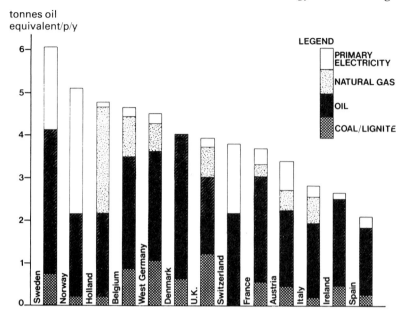

Fig. 1.1. Energy consumption in European countries (tonnes oil equivalent/p/y).[1] (Primary electricity is that derived from nuclear power and hydroelectric plants.)

done there we have tried to bridge the oceanic gap and draw extensively on developments in the US. From another point of view, characteristics of the European built environment such as higher densities and lower acceptable space heating temperatures are of relevance to energy use across the Atlantic.

We have not dealt with numerous topics ranging from alternative sources of energy such as wave power to feed the national grid to the legal aspects of alternative sources of energy, for example overshadowing of solar collectors by neighbouring buildings. In certain cases we have provided references to specialist topics not included in the text.

1.2 Energy and the built environment – past and present

It is worthwhile recalling that, historically, oil production is negligible. On a broad timescale the oil age presently coming to its conclusion is rather insignificant, although all of us have been profoundly marked by its effects. An alternative solution to energy supply and use must be found based on renewable sources of energy.

Fig. 1.1 shows *per capita* consumption of energy in various European countries and the energy sources that contribute to the figures.

Europe on the whole is a net fuel consumer but the UK is in the unusual and fortunate position of being a net producer. For 1981, estimated total consumption of primary energy was 317 MTCE compared to a production of 350 MTCE.[2] (Primary or gross energy is the calorific value of the raw fuel, for example oil, coal, natural gas, nuclear and hydroelectricity. Delivered, or net, energy is the calorific value of the fuel actually received by the consumer.)

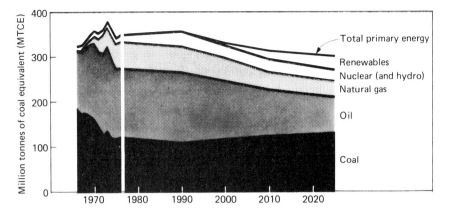

Fig. 1.2. Primary energy projections for a low-energy scenario.[4]

Coal is of major importance in the UK energy supply. Coal supplies are well distributed throughout the UK and estimates of technically recoverable reserves are in the order of 45 000 MTCE.[3] By comparison, recoverable UK North Sea oil reserves are estimated at about 6000 MTCE.[4] During the 1960s, pressure from oil led to a drastic drop in production to 110 Mt/y. In 1974 after a miners' strike a plan was developed to increase production to between 125 and 150 Mt/y by 1986. In 1981 annual output was about 128 Mt/y.[2]

In 1979 (when UK energy consumption was 356 MTCE and production only 329 MTCE), it was estimated that if a 'low-energy' strategy was adopted, in the year 2000 the UK could be entirely self-sufficient on North Sea oil and gas and indigenous coal supplies (renewable energy sources were assumed to contribute less than 2% of the total). Fig. 1.2 shows the makeup of energy sources in such a scenario, which aims for a decrease in energy consumption through a combination of conservation and increased efficiency in industry, the domestic sector, the commercial and institutional sector, transport and energy-supply industries.

Essential to any low-energy strategy is a change in the energy use in buildings. Before considering energy, though, it is worthwhile taking a quick look at population patterns and built forms. Europe is much more densely populated than the US. To the average American there was a hint of this in the absurdly small sizes of imported cars – but such cars are perfectly in scale with road widths and the pattern of buildings.

The UK has a population of about 60 million and an area of 245 000 km² – a density of about ten times that of the US. As one observer, commenting on the dangers of nuclear power plants and reflecting on the accident at Harrisburg, said: 'Ours is a small, densely populated island. It is not possible to run very far or very fast.'[5]

Housing densities reflect this with average residential developments in the UK having about 125–175 p/ha.[6] A recent scheme for energy-conscious housing uses passive solar gain and a density of about 40 houses/ha. Looking back a bit, one of the present authors has a home in an area of nineteenth-century, urban two-storey terraced houses which has a present density of about 300 p/ha. The approximate breakdown of the UK housing

Fig. 1.3. Estimated gross energy consumption of buildings in the public-service and miscellaneous categories.[11] (BRE, Crown Copyright, HMSO.)

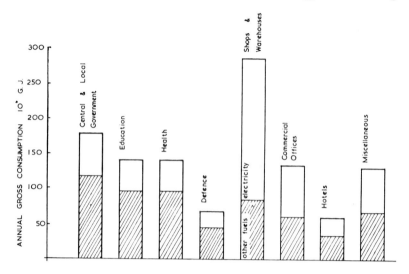

stock is semidetached houses, 31%; terraced houses, 30%; detached houses, 16%; purpose-built flats, 14%; converted flats, 8%, and miscellaneous, 1%.[7]

It has been estimated that building services account for 40–50% of the national consumption of primary energy. By sector, industry uses 41% of the energy input; households use 29%; transport, 16%; and other users, 14%.[8] Of the mean net energy consumption per household of 81 GJ/y the breakdown has been estimated as space heating, 64%; water heating, 22%; cooking, 10%; and TV, lighting, etc., 4%. The comparable mean gross energy-consumption figure is 138 GJ/y and the respective percentages 54, 22, 11 and 13.[8] In the domestic sector in the UK and in much of the rest of northern Europe there is no requirement for air conditioning. This is, of course, in contrast to a large part of the US.

During the space heating season it has been estimated that 41–61% of households are intermittently occupied.[9] It is now common for occupied living rooms to be kept at about 20 °C.[10] This represents a rise of about 1 °C per decade from a post-World-War II level of about 17 °C. The most likely reason for this trend was reported to be that, given the choice, people now prefer to wear less clothing in a warmer room than more clothing in a cooler one, although their thermal comfort would be the same in both. (See Chapter 3.) In the interest of energy conservation for society as a whole, a number of countries have taken steps to limit temperatures in spaces. In Britain, for example, temperatures in non-domestic buildings should not exceed 19 °C.

In non-domestic buildings the pattern of energy use varies greatly. Fig. 1.3 shows energy-consumption estimates.

In the non-domestic sector lighting, and hence electricity, plays a much greater role. Cooling loads, often linked to lighting as an unwanted heat source, are also often important.

Table 1.1. *Approximate delivered energy demands (for all energy requirements)*

Building type	Delivered energy demand (MJ/m^2 floor area/y)
(1) Commercial glass houses[a]	2500
(2) Large office buildings and large shops[b]	1700
(3) 1975 schools[c]	1280
(4) 2000 low-energy school target[c]	600
(5) 1982 house[d]	610
(6) Extremely well-insulated house of the future[e]	300

[a] Ref. [12]
[b] Ref. [4]
[c] Ref. [13]
[d] Authors' calculation updating[14] for 1982 Building Regulations.
[e] Based on the Wates House at the National Centre for Alternative Technology.[15]

An approximate idea of the energy consumption in a range of building types is given in Table 1.1.

1.3 Energy and the built environment – the future

The built environment changes slowly – a commonly cited figure for the lifetime of buildings is 60 years although, of course, this varies greatly. Office buildings are likely to last 30–70 years and houses 60 years or more. Servicing systems change more rapidly, with innovation and replacement occurring, say, every 10–30 years. In the last 40 years in the home of one of the authors the heating system has changed from open coal-burning fires to a mixture of gas room heaters and electric resistance heaters to gas central heating with an open fireplace for amenity. Each change has represented an increase in comfort and, because of this, probably in primary energy consumption, although the efficiency of use, particularly from open coal-burning fires to central heating, has been improved.

Will the next change be to solar energy? Probably not or at least not in the immediate future. Solar energy influenced both Greek and Roman urban planning and house design and in the latter civilization by the end of the first century the use of the sun for heating had become so important that laws were enacted to protect a structure's exposure to the sun.[16] For the next 2000 years solar energy applications were limited but diverse and included lenses for heating copper (Leonardo da Vinci) and coupled copper tanks partly filled with water and air to produce pressurized air to play organ pipes (de Caus).[17]

In the nineteenth century the foundations of widespread use of solar energy were laid with experiments in steam generation to drive an engine and pump (1876), generating electricity by focussing sunlight on selenium—platinum junctions (1876), and theoretical studies on solar radiation intensity and transmission through glass.[17] In the early twentieth century, flat-plate collectors were being used for domestic hot-water heating in California. Since then, numerous attempts have been made to use solar energy actively and passively in building design and services, but until the 1970s the ready availability (for part of the world's population) of oil delayed research and development for alternative sources of energy and meant that those applications available, as well as a number of energy conservation measures, were not cost-effective.

In the last decade, increasing prices and a realization that fossil-fuel resources are limited encouraged solar energy work. One UK project on which the present authors participated was that of the Cambridge University Autarkic House,[18, 19, 20] a plan for a house that would be completely independent of mains services, that would rely solely on solar and wind energy for heating and power and that would recycle water and wastes to make the best use of scarce resources. If ever a project was a child of its time, this was it, but unfortunately it was a child of the world and an orphan in the UK with this country's North Sea oil and gas and its 300-year supply of coal. The project was not continued but it did stimulate enormous interest. Unfortunately, many of the problems it had hoped in part to alleviate remain.

For example, the use of solar energy for space heating is environmentally sounder than the use of coal. Coal is a source of pollution and perhaps, on a large scale, a major danger to our climate. One immediate problem with coal (and oil and natural gas) is the so-called 'acid rain' produced when sulphur dioxide and other contaminants from power stations combine with vapour in the atmosphere to form sulphuric acid. Upon falling to earth as rain it seeps into the soil, releasing aluminium and manganese, and poisoning trees. It has been estimated that half of Germany's trees are suffering from this pollution.[21] In Norway, the effect of Britain's power stations has been to destroy fish life in more than half of the country's rivers.[21]

In the long term, fossil-fuel burning affects the carbon dioxide content of the atmosphere and this is presently a subject of great concern. The danger is that the increasing CO_2 content could lead, via the 'greenhouse effect', to a retention of heat that would otherwise escape from the lower atmosphere and so warm the earth's surface and perhaps drastically affect climate. More knowledge of the carbon cycle is required but, in the meantime, solar energy should not be dismissed as uneconomical or unnecessary because of abundant coal. In both the short and long terms we should be striving for environmentally safe solutions to energy supply and, indeed, ultimately we must find them.

Some results of the research into the potential of solar energy have been encouraging. For example, one recent study[6] of passive solar heating of houses showed that a number of measures, including lean-to conservatories and glazing the south-facing aspect of a roof so that it may be used as a warm-air plenum, are likely to be cost-effective in the not too distant future. Reports on active solar heating systems for domestic hot water vary. The Building Research Establishment (BRE) finds them uneconomic at present but estimates that costs might be reduced to about 50% of the current figure for a 4.5 m^2 system by attention to collector design (with significant potential for high-performance collectors) and to satisfactory installation on buildings.[22] However, at least one report has indicated that domestic solar hot-water heating systems will soon become economically viable.[23] One difficulty facing all ambient energy sources is that they are judged by different economic criteria than are conventional and nuclear sources. Similarly, they are not given the generous tax allowances that these latter often receive. Thus, for example, a comparison between solar energy and energy from coal, where the UK Central Electricity Generating Board is heavily subsidized, is necessarily biassed.

A general consensus seems to be that, given present trends, which is another way of saying if we bumble along and continue to spend phenomenally more on nuclear power than alternative sources of energy, the 'alternatives' may be contributing something under 5% to the UK primary energy requirement in the year 2000.[4, 11, 24]

The most likely development in the field of energy and buildings and one which is to be welcomed is greater emphasis on energy conservation. Ever since cow dung was applied to the walls of dwellings by our ancestors, the benefits of increased insulation and greater thermal comfort have been appreciated. Perhaps because it is one of the crudest and easiest ways to conserve energy, as well as one of the most cost-effective, greater insulation has been widely adopted. The dramatic trend in insulation of domestic buildings is shown in Table 3.11.

There are, of course, numerous other ways to conserve energy. Good management can be exceptionally effective – particularly in existing buildings that have undergone numerous transformations; over the years the scope for conservation is enormous. Techniques now available such as heat recovery on extract air and activated carbon filters, which permit a greater proportion of recirculated to fresh air in controlled ventilation systems, can be expected to reduce energy consumption in the near future. Other means are only in their infancy – controls are likely to be one of the greatest areas for development. On the domestic scale, for example, one of the present authors is currently developing a home-energy management system linked to a microcomputer which will sense temperatures throughout the home and outdoors, sense solar radiation and wind speed, allow for the thermal response time of the house and provide desired comfort conditions according to as complex a schedule as the occupants desire – all they will have to do is vary the program at the computer console which may well be linked to their own television for display. Additional features include control of the electrical loads, an energy-accounting system for recording consumption and paying bills and a home-security system.

The BRE has estimated that a combination of energy-conservation measures and the use of ambient energy sources could save 15% of the primary national energy consumption.[11] The programme would include combined generation of heat and electricity in which the waste heat of electrical generation would be used in district heating schemes, widespread employment of heat pumps, increased thermal insulation, controlled mechanical ventilation, solar collectors to supplement the hot-water supply of existing dwellings and a limited use of aero generators for space and water heating.

In the following chapters, we have tried to present a systematic approach to energy conservation and the use of ambient energy in buildings and to provide information on a variety of techniques, not all of which are presently cost-effective, but which we expect to be in the short- to medium-range future.

The question of costs is a vexed one, in part because of its relation to the assumed future real cost of energy. The uncertainty of energy costs affects everyone and large organizations such as the Central Electricity Generating Board are attempting to develop flexible, step-by-step strategies which try to ensure that, whatever the future vagaries of energy supply and demand, the requirements of customers for economical and reliable supplies of electricity are met. Predictably enough, one of the central strategies is the continued development of nuclear power. Another is to explore better ways of burning coal and a third is to identify and develop renewable sources of energy which may be useful for electricity generation.[25]

An approach to cost-calculations, which is gaining in favour, is lifecycle costing in which assumptions about the future, such as those for fuel costs, are made explicit. When energy-conservation measures are assessed on this basis they are often much more attractive than on the simple payback method commonly used. Nevertheless, there is no clear consensus as to what is to be done and in what order. As a preface to the work which follows we present the results of a number of studies of the comparative value of energy-saving measures. One of the present authors made one of the first detailed studies and the results are given in Table 1.2. (Because of the context of the study, passive solar measures were not included.)

Fig. 1.4 shows the Electricity Council Research Centre's results for a number of conservation measures based on dividing the potential annual energy savings by the additional capital cost (above that of an electric warm-air heating system).

Fig. 1.5 shows the results of a study of energy conservation in the Felmore Housing Scheme at Basildon, a group of 430 houses supplied by a district heating scheme.

Of course, there are many considerations other than cost involved in energy conservation — these vary from providing resources for future generations and achieving self-sufficiency in energy in order to escape political dependence on countries controlling the energy supply, to individual decisions about comfort and a desire to have the most modern technology available in one's home.

We believe that we are moving towards a community which uses energy and resources both more efficiently and in closer harmony with nature — but progress is, and will continue to be, slow. There is a great need for research into materials, equipment, techniques and attitudes. The vicious self-fulfilling argument that 'since the renewable sources of energy have

Energy and the built environment – the future

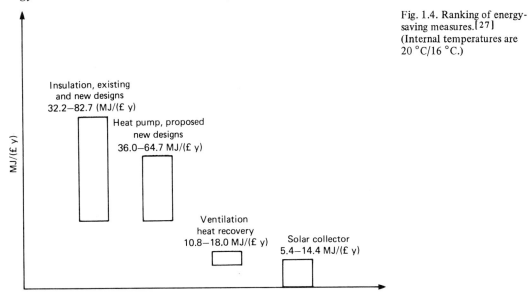

Fig. 1.4. Ranking of energy-saving measures.[27] (Internal temperatures are 20 °C/16 °C.)

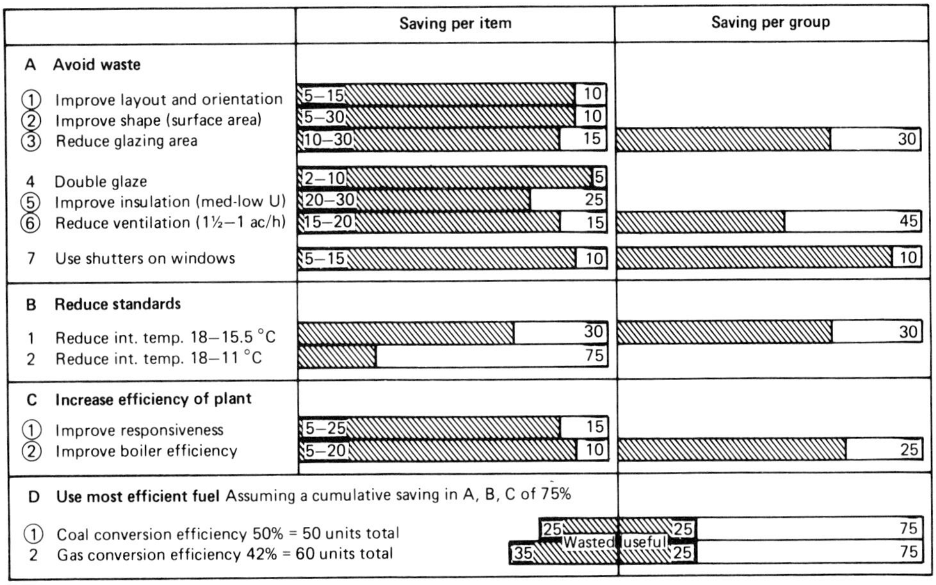

Fig. 1.5. Aspects of energy conservation at the Felmore Housing Estate in Basildon.[28] (The main design effort was placed on items marked by a circle.)

not been proven to be reliable and capable of contributing significantly to our energy supplies, the capital allocated to their research and development must be severely limited' will have to be attacked.

It seems reasonable to expect the following for the near future in the field of energy and buildings:

(1) The building stock will not have changed greatly; design professionals will be working largely on existing buildings; both new and existing buildings will be vital to improving the national use of energy.

Table 1.2. *Ranking of energy cost-effectiveness for measures of conservation and alternative energy supplies*[26]

			Annual GJ/£[a,b,c]	
Measure			Break-even	9% dividend
Roof insulation	50 mm	DIY	3.8	1.2
Roof insulation	100 mm	DIY	2.4	0.7
Roof insulation	150 mm	DIY	1.8	0.5
Roof insulation	50 mm	Installed	1.4	0.4
Double glazing	—	DIY	1.5	0.5
Roof insulation	200 mm	DIY	1.4	0.4
Roof insulation	100 mm	Installed	1.2	0.4
Roof insulation	250 mm	DIY	1.2	0.3
Roof insulation	150 mm	Installed	1.0	0.3
Wall insulation (cavity)	100 mm	Installed	0.9	0.3
Roof insulation	200 mm	Installed	0.8	0.3
Double glazing	—	Installed	0.8	0.2
Double glazing with heat-reflecting coatings	—	Installed	0.8	0.3
Heat recovery from hot water	—	Installed	0.7	0.2
Wall insulation (cavity)	150 mm	Installed	0.7	0.2
Roof insulation	250 mm	Installed	0.7	0.2
Double glazing with heat-reflecting coatings and krypton	—	Installed	0.7	0.2
Single glazing with foam blinds	—	Installed	0.6	0.1
Wall insulation (cavity)	200 mm	Installed	0.6	0.2
Electric heat pump in new house	—	Installed	0.5	0.1
Wall insulation (solid)	50 mm	Installed	0.4	0.1
Wall insulation (solid)	100 mm	Installed	0.4	0.1
Wall insulation (solid)	150 mm	Installed	0.4	0.1
Conversion from oil to electric heat pump	—	Installed	0.4	0.1
Conversion from electric resistance heating to electric heat pump	—	Installed	0.3	0.1
Wall insulation (solid)	200 mm	Installed	0.3	0.1
Heat recovery from warm air	—	Installed	0.3	0.1
Solar collectors for water heating on wall of:				
New house			0.3	0.1
Existing house			0.3	0.1
Solar collectors for water heating on roof of:				
New house			0.3	0.1
Existing house			0.3	0.1
Solar collectors for space and water heating without long-term storage on wall of:				
New house			0.2	0.1
Existing house			0.2	0.1
Solar collectors for space and water heating without long-term storage on roof of:				
New house			0.3	0.1
Existing house			0.2	0.1
Solar collectors for space and water heating with long-term storage on a house of floor area 100 m^2			0.1	0.0
Aerogenerator			0.1	0.0

Energy and the built environment – the future

(2) Energy sources will be varied; there will be greater reliance on solid fuel; renewable energy sources are still in their infancy; energy conservation will change the demand profile.

(3) The construction industry is unlikely to have evolved greatly and, consequently, the widespread use of techniques of energy conservation requiring a high degree of skill on-site is likely to prove difficult. However, the introduction of components such as prefabricated and preinsulated panels for timber-frame housing, which are common in Scandinavia, may challenge traditional building methods on cost grounds.

(4) The preference for houses rather than flats is unlikely to change; more one- and two-person households may be expected; dwelling size will continue to be small although smaller household sizes will mean more space per person.[29]

(5) Houses will be better insulated and will be more airtight.

(6) Heating systems will become more sophisticated with microelectronic-based control systems contributing to energy conservation.

(7) Occupants will become more interested in energy conservation and more adept at controlling their environment.

In many ways the situation is analogous to that of buildings in New York City just before the widespread introduction of lifts. The more far-sighted designers realized the potential of lifts, and at relatively little cost to their clients provided a lift shaft which could be used later according to the capital available. We shall argue for the provision of a large number of 'shafts' and a smaller number of 'lifts'.

We believe that today's buildings should be considered by their designers and users to be worth preserving for at least the next 50 years and hope that in the remaining chapters the reader will find advice that will help to make them so.

Notes to Table 1.2

[a] Break-even value = $\dfrac{\text{Annual energy (GJ) saved by the measure}}{\text{Annual cost of installing the measure (i.e. mortgage interest plus capital repayment)}}$

9% dividend value = $\dfrac{\text{Annual energy (GJ) saved by the measure}}{\text{Mortgage interest + capital repayment + 9\% interest on capital}}$

The 9% dividend figure is included for comparison since there is a school of thought which says that when deciding, for example, to buy insulation one should compare the savings with the return one could expect if, instead, one invested the money. This approach is rather stringent since energy conservation measures will normally increase the value of the property.

[b] Figures in GJ/£ have been converted from the original figures which were in kWh/£ rounded to the nearest 10.

[c] Capital and the following fuel costs, have been left at 1976 levels. The GJ of fuel which could be purchased for £1 are as follows, and they are quoted as mean costs over a 20-year period assuming either 0% or 10% inflation of fuel price per year.

Electricity (0% inflation) = 0.18 GJ/£
Electricity (10% inflation) = 0.07 GJ/£
Gas (0% inflation) = 0.50 GJ/£
Gas (10% inflation) = 0.18 GJ/£

Thus an occupant who uses gas and who assumes that the price will rise by 10%/y should carry out all the measures in Table 1.2 with an index of more than 0.50 GJ/£. This implies all the measures above 'electric heat pump in a new house'. If the more stringent criterion of a 9% return is demanded, then all measures above 'DIY double glazing' qualify.

References

[1] Anon. (1979). 'France Bottom of Energy Supply League'. *Building Services and Environmental Engineer*, April, p. 22.

[2] Anon. (1982). *Department of Energy Press Notice.* Reference No. 51, March, p. 31.

[3] Anon. (1976). *Coal for the Future Progress with Plan for Coal and Prospects for Coal to the Year 2000.* Department of Energy.

[4] Leach, G. (1979). *A Low Energy Strategy for the United Kingdom.* London: IEED.

[5] Shayer, B. (1979). 'Harrisburg here?' *Original Equipment Manufacture Design*, May, p. 3.

[6] Turrent, D., Doggart, J. & Ferraro, R. (1980). *Passive Solar Housing in the UK.* London: Energy Conscious Design.

[7] Romig, F. & Leach, G. (1977). *Energy Conservation in UK Dwellings: Domestic Sector Survey and Insulation.* London: IIED.

[8] Anon. (1976). *Energy Consumption and Conservation in Buildings.* BRE Digest 191. Garston, Watford: BRS.

[9] Desson, R.A. (1976). *Energy Conservation: The Intermittent Occupation of Dwellings and Domestic Energy Consumption.* BRE CP 37. Garston, Watford: BRS.

[10] Hunt, D.R.G. & Steele, M.R. (1980). 'Domestic temperature trends'. *The Heating and Ventilating Engineer*, April, pp. 5–15.

[11] Bakke, P. (Chairman) (1975). *Energy Conservation: A Study of Energy Consumption in Buildings and Possible Means of Saving Energy in Housing.* BRE CP 56/75. Garston, Watford: BRS.

[12] Winspear, K. (1980). 'Energy and UK glasshouse crop production'. In: *Energy Conservation and the Use of Renewable Energies in the Bio-Industries*, Vogt, F. (ed.). Oxford: Pergamon.

[13] Romig, F. & O'Sullivan, P. (1979). 'Buildings and the energy future'. In: *Buildings: the Key to Energy Conservation*, Kasabov, G. (ed.). London: RIBA.

[14] Anon. (1976). *Heat Losses from Dwellings.* BRE Digest 190. Garston, Watford: BRS.

[15] Todd, R. (1977). 'Low energy housing at the National Centre for Alternative Technology'. *University of Nottingham Conference on Ambient Energy and Building Design.* Welwyn: Construction Industry Conference Centre.

[16] Butts, K. & Perlin, J. (1978). 'The middle age of solar energy use'. *Solar Age*, 3 (3), 24–5.

[17] Kemper, J.P. (undated). *Pictorial History of Solar Energy Use.* University of Capetown, South Africa.

[18] Pike, A. (1972). 'Cambridge studies'. *Architectural Design*, 42 (7), 441–5.

[19] Pike, A. (1974). 'The autonomous house'. *Architectural Design*, 44 (11), 681–9.

[20] Littler, J.G.F. & Thomas, R.B. (1977). 'Solar energy use in the Autarkic House'. *Transactions of the Martin Centre for Architectural and Urban Studies*, 2, 93–110.

[21] Girling, R. (1982). 'Ten billion dying trees'. *The Sunday Times*, 4 April, p. 20.

References

[22] Wozniak, S.J. (1981). *Cost and Performance of Solar Water Heating Systems in the UK*. BRE IP 14/81. Garston, Watford: BRS.

[23] Soldatos, P.G. (1975). *UK–ISES – Midlands Branch Conference on Solar Energy in the 80s – Technical and Economic Viability*.

[24] Long, G. (1975). 'Solar energy: its potential contribution within the United Kingdom'. *Department of Energy Paper No. 16.* London: HMSO.

[25] England, G. (1980). 'Planning for uncertainty'. Lecture, 4 December. Central Electricity Generating Board.

[26] Littler, J.G.F. (1976). 'Priority in selecting energy conservation methods and solar or wind heating in houses'. *Autarkic Housing Project Working Paper No. 27.* Cambridge: University of Cambridge, Department of Architecture.

[27] Siviour, J.B. (1979). 'Prospects for energy savings in home heating'. *CIBS Annual Conference.*

[28] Anon. (1979). 'Felmore Housing, Basildon'. In: *Buildings: the Key to Energy Conservation*, Kasabov, G. (ed.), London: RIBA Energy Group.

[29] Hitchin, E.R. & Cheyney, B.S. (1980). 'Building trends and their effect on appliance and heating-system design'. *Building Services and Environmental Engineer*, 2 (10), 6–9.

2

Site planning and analysis

2.1 Introduction

Traditional site planning includes evaluation of the aesthetics of a site, population densities, land-use patterns, slope, drainage, soil characteristics, incident solar radiation, daylighting, exposure to wind and numerous other considerations which are treated in standard works.[1, 2] In this chapter these subjects will be discussed only when specifically applicable to the use of ambient energy sources or the opportunity for energy conservation in buildings. In the more recent past the attitude of many designers has been one of ignoring both the natural characteristics of the site and the potential of solar and wind energy. Instead, they concentrated merely on avoiding potentially deleterious effects such as summertime overheating. Important exceptions to this way of thinking include Olygay & Givoni who wrote classic works on climate and architecture.[3, 4] In an age of rapidly dwindling fossil-fuel reserves, though, it is important to use a site to best advantage. Fortunately, much can be done to conserve energy merely through good design, on both the large and small scale. In the former category especially, the possibilities depend on social conceptions of work, home and leisure but in the future we may see a closer integration of places of work and residence to reduce transportation energy. This could be encouraged by a gradual renovation of cities, resulting in their increasing attractiveness as places of residence and thus reducing the tendency towards suburban sprawl. For example, near the city home of one of the present authors, a former warehouse is being converted into flats. Among the results are a higher density and, for the occupants, less dependence on vehicular transport.

If urban densities increase in such a socially acceptable way it becomes easier and more economical to implement technical measures such as energy and materials recovery on sewage and solid waste.

Quantitative work[5, 6] has begun on some of these questions, particularly for new developments, and several years hence we can expect to discuss issues such as whether an upper limit should be imposed on densities to permit extensive exploitation of solar energy for space and water heating or whether such a limit might be too low to permit combined heating and power to be provided economically.[7] Or perhaps there is a range of options. The real difficulty, as suggested previously, is attempting to anticipate what our sources of energy will be and what they will cost during the next 50–100 years.

Introduction

In assessing a site, the meteorological scale of interest must be determined. A convenient classification[8] is
(1) macroscale including the regional scale;
(2) local scale (town and site conditions);
(3) microscale (around, on and within buildings).

Lacy[8] has also provided a very useful discussion of the way in which 25 principal meteorological elements (solar irradiance, temperature, humidity, pollution content, wind speed, etc.) may be used in dealing with problems of town planning, design (of individual structures), construction and building maintenance and running costs.

The macroscale is a 'given' element and on some of the colder, wetter, greyer days in the UK, one is reminded of the apophthegm that the only problem is architects' insistence on continuing to build out-of-doors. On the local scale the designer must take account of:[5]

(1) The natural drainage and transpiration — these often differ from that of the surrounding countryside and so can affect humidity.

(2) Smoke and waste gases produced by factories — gases and suspended solids can screen localities from sunlight; in the past Manchester and London lost nearly an hour of daylight because of this.

(3) Reduction of wind speed due to the increased surface roughness over cities (see Section 2.3) — surface wind speeds also tend to be variable and unpredictable due to the channelling effect of the buildings.

(4) More-effective absorption of solar radiation in city areas — this combined with combustion of fuels and reduced wind speeds creates an area of higher temperature, a so-called 'heat island'. The temperature of a large industrial city can be $1-2$ K higher than that of its surroundings. The Swedish Council of Building Research, in its reporting format for solar energy buildings, requires information on atmospheric clarity and for one house at Linköping, for example, has called for an investigation of radiation losses due to local cloud formation produced by condensation trails from intensive air traffic overhead.[9]

It is the microscale especially that interests us here because it lends itself best to artificial changes and hence to an element of control. Judicious selection of a site, based on a knowledge of existing or potential variations in microclimates, may also permit the designer to overcome climatic factors, as well as to exploit to the utmost the use of ambient energy sources. The principal elements to consider are solar energy, the wind, ambient temperature, soil and vegetation.

Fig. 2.1 summarizes the typical energy gains and losses in winter and summer for any structure. The sun will, of course, determine the solar radiation loads and the wind will affect air movement through the structure, as well as changing the external surface resistance (see Appendix 2).

Siviour, in a study of the effects of weather on house-heating requirements,[11] listed the following parameters as those expected to be of importance:
  outside air temperature;
  inside air temperature;
  outside radiant temperature;

Fig. 2.1. Typical energy gains and losses in winter and summer for any structure.[10]

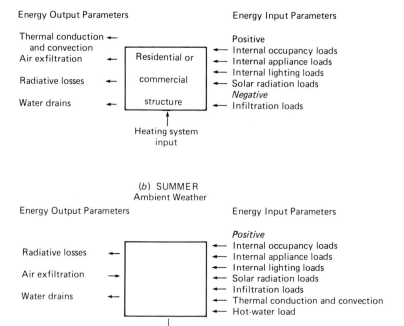

    wind speed;
    cloud cover;
    rainfall;
    solar radiation;
    vapour-pressure difference;
    orientation.

For the unoccupied semidetached houses he studied he found statistically that outside air temperature and solar radiation were the most important weather variables governing heating requirements. Other variables such as wind, cloud and rain had only a small effect. It is important to note, however, that the test houses were unoccupied — occupied ones will have all too human complications such as people opening windows with the heating system on — and that the results are a function of the building. A well-insulated home designed with large areas of south-facing glazing for passive solar gain will show a marked performance difference if its orientation is changed by 180°.

2.2 Solar radiation

The evaluation of the solar energy incident on building facades has been the subject of considerable scientific and practical work[12, 13, 14, 15] because of its importance for interior comfort conditions for daylighting

Solar radiation

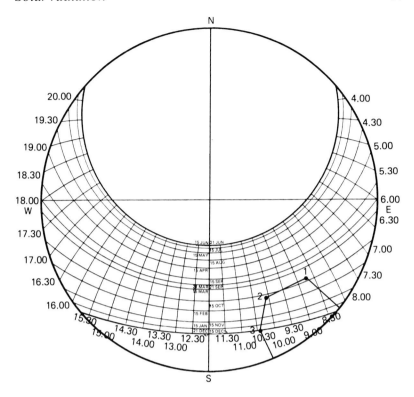

Fig. 2.2. Stereographic sunpaths for latitude 53° north with shading mask superimposed.[17]

and, commercially, for air-conditioning loads. Appendix 1 provides data on solar energy in the UK — here we will examine some practical consequences and applications.

Suppose we have a house with a south-facing bay window (*P*) and wish to determine whether a neighbouring building to the south-east will significantly reduce the direct passive solar gain into the window. We start with a stereographic sunpath as shown in Fig. 2.2. Then, as in Fig. 2.3, we draw a section and plan through point *P* in order to calculate the vertical and horizontal angles subtended at *P* by the obstruction.

These are marked on a shading mask and an outline of the mask is superimposed on the stereographic sunpath diagram as shown in Fig. 2.2. From this it is possible to see, for example, that point *P* will be shaded from the sun's direct radiation between approximately 8.45 a.m. and 10.15 a.m. on 15 October. (A more impressive and immediate method is to use a device[16] which is similar to a sextant in which a tiny telescope mounted on a level platform views the obstructions through a grid appropriate to the latitude. The observer stands at the proposed site of the building and can see through the grid an indication of the times at which any object casts a shadow on the site. There is no need to measure the heights of obstructions and there is the further advantage that the image may be photographed.)

Then, with data for the average distribution of direct solar radiation throughout the day in question,[13] we can assess the importance of

Fig. 2.3. Determination of overshadowing by neighbouring buildings. (*a*) Section. (*b*) Plan. (*c*) Construction of a shading mask.

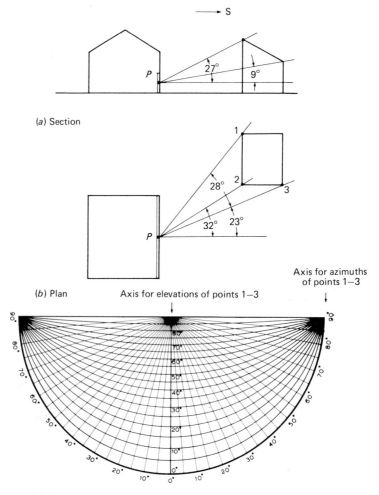

(*a*) Section

(*b*) Plan

(*c*) Construction of a shading mask

shading. A more refined calculation would also take diffuse radiation into account. A simple model of diffuse radiation would probably assume an isotropic sky — more-complex models[15] will refine this assumption to allow for a circumsolar diffuse component and an isotropic diffuse component.

Similar calculations for other days during the heating season facilitate the estimation of the loss of passive solar gain. (Chapter 4 discusses passive solar buildings in detail.)

The obvious next step is to consider how houses should be grouped to best use the sun's energy. Pennyland, a local-authority scheme for 177 houses in Milton Keynes, was designed to take advantage of passive solar gain. The planning density is, very roughly, about 100 p/ha of the total site area, a figure which gives considerable scope for selective orientation. Ideally, for maximum passive solar gain, houses should be orientated towards the south (see Chapter 3) but this is, of course, not always

Solar radiation

possible and, indeed at times, not desirable for aesthetic or other reasons. In Pennyland[18] it was necessary to consider, in addition to passive solar gain,

(1) the need to establish a movement network with special emphasis on footpaths and cycleways to reduce mobility energy requirements;
(2) the shape and contours of the site, existing hedgerows and trees, land use, density and tenure policies; and
(3) the need to relate to the development in adjoining areas.

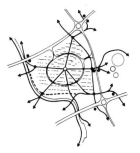

Fig. 2.4. South-facing housing option for Pennyland.[18]

One of several preliminary studies investigated orientating all houses due south, as shown in Fig. 2.4, but this resulted in a conflict between the housing groups and the footpath network. Similar conflicts occurred as the orientation constraint was relaxed further until the eventual solution, which has now been constructed, of 75% of the houses within ± 30° of due south, 20% within ± 45° of due south and 5% aligned east–west (see Fig. 2.5). The final effect is incontestably more pleasing and should appeal to designers concerned about energy constraints encouraging rigid and monotonous developments.

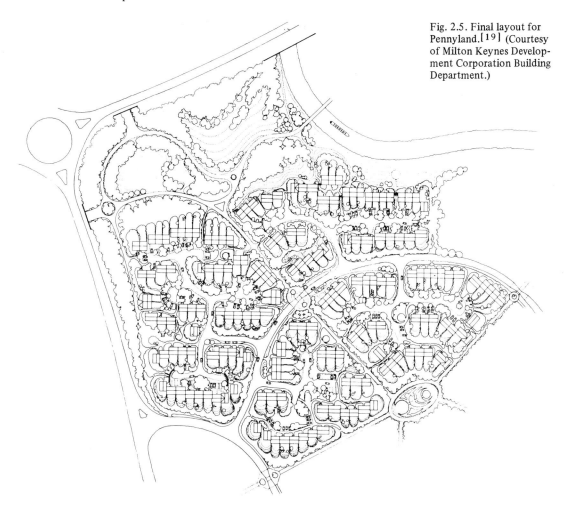

Fig. 2.5. Final layout for Pennyland.[19] (Courtesy of Milton Keynes Development Corporation Building Department.)

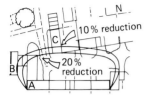

Fig. 2.6. Use of a solar shading calculator. (Building group A reduces radiation on B by less than 10% and on C by 10–20%.) (Courtesy of B. Everett.)

Fig. 2.7. View of Basildon Northlands 1 Housing showing insolation of south facades over lower buildings.[21] (Courtesy of Antoine Rafaul.)

The mean distance between approximately parallel rows of south-facing houses is about 23 m. The extent of overshadowing of one group of houses by another for the design options was estimated using specially developed solar shading calculators (see Fig. 2.6).

Similar considerations influenced the Northlands 1 Housing at Basildon.[20, 21] All houses face within ± 22° of south and overshadowing is avoided by a spacing of about 20–30 m between adjacent rows of two-storey buildings (see Figs. 2.7 and 2.18).

Turrent *et al.*[18] provide the following rules of thumb for the spacing distances between two-storey terraces assuming a flat site:

daylighting   6–10 m;
sunlighting   10–16 m;
privacy   15–18 m;
passive solar 20–23 m;

A separation of 20–23 m between buildings of a height of 6.2 m corresponds to a vertical shadow angle of 14.6–16.7° and since the solar altitude falls to 17° at latitude 50° N on 22 December at noon there will still be some overshadowing of ground-floor windows throughout part of the day. However, solar radiation in the depth of winter is not very significant (see Appendix 1) and so a degree of overshadowing may be acceptable. At Basildon, for the particular design of two-storey terraces houses used, and taking a ventilation rate of 1 air change per hour and an internal temperature of 21 °C, it was shown that substantial overshadowing result-

Table 2.1. *Density, plot width and energy loss*[24]

Density (p/ha)	Number of storeys	Plot width[a] (m)	10% energy loss	20% energy loss
100	1	5	P[b]	P
		6	P	P
		7	P	P
125	2	5	P	P
		6	?	?
		7	F	F
180	3	5	F	P
		6	F	F
		7	F	F

[a] The cases shown are for orientations of due south.
[b] P (pass) indicates that at the given density, a plot width is compatible with no more than the energy loss shown; F (fail) represents incompatibility and (?) represents an inability to make confident predictions with the mathematical model used for the analysis.

ing from a spacing between rows of about 10 m increased the energy demand by 6% over that of the unshaded house.[22, 23]

Plot width is another element in the solar energy–land-use matrix. Ó Catháin and Jessop at the Martin Centre in Cambridge have studied some of the theoretical aspects of layout for the Milton Keynes Development Corporation. Using several proposed density figures and the solar shadow prints of Fig. 2.6 they were able to analyse which combinations of plot width, building height and density were consistent with 10 and 20% energy losses due to overshadowing. Table 2.1 summarizes their results.

Narrow plot widths (with correspondingly deeper plots) reduce energy losses, and at higher densities it becomes more difficult to make maximum use of passive solar gain. Turrent *et al.*,[18] however, believe that with 'some design ingenuity' it should be possible to increase the maximum compatible density up to 180–200 p/ha. To put this figure in context they cite 1978 public-sector housing-scheme figures of 125 p/ha (38%), 150–200 p/ha (38%), and above 200 p/ha (24%); densities in the private sector are generally lower at 100–150 p/ha.

Vegetation has been associated with housing for millennia by both native and professional designers attempting to create an aesthetically attractive and a functionally well-tempered environment. The influence of natural vegetation on the microclimate near the house can include reduced solar radiation, lower wind speeds, cleaner air due to the filtering of pollution products, reduced noise levels and lower temperatures caused by evapotranspiration.

The reduction in solar radiation due to deciduous trees is well known and designers often include them in plans without perhaps always considering the question in adequate depth (or, more appropriately, height). There seems to be a certain belief that once the leaves fall, so does the tree, or at the least that the tree turns transparent. Careful observation

Fig. 2.8. A horse chestnut tree in Cambridge in July and December.

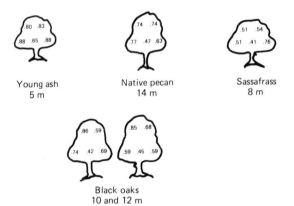

Fig. 2.9. Relative light penetration at five sampling sites for North American barren trees.[25] (Heights indicated are approximations.)

quickly reveals, however, that this is not so. Fig. 2.8 shows a horse chestnut tree in July and December and, obviously, even such a graceful young tree is not to be ignored in the winter.

Fig. 2.9 shows the results of one of the rare scientific studies of sunlight penetrating barren trees (in the US). The variability among species is particularly noteable.

Other variable aspects are the times of leaf initiation and leaf fall. While these will vary from year to year it would be interesting to see a study comparing the periods when certain trees are bare with a measure of the space heating requirement, such as degree days.

Mature height also depends greatly on species. To the south of houses we would suggest smaller trees. Depending on the distance from the house, trees in the 5–10 m range are recommended if passive solar gain is to be maximized. The following are just a few of the suitable species: maple

Solar radiation 23

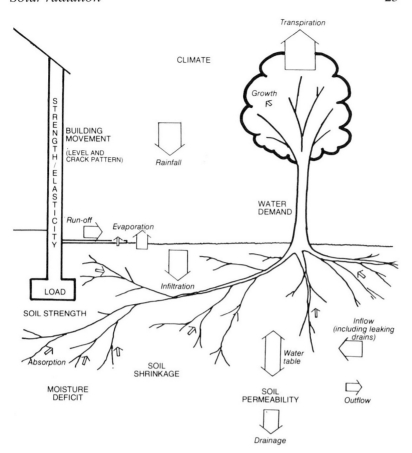

Fig. 2.10. Main water movements and factors for consideration when investigating subsidence damage.[27] (Courtesy of P.G. Biddle.)

(*Acer capillipes*), dogwood (*Cornus kousa chinensis*), magnolia (*Magnolia spp*), weeping elm (*Ulmus glabra*) and fruit trees.

Numerous other considerations, both aesthetic and technical, may enter when siting trees, and a good standard work[26] will deal with many of these. Suffice it to say here, because of the possibility of underground thermal storage being used in the house (see Chapters 5–7), attention must be paid to possible foundation damage. Fig. 2.10 shows the main water movements and factors for consideration when evaluating possible subsidence damage.

For buildings in non-shrinkable soils, that is, chalks, gravels, sands and loams, there is no threat of damage from trees and hence no need to specify a minimum distance of trees from buildings, according to Flora.[27] However, on certain shrinkable clay soils and with the rooting systems of some trees, damage is possible[28] although experts disagree on the extent of the problem and its implications for tree siting: on the one hand, there is a rule-of-thumb, based on the idea that the radius of root spread of a tree generally equals its height, that single trees should be located a distance equal to the height of the tree at maturity away from a building.[27] On the other hand, Flora, who disagrees with previous recommendations, maintains that 'any building which is adequately founded to cope with all problems normally associated with building in shrinkable or expansive clay soils, should present no problems when trees are planted'.[27] The thumb rule cited corresponds to an angle of 45° if the tree is in line with the sun at noon and a collecting surface such as a window, thus it would normally have to be moved back (to about 3.8 times the mature height) anyway to completely avoid interfering with passive solar gain. Nonetheless, attention should be paid to both loss of solar radiation and possible structural damage in shrinkable clay soils when locating trees, particularly those displaced to the east or west of the line due south from the window.

2.3 Wind

In temperate climates and particularly where summer cooling problems are not significant, as is generally the case in northern Europe, site design usually centres on reducing the wind speed around the building to minimize heat loss by infiltration and conduction—convection at the surfaces (see Fig. 2.1). The potential conflict between such a situation and one where a windmill is sited near the building is evident; additionally the means of reducing the wind speed, which are usually vegetation or fences, may interfere with passive solar gain. We will first examine the more typical case where there is no windmill.

Avoiding exposed sites is a commonsense first rule — annual mean wind speeds at the summits of hills can be from 30 to 100% higher than the annual mean wind speeds in the surrounding districts.[29] The BRE have devised a system for assessing wind loads, which accounts for local topographic influences and for the surface roughness of the environment[30] (see Chapter 8). For the latter the categories used are

(1) open country with no shelter;
(2) open country with scattered windbreaks;
(3) country with many windbreaks, small towns, outskirts of large cities;
(4) surface with large and numerous obstructions, for example, city centres.

The effect on the wind speed varies with height: at 3 m or less the BRE factor used for determining the design wind speed for the first category is about 50% higher than for category 4; at a height of 50 m the comparable figure is 10%.

The second rule is to lay out houses to reduce wind speed. Fig. 2.11 shows some simple schematics of air flow around buildings.

On the windward wall there is a build-up of pressure and on the leeward walls a reduction of pressure, that is, suction; the roof can experience either pressure effect depending on such factors as slope and angle relative to the wind direction.[30, 32] The wind-affected infiltration is generally an air flow through the structure from windward to leeward wall and roof surfaces.

Reducing the wind velocity at low level cuts infiltration losses and provides a more comfortable environment with fewer draughts, as well as a building that consumes less energy and, although this has long been known, little quantitative work has been undertaken to assess how to

Wind

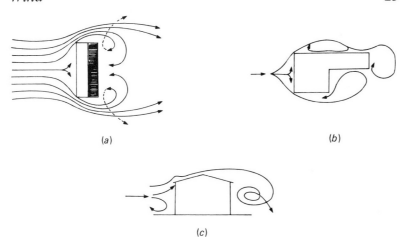

Fig. 2.11. Schematics of air flow around buildings.[30, 31]

reduce the wind speed most effectively. At the Princeton Centre for Environmental Studies, research has included an analysis of the relative benefits of fences, houses and trees as wind barriers.[32] Fig. 2.12 shows the wind flow, as represented by smoke patterns generated in a wind tunnel, for an isolated house and for houses sheltered by fences, upwind houses and simulated trees.

In Fig. 2.12(a) the wind flow is basically that of an oncoming stream stagnating against the windward wall of the house and then flowing up and over the windward roof to the peak from which it separates. Where the external pressure is higher as on the windward wall, flow tends to be into the house through poor seals around windows, cracks around doors and so forth; where the external pressure is lower, as in the visible wake on the leeward side, air flows from inside to out through any available conduit. The infiltration loss can be reduced by blocking the paths available for flow — through better construction, draught sealing, etc., as described in Chapter 3 — and by sheltering the house from the deleterious effects of the wind. Interestingly, a series of tests on the unsheltered house indicated that infiltration was worst when the wind blew towards the house at a 45° angle. This was due mainly to changes in the pressure produced by the wind angle on the roof and leeward wall surfaces.

Fig. 2.12(b) shows how a solid fence in particular affects the flow pattern on the windward side. Experimental results indicated that a fence one-fifth the height of the house and approximately one house height to the windward side reduces air infiltration by about 30% when the wind is perpendicular to the house (this falls to about 15% when the wind angle is 45°). It should be noted that this was in unheated models. In heated buildings infiltration will also depend greatly on the stack effect (see Chapter 3).

In Fig. 2.12(c) the much more balanced pressure distribution produced over the sheltered house is apparent. As it was not the object of the particular study to assess spacing effects, quantitative results were not provided but the authors did indicate that house sheltering should be far more effective than fence sheltering. For the windward boundaries of com-

Fig. 2.12. Flow patterns over houses.[32] (a) Isolated housing – no sheltering. (b) Solid fence provides sheltering. (c) Upwind house provides sheltering. (d) Simulated tree provides sheltering. (Courtesy of G.E. Mattingly.)

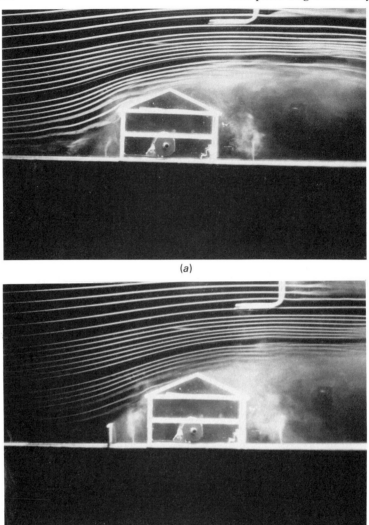

(a)

(b)

munities of such houses they suggested that both deciduous and evergreen trees might be used.

Fig. 2.12(d) shows the flow pattern resulting from sheltering by an evergreen tree. Of particular note is the relatively calm wake similar to the effect produced by an upwind house. Wind-tunnel results indicated that a single row of trees approximately the height of the house and planted 1.5–2.5 house heights upwind from the house can reduce the air infiltration by 40%. A combination of fence and trees results in a reduction of about 60% compared to the case of an unsheltered house.

It was then estimated that, given that air infiltration was responsible for about one-third of a winter heating bill, fence and tree sheltering could result in a 20% reduction in total fuel consumption, depending upon the prevailing wind direction and the outside temperature. This estimate may be compared with a fuel reduction of 40% in a house in South Dakota

Wind

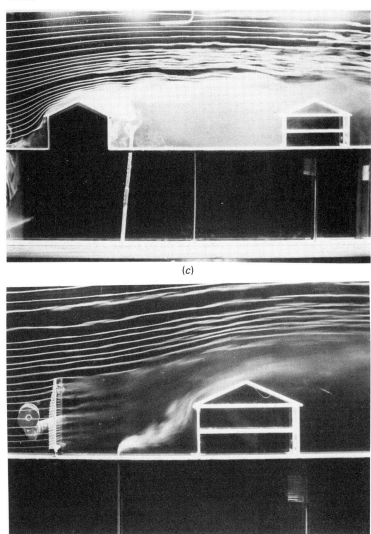

(c)

(d)

with plants on three sides, compared to an identical one with no windbreaks.[33]

Landscape elements such as trees and hedges are thought to be more effective than solid barriers, such as fences, by several authorities.[32, 34] Branches, leaves and evergreen needles tend to 'filter' the wind and break down the organized eddy patterns which prevail in the wakes of solid barriers. This sifting effect reduces wake pressures, giving a more balanced pressure distribution on windward houses.

The suggested separation of trees from house is of particular interest. A gap of 2.5 house heights corresponds to an angle of about 22° and so means that on a clear day at noon in December there will be some overshadowing of the ground floor and so some loss of direct passive solar gain. There will be, of course, a continuous obstruction of a small amount of diffuse radiation over the course of the heating season – the effect should

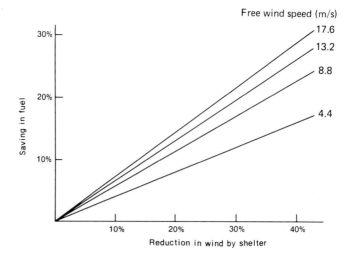

Fig. 2.13. Reduction of wind by a shelterbelt.[35] (Courtesy of G.E. Sheard.)

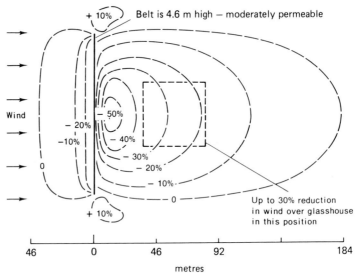

Fig. 2.14. The effect of shelter on greenhouse fuel consumption (measurements 3.7 m above ground level). (Courtesy of G.E. Sheard.)

not be very great but it may nevertheless be worthwhile to quantify the loss for a specific site. It should be noted too that a shelterbelt at even 1.5 house heights separation affords little shading in the summer and if overheating is a problem another means of protection must be sought.

Greenhouse researchers have also been interested in the use of natural or artificial shelters to reduce windspeed. Sheard[35] cites a rule-of-thumb that windbreaks should have a minimum height equal to half the height of the ridges of the glasshouses and be sited about ten times their height away from the glasshouses. He gives a diagram, reproduced as Fig. 2.13, showing the reduction of wind by a shelterbelt, as well as an estimate of the resultant fuel savings (Fig. 2.14). As for homes, the loss of passive solar

Wind

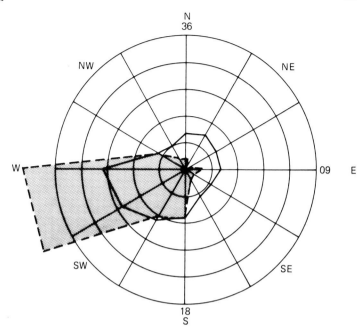

Fig. 2.15. Wind diagram for Malmslätt, Sweden.[9] (The diagram gives the relative frequency (%) of winds from varying directions — each ring represents 5%. Observations were over a 24 h period for 12 months and were made during all weathers. The solid line is for wind speeds of 1–99 knots (17 knots = 8 m/s) and the broken line for wind speeds of 17–99 knots. Thus, for example, approximately 7% of the time, winds in the range 1–99 knots come from the direction of due north to 15° east of north.

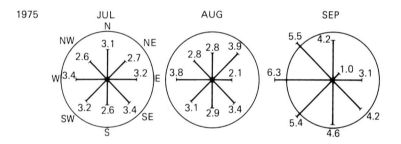

Fig. 2.16. Typical monthly (in 1975) wind roses for Malmslätt, Sweden (mean wind speed in m/s).[9]

gain must be remembered when evaluating the fuel savings. (It is also possible that the crop will be of lower value due to lack of light for photosynthesis and thus slower growth.)

To aid the siting of windbreaks for any particular application, wind diagrams are used. Fig. 2.15 shows the relative frequency of wind direction in Malmslätt, Sweden, used in describing the ambient environment of a solar house.[9] Monthly wind roses (see Fig. 2.16), if available, are ideal because they allow design for a specific heating season. Wind roses for a number of locations may be found in the Chartered Institution of Building Services (CIBS) Guide.

Wind speed and direction can only be obtained accurately by measurement but often, on first examining a site, a very rough quantitative idea of wind conditions is adequate and useful. For this reason we include here the Beaufort scale (Table 2.2).

At least one UK project has investigated and employed shelterbelts to reduce heat losses. The development at Basildon, mentioned previously, incorporated shelterbelts (see Fig. 2.17) which run predominantly north–

Table 2.2. *The Beaufort scale*[36]

Beaufort number	Description	Wind speed (m/s)
0	Calm; smoke rises vertically	0.3
1	Direction of wind shown by smoke drift but not by wind vanes	0.3–1.5
2	Wind felt on face; leaves rustle; ordinary vane moved by wind	1.6–3.3
3	Leaves and small twigs in constant motion; wind extends light flag	3.4–5.4
4	Raises dust and loose paper; small branches are moved	5.5–7.9
5	Small trees in leaf begin to sway; crested wavelets form on inland waters	8.0–10.7
6	Large branches in motion; whistling heard in telegraph wires; umbrellas used with difficulty	10.8–13.8
7	Whole trees in motion; inconvenience felt when walking against wind	13.9–17.1

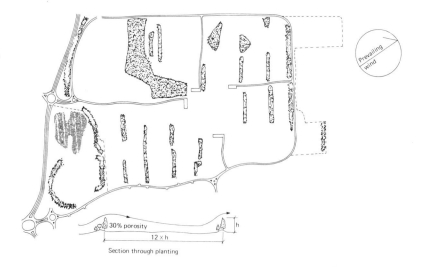

Fig. 2.17. Shelterbelt pattern and recommended spacing.[23] (Courtesy of Ahrends, Burton & Koralek.)

south (thus effectively countering the prevailing wind), while the housing runs east–west (see Fig. 2.18), so conflict between passive solar gain and vegetation for shelter is fortunately minimized. Ash, grey alder, red oak and wild cherry are among the principal trees in the windbreaks and the mature height is expected, on average, to be about 25–30 m. With the trees in leaf the estimated porosity is 30%.[37]

Lest it be thought that consideration of solar energy and wind patterns on site is relatively new, consider Henry Niccolls Wright's 'heliothermic' site planning for the American suburbs in the 1930s.[38] Fig. 2.19 shows

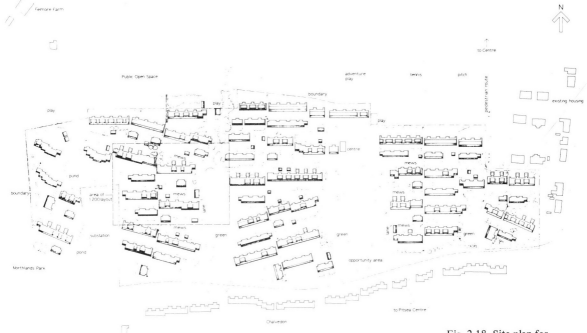

Fig. 2.18. Site plan for Felmore Housing at Basildon.[22] (Courtesy of Ahrends, Burton & Koralek.)

his plans for a development designed to be cool in summer and warm in winter. The plan of the houses, their orientation and location on site all contribute to using winter insolation to best advantage and to reducing summer insolation. Similarly, winter winds are minimized to reduce heat losses and summer breezes maximized for cooling.

Now let us examine the case where a windmill is to be located on-site to provide either electricity or space heating. Chapter 8 deals with windmills in detail but, for the moment, we must introduce some idea of the approximate size of the structure under consideration. In a relatively sheltered area such as Kew, London, a wind turbine with a swept area of 24 m^2 (4 m blades on a 6 m cross-member, see Fig. 8.18) is likely to provide (at the alternator output) approximately 15.8 GJ/y and so could make a significant contribution to the space heating load or could meet the entire electrical demand of a home (excluding the domestic hot-water demand).[39]

Normal practice would place such a wind turbine on a tower at least 10 m high to take advantage of the increased wind speed (see Chapter 8).

As for siting, precise information is unfortunately scarce. Manufacturers' guidelines, which tend to be based on experience, are often qualitative exhortations, for example to avoid areas where obstructions can either prevent the wind from actually reaching the turbine or where they create turbulence which decreases the efficiency of the machine and increases vibration. Most manufacturers of small- to medium-sized turbines (say 1–5 m in diameter) suggest towers of at least 5–6 m (Dunlite mills were sold with 12–18 m towers, and Golding[29] recommends at least

Fig. 2.19. Heliothermic site planning.[38]

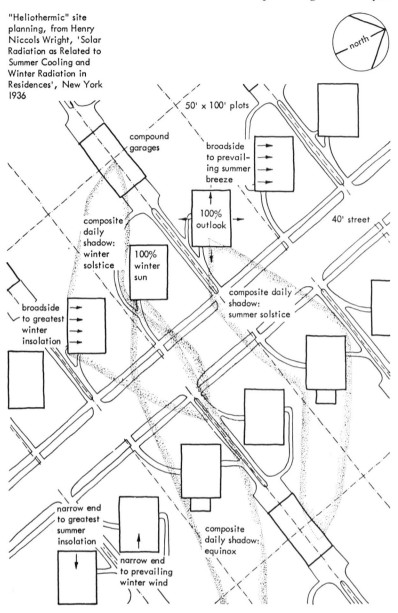

9–12 m) and, furthermore, add that the centre of the turbine should be 4–5 m above buildings and clear of vegetation within 200–300 m.

It is obviously quite difficult in many situations to meet such constraints and it is to be regretted that, at present, the loss of available energy as a function of distance of a small windmill from a configuration of obstructions is not known. One step towards relaxation is to site the turbine where it at least has access to the dominant winds, using a wind rose for this purpose, and to estimate the loss of energy.

Sheltering a house from the wind and siting a windmill which will be

Soil   33

used to supply energy to the house will thus pose a complex problem. If transmission losses from the turbine to the energy store or point-of-use were negligible the situation would be simplified but, as Chapter 8 explains, this is not necessarily so. As Fig. 2.12(c) and (d) suggests, if a shelterbelt, be it vegetation or other houses, is used, a wind turbine must be sited either at a considerable distance from it or on a tower sufficiently high to avoid the effects of turbulence.

One possibility that was considered by the Cambridge Autarkic Housing Project, as well as other groups, was to site the wind turbine on the roof of the house. This has a number of disadvantages: Golding[29] cites the possibility of the usually turbulent wind flow over the roof producing enough thrust on the turbine to damage the roof in high winds, but there may also be adverse reactions from the occupants due to noise and vibration. The Autarkic Housing Project viewed the mounting of a turbine on the roof as a research project and hoped that by using the house structure to support the turbine, the cost of the tower could be avoided; the savings were expected to outweigh the added costs incurred in strengthening the house structure to support the turbine. Precise expenses were not determined but a preliminary cost-analysis indicated that the cost of incorporating the tower would in fact be quite high. This is not surprising given the size of the wind turbine compared to the house as shown in proper scale in Fig. 12.3.

2.4 Soil

Aside from the traditional considerations of foundations, thermal energy storage and the use of the soil to reduce heat losses are the two principal points of interest when evaluating the soil of a site for a low-energy building. The effect of soil type on the choice of a waste-disposal system, which may also be important, is covered in Chapter 10. Soil as a building material is unlikely to play a major role in low-energy European homes of the near future and is not discussed here. (For those who wish to pursue the subject, a number of people, especially architects interested in the use of natural materials, have looked at the possibilities available.)[40, 41, 42]

The pros and cons of aboveground and underground storage of energy are presented in Chapter 7. In a variety of situations, for reasons of space especially, underground storage will be the sole viable solution and if the water table is low this need not be more than a routine difficulty.

However, in much of the UK and many areas of northern Europe the water table is quite high. Fig. 2.20, developed for agricultural purposes, shows how large an area is seasonally or permanently wet as judged by examining the top 40 cm of soil[43] — a deeper sampling depth, as would be needed for underground storage of energy, would of course increase the wet acreage.

For interseasonal storage of 10–20 GJ solar energy for a house, it would not be unusual to have to excavate, say, 100 m^3 to allow for both a water store and its insulation.[44] A concrete basement is the likely structural solution (see Chapter 7). It will have to resist the lateral pressure of

Fig. 2.20. Types of soil water regime in southeast England. After [42]

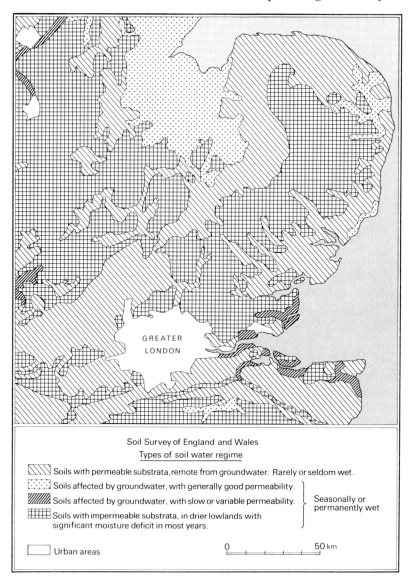

the earth and, additionally, if the basement is submerged partly or wholly in water, the hydrostatic pressure of the water. For dry soils the lateral pressure is about 4.7 Pa/m of depth; the hydrostatic pressure of water is 9.8 Pa/m of depth.[45] An example of the loading for a basement at a depth of 3 m with a water table at 1 m is shown in Fig. 2.21.

If, as is likely, water is used as the storage material in the heat store in the basement, its weight cannot be relied upon to keep the structure from rising majestically out of the earth. This is because allowance must be made for the basement being empty initially, during inspections and in the event of leaks. Furthermore, a high water table may mean that even with backfilling, cohesion between the soil and the basement walls is negligible

Soil

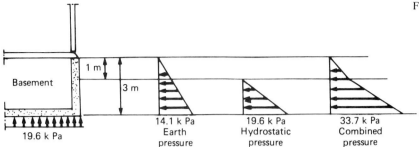

Fig. 2.21. Basement loading.

and so may not contribute to resisting flotation. The costly consequences are likely to be a need for additional concrete in the form of a 'lip' for the basement slab or thickened walls and slab, or some combination of these.

Finally, the tendency of the clay soils common to England to shrink and expand with changes in moisture content (which affects the siting of vegetation as mentioned earlier) can result in differential movements and lead to cracking. If moisture were able to penetrate and soak the thermal insulation in the store the effect on the thermal performance would be catastrophic. Consultation with a structural engineer is recommended to determine the appropriate foundation to minimize such risks.

More optimistically, but in more favourable circumstances, a number of full-scale experiments are underway to evaluate storage of solar energy in the soil itself. In France, near Grenoble, at a site with a water table 9 m below the surface, heat from solar collectors is being stored in a clay soil.[46] The boundary sides of the store are insulated and on top the soil is covered with a 15 m × 10 m slab of 20 cm thick polyurethane. On Prince Edward Island, Canada, water storage is expensive and rocks are scarce but an abundant supply of clay has encouraged work on the use of wet dirt or 'mud' as a storage medium.[47]

Soil may be used to reduce the heat loss from a building by delving to an extreme and locating the structure in the ground or by creating a berm around the ground-floor storey. Sterling and his colleagues[33] have covered the topic in depth. The principal advantage claimed is substantial energy savings due to reduced conduction and ventilation heat losses. Fig. 2.22 shows the soil temperature compared with the air temperature at Cambridge (UK). At greater depths the amplitude of the mean temperature fluctuation decreases. Sterling[33] notes that in the Minneapolis–St Paul area, at 5–8 m the temperature is almost constant at 10 °C.

The relatively high soil temperatures during the heating season afford an opportunity to reduce the heat losses, and sheltering the structure with the earth both minimizes the effect of the wind and provides insulation. Fig. 2.23 shows the Ecology House by architect John Barnard[50] at Osterville, Massachusetts, which is completely underground and features an internal courtyard; solar collectors are combined with a forced-air furnace for heating. A reduction of 75% in the energy demand for heating and cooling has been claimed.

A subsequent development, likely to be more practical in northern

Fig. 2.22. Average temperatures at Cambridge.[48, 49]

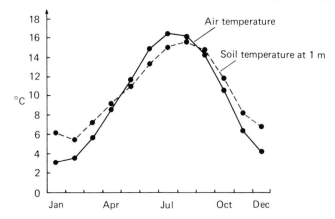

Europe (by the same architect), combines an earth surround and covering with an exposed south facade to take advantage of passive solar gain. Fig. 2.24 shows the Ecology House in Stow, Massachusetts. Heating is by forced air with a heat pump and energy use is 50–60% of that of a normal house of the same size.

Additional loading due to earth, groundwater and frost; avoidance of moisture penetration (from surface run-off, a high water table, during heavy rainfalls, etc.) and provision of adequate ventilation and light are some of the factors that must be evaluated carefully before opting for an underground or semiunderground structure. (Metz[51] provides a useful discussion of some of the severe problems that arise from the need to waterproof the underground structure.) That these potential difficulties can be overcome is being illustrated in the US, but for northern Europe, at least for the time being, there is insufficient information on such buildings to recommend their use in other than exceptional circumstances.

An effective partial measure which may easily be taken is the use of built-up earth berms. These avoid, in particular, the problems associated with a high water table, while lowering the heat loss due to conduction and reducing air infiltration; the energy savings resulting will, of course, depend on the amount of wall area covered, among other factors. In houses incorporating underground thermal storage, berms may be the most suitable and inexpensive way of using the excavated soil. It must be remembered, however, that berms are not a panacea. Earth is not a very good insulator and the soil temperature will normally be below that of the house so there is always a case for including insulation between the soil and the house. This is particularly true when the berm is exposed to the weather.

An additional consideration when examining the soil of a site is its potential for producing a source of fuel. In most soils of northern Europe there should be little difficulty in growing some variety of tree which, as wood, could contribute somewhat to the fuel supply, as well as providing a windbreak. For example, by coppicing certain species of hardwoods (such as sycamore, poplar, alder), it is possible to get higher yields than by traditional methods. The UK Department of Energy[52] cites yields in

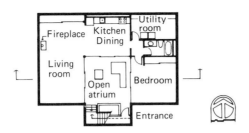

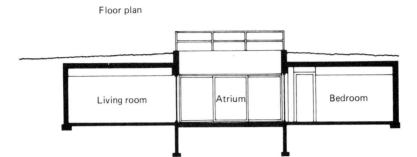

Fig. 2.23. Ecology House at Osterville, Massachusetts.[33] (Courtesy of J.E. Barnard Jr.)

Fig. 2.24. Ecology House at Stow, Massachusetts.[32] (Courtesy of J.E. Barnard Jr.)

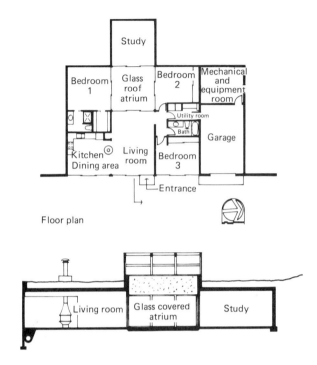

References

Pennsylvania of up to 1 kg/(m² y), or approximately 18 MJ/(m² y) (see Table 6.5). If we use a combustion efficiency of 65%, an area of 50 m² would provide about 1.6% of an annual space heating load of 36 GJ.

References

[1] Lynch, K. (1971). *Site Planning*. Cambridge, Mass.: MIT.
[2] McHarg, I. (1971). *Design with Nature*. New York: Doubleday.
[3] Olgyay, V. (1963). *Design with Climate*. Princeton: University Press.
[4] Givoni, B. (1976). *Man, Climate and Architecture*. London: Applied Science.
[5] van Straaten, J. (1967). *Thermal Performance of Buildings*. Barking: Elsevier.
[6] (1979). *Built Environment*, 5 (4).
[7] Owens, S.E. (1979). 'Energy and settlement patterns'. *Built Environment*, 5 (4), 282–6.
[8] Lacy, R.E. (1972). *Survey of Meteorological Information for Architecture and Building*. CP 5/72. Garston: Building Research Station.
[9] Ahlstrom, B., Hedeman, E. & Dättermark, B. (1977). 'Solar energy house in Linköping'. *Document D9*. Swedish Council for Building Research.
[10] Kovach, E.G. (ed.) (1973). *Technology of Efficient Energy Utilization*. Brussels: NATO.
[11] Siviour, J.B. (1974). *The Effects of Weather on House Heating Requirements – An Interim Report on Unoccupied Houses*. ECRC/M710. Capenhurst, Chester: Electricity Council.
[12] (1971). *Sunlight and Daylight Planning Criteria and Design of Building*. Department of the Environment Welsh Office. London: HMSO.
[13] *CIBS Guide* (1975). Section A6, 'Solar data'. London: CIBS.
[14] Page, J.K. (1978). 'Methods for the estimation of solar energy on vertical and inclined surfaces'. *International Report BS46*. Department of Building Science, University of Sheffield.
[15] (1979). 'Meteorology for solar energy applications'. *UK – ISES Conference* C18.
[16] *Solar Site Selector*, Lewis and Associates, Grass Valley, CA.
[17] Petherbridge, P. (1965). *Sunpath Diagrams and Overlays for Solar Heat Gain Calculation*. BRE CP 39. Garston, Watford: BRS.
[18] Turrent, D., Doggart, J. & Ferraro, R. (1980). *Passive Solar Housing in the UK*. London: Energy Conscious Design.
[19] Seed, J.L. (1982). 'Milton Keynes Development Corporation'. Private communication.
[20] Berry, J., Emerson, R., Harrison, J.W. & Kasabov, G. (1977). 'Conservation of energy in housing'. *Building Services Engineer*, 2 (45), 288–93.
[21] Anon. (1978). 'Energy conscious housing layout'. *RIBA Journal*, 85 (6), 229–30.
[22] Hermsen, J., Nelson, G. & Witham, D. (1979). 'Energy-conscious housing at Basildon, Essex', *AJ*, 41 (170), 758–72.

[23] Anon. (1975). *Northlands–Felmore Report.* Northlands–Felmore Design Team.
[24] Ó Catháin, C. & Jessop, M. (1978). 'Density and block spacing for passive solar housing'. *Transactions of the Martin Centre*, 3, 137–63.
[25] Holzberlein, T.M. (1979). 'Don't let the trees make a monkey of you'. *Proceedings of the Fourth National Passive Solar Conference.* Newark, Delaware: ISES – American Section.
[26] Marlowe, O.C. (1977). *Outdoor Design.* London: Crosby, Lockwood and Staples.
[27] Flora, T. (1977). 'Trees and building foundations'. In: *Trees and Buildings*, Hall, T.H.R. (ed.). Westbury, Wiltshire: Arboricultural Association.
[28] Biddle, P.G. (1977). 'Tree-root damage to buildings – an arboriculturalist's experience'. In: *Trees and Buildings*, Hall, T.H.R. (ed.). Westbury, Wiltshire: Arboricultural Association.
[29] Golding, E.W. (1977). *The Generation of Electricity by Wind Power.* London: Spon.
[30] Anon. (1970). *The Assessment of Wind Loads.* BRE Digest 119. Garston, Watford: BRE.
[31] Arens, E.A. & Williams, P.B. 'The effect of wind on energy consumption in buildings'. *Energy and Buildings*, 1, 77–84.
[32] Mattingly, G.E. & Peters, E.F. (1977). 'Wind and trees: air infiltration effects on energy in housing'. *Journal of Industrial Aerodynamics*, 2, 1–19.
[33] Sterling, R. (undated). *Earth Sheltered Housing Design.* University of Minnesota: The Underground Space Center.
[34] Woodruff, N. (1954). 'Shelterbelt and surface barrier effects on wind velocities, evaporation, house heating and snow drifting'. *Technical Bulletin No. 77.* Agricultural and Engineering Station, Kansas State College of Agriculture and Applied Science.
[35] Sheard, G.F. (1975). *The Effects of Wind on Glasshouse Production.* Littlehampton Glasshouse Crops Research Institute.
[36] Anon. (1939). *The Meteorological Glossary.* Littlehampton. London: HMSO.
[37] Price, C., Ahrends, Burton & Koralek. (1981). Private communication.
[38] Wright, H.N. (1936). *Solar Radiation as Related to Summer Cooling and Winter Radiation in Residences.* New York: J.B. Pierce Foundation, 40 W 40th St, NYC.
[39] Littler, J.G.F. & Thomas, R.B. (1979). 'Wind power for domestic use in the United Kingdom'. *Proceedings IEE International Conference on Future Energy Concepts.* London.
[40] Vale, B. (1973). 'A review of the Ministry of Agriculture's earth houses'. *Autonomous Housing Study Working Paper 17.* University of Cambridge, Department of Architecture.
[41] Weeks, C.T. (1978). 'Method in madness'. *Building Design*, p. 19.
[42] Fitzmaurice, R. (1958). *Manual on Stabilized Soil Construction for Housing.* New York: UN Technical Assistance Programme.
[43] Thomasson, A.J. (1975). *Soils and Field Drainage.* Soil Survey Technical Monograph No. 7. Harpenden: Soil Survey.
[44] Littler, J.G.F. & Thomas, R.B. (1980). 'Thermal storage in the

References

Autarkic House'. University of Cambridge, *Transactions of the Martin Centre*, 4, 139–56.

[45] Salvadori, M. & Levy, M. (1976). *Structural Design in Architecture.* New York: Prentice-Hall.

[46] de la Casinière, A., Person, J.-P. & Vachaud, G. (1980). 'Field experimentation of the "soil therm" interseasonal storage system of solar energy in the subsoil application to bio-industries'. In: *Energy Conservation and Use of Renewable Energies in the Bio-Industries.* Vogt, F. (ed.). Oxford: Pergamon.

[47] Caffell, A. & MacKay, K.T. (1980). 'Mud storage: a new concept in greenhouse heat storage'. In: *Energy Conservation and Use of Renewable Energies in the Bio-Industries.* Vogt, F. (ed.). Oxford: Pergamon.

[48] Anon. (1976). *Averages of Temperatures for the United Kingdom.* Meteorological Office, London: HMSO.

[49] Data from the University Botanic Garden, Cambridge.

[50] Barnard, J.E., AIA, Marston Mills, Mass.

[51] Metz, D. (1982). 'Keeping dry underground'. *Solar Age*, 7 (5), 24–31.

[52] Long, G. (1976). 'Solar energy: its potential contribution within the United Kingdom'. Department of Energy. *Energy Paper No. 16.* London: HMSO.

3

Building design

3.1 Introduction

Le Corbusier[1] said that a house was a machine for living in — in this chapter we attempt to describe some of the machine's components. Unfortunately, the state of the art is such that, for a given set of inputs, say environmental and material, one isn't quite certain how the machine or building will perform. The heart of our present dilemma is that we cannot precisely say even how individual building components respond in situations outside laboratories. To take but one example, infiltration losses are difficult to determine because of the effect of workmanship, condition of seals, extent of exposure to wind and so forth. And, remembering the old adage: 'The best laid plans of mice and men . . .', if we consider that, even in a well-designed low-energy house, the occupants will undoubtedly use the house in ways that would amaze the designers, the description of a building's performance is a hazardous task indeed. What is even more difficult though is to make accurate recommendations about how to design different, more-energy-efficient buildings. Of course there are the banal (but worthwhile) solutions such as increasing the insulation level and decreasing the ventilation rate but, as we will see, the latter can only be done to a certain point which is well short of the one where occupants and designer breathe uneasily. For more-complex questions such as choosing between lightweight and heavyweight buildings our information is imperfect. And as for other areas which may become the technology of tomorrow, evacuated walls, say, we know almost nothing.

We start with a brief introduction to some of the broader physical principles of design, such as materials and environmental conditions, before examining the detailed questions of form, layout and constructional elements. The reader will note that we have ignored some aspects entirely. For example, we have left aesthetics to designers and clients but, or perhaps, since, we feel that there need be no conflict between low-energy buildings and attractive ones.

3.2 Energy demand and thermal response

Fig. 3.1 shows the components of the seasonal heat loss in a typical house built to the standard of the 1982 Building Regulations[2, 3] (U-values (thermal transmittances) in W/(m^2 K): windows 5.6; walls 0.6; floor 0.6

Energy demand and thermal response

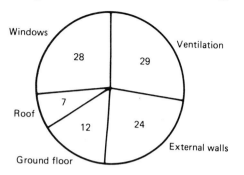

Fig. 3.1. Percentage composition of heat loss in a typical semidetached house built to 1982 Building Regulations Standards.

(assumed); roof 0.35; ventilation rate = 1 air change/h = 225 m³/h (assumed); average internal to external temperature difference (assumed) 7 K; heating season of 5544 duration (assumed)).

Such a house has an estimated annual gross space heating requirement of 36.3 GJ; Siviour suggests that this would be reduced to about 13 GJ (assuming 70% overall utilization of 'free heat' from occupancy, solar heating and losses from the hot-water system).[4] (In related work, Siviour has shown that due to the 'free heat' from the sun, typical houses function as solar collectors with efficiencies of 6–7% for space heating.)[5]

Siviour noted that problems arise in using 100% of the free heat because of the variability of input (some of course occurs when there is no space heating demand) and its association with steam and smells. His estimate of 70% utilization followed his detailed analysis shown in Table 3.1.

Table 3.2 shows the heat emission from the adult male human body — for women and children correction factors of 85 and 75%, respectively, may be used.

Siviour has compared a number of published estimates of internal free-heat gain. Table 3.3 summarizes his results.

The CIBS estimates that the radiation heat gain during the heating season per square metre of glass will be approximately as follows: south-facing window, 670 MJ; east- and west-facing windows, 400 MJ; north-facing window 240 MJ.

Obviously, utilization of free heat will vary from building to building. One of the objects of energy efficient design is to use as much of it as is feasible and economical. (Appendix 2 discusses how the design peak and annual heating requirements may be calculated.)

An as yet unresolved point but one of some interest is whether lightweight or heavyweight buildings use energy more efficiently. The terms themselves are relative. The old Institute of Heating and Ventilating Engineers (IHVE) Guide[8] described as heavyweight, buildings with a construction of solid internal walls and partitions and solid floors and ceilings. Lightweight buildings were characterized by lightweight demountable partitions with suspended ceilings and floors which were either solid with a carpet or wood block finish or were of the suspended type. For houses it may be convenient to think of a heavyweight house as one with a double wall of, say, brick on the outside and concrete block inside and a lightweight house to be of single-brick construction with, say, expanded

Table 3.1. *Analysis of free-heat gains*[4]

Source	Degree of availability	Comments
(1) Occupancy		
a. People	Good	Heat from people is where it is required
b. Appliances	Fair	Heat is sometimes where required, e.g. television
c. Lights	Good	Heat is where required
d. Cooking	Poor	Concentrated in time and space and accompanied by steam and smells
e. Drawing curtains	Good	Heat saving can be throughout the house
(2) Domestic hot water		
a. Storage vessel	Fair	Concentrated but continuous
b. Pipe losses	Fair	Dispersed but intermittent
c. Point of use	Poor	Concentrated and accompanied by moisture
(3) Solar heating	Fair	Varies from day to day; during the day from room to room

Table 3.2. *Heat emission from the adult male human body at 20°C dry-bulb temperature*[6]

	Heat emission (W)		
Degree of activity	Sensible	Latent	Total
Seated at rest	90	25	115
Light work	100	40	140
Walking slowly	110	50	160
Light bench work	130	105	235
Medium work	140	125	265
Heavy work	190	250	440

polystyrene adjacent to the brick or to be of timber-frame construction (with insulation between the studs).[9]

The most recent CIBS Guide[10] provides a more rigorous interpretation of heavyweight and lightweight in terms of the response factor given by

$$f_r = \frac{\Sigma (AY) + (NV/3)}{\Sigma (AU) + (NV/3)}$$

where

A = area of a structural element (m^2);
Y = admittance of a structural element (see Appendix 2) (W/(m^2 K));
N = number of room air changes per hour;
V = room volume (m^3);
U = thermal transmittance of a structural element (W/(m^2 K)).

A thermally lightweight building will have a response factor of about 2.5 and a heavyweight building one of about 6.

Table 3.3. *Assessments of internal free heat gains in domestic premises during the winter*[7]

Source	MJ/day
Occupants[a]	14.4–21.6
Lighting, appliances, cooking	29.5–63.3
Hot water[b]	12.2–54.0

[a] Family size varies from 2.9 to 4; note that only sensible-heat production should be accounted for.
[b] Assumed to be 50% of total energy input to water heating.

Table 3.4. *Specific heats of selected materials*[11]

Item	Specific heat (J/(kg K))	Item	Specific heat (J/(kg K))
Granite	330	Polystyrene	1250
Copper	376	Standard hardboard	1250
Mild steel	502	Insulating fibreboard	1400
Glass	830	Perspex	1460
Concrete	880–1040	Timber	1500
Aluminium	920	Polythene	2300
PVC	1040	Water	4187

Heavyweight buildings will have high thermal capacities. Thermal capacity is the ability of a body to store heat and is given by the product of its mass and its specific heat. Specific heat is the quantity of heat required to raise a unit mass of a material through a unit degree of temperature. Table 3.4 shows the specific heat of some common materials (see also Table A 2.3 of Appendix 2).

In steady-state conditions the same thermal transmittance, or U-value, can be provided by a thick heavyweight wall, say that of a stone castle, or a thin lightweight wall, say a timber-frame house with insulation. However, when sudden temperature changes occur, as they do continually, a thick wall of high thermal capacity will only warm up, or cool down, slowly; the converse is true for thin lightweight walls. This response will affect the energy consumption of buildings.

For large buildings, Burberry[12] has stated that, if there is any intermittency of use or if close control of temperature is appropriate, lightweight structures offer better economy of heating. (On the other hand, heavyweight construction, in part because it resists summertime peak temperatures better, may provide a better economy of air conditioning.)[13]

For small buildings such as houses, opinion is divided on the merits of lightweight, rapid-response structures and heavy, longer response times. One important point to consider pattern. Although this is often difficult to predict fact, be designed to accommodate a variety of patterns during its life, if it is known that the house will be continuously occupied, for example housing for the elderly as opposed to that for young, single people, a heavyweight structure is likely to offer advantages. Where use is intermittent a lightweight structure which warms up quickly will tend to save energy. Another consideration is the importance of passive solar gain in the design but even here opinion varies. At Basildon, in the estate described in Chapter 2, the houses were designed to take advantage of passive solar gain by attention to orientation and larger glazed areas on the south walls. The design study recommended a lightweight structure because of cheaper installation costs and, with an intermittent heating regime, lower heating costs.[14] The subsequent report to the client, the Basildon Development Corporation, incorporated the study and went on to discuss the importance of heavyweight solid party walls in terraced housing. Because of such walls, the difference between heavyweight and lightweight external constructions is less distinct. Similarly, such unknowns as the occupants' choice of furnishings will affect thermal response. Heavy sofas and chairs may store a fair amount of energy, thus tending to make the house more heavyweight. On the other hand, a carpet over a solid floor will reduce the response time, just as tapestries on mediaeval walls did. The final construction, shown in Fig. 3.21, was a compromise with cavity brickwork on the ground floor and timber-frame construction on the first.

Another hybrid approach was used in a development of 29 two- and three-person solar heated flats in the London Borough of Lewisham.[15] There, a heavyweight floor with embedded coils (see Chapter 6) to provide background heating was used in conjunction with highly insulated walls which also had a low admittance. The construction technique was dry lining on plaster dabs, 150 mm aerated concrete blocks, 50 mm cavity and 115 mm brick, giving a U-value of 0.76 W/(m^2 K) and an admittance of 2.0 W/(m^2 K) (compared to, say, 4 for a more traditional construction).[16]

(In addition to the greater insulation afforded by 150 mm blocks, they also add strength to the construction and help overcome the problem, often encountered with 100 mm lightweight blocks, of their being too weak to carry the structure's weight.)

An example of heavyweight construction deliberately used in conjunction with passive solar gain may be found in a group of 14 houses for the elderly in Bebington (see also Chapter 4).[17, 18] Occupancy there is continuous and to increase the effect of the thermal capacity of the structure the heavyweight walls are insulated externally. The principle may be extended even further and special storage elements incorporated in the building to 'soak up' excess solar energy for later use — examples are given Chapter 4. The approach depends on whether the designer thinks there be solar energy available for storage and whether his or her design can ke advantage of such heat and avoid overheating. If overheating example when the building fabric and furnishings cannot h energy and if the heating system cannot be turned down

The internal environment 47

quickly enough during a period of intense passive solar gain, there is no energy saving but there will be discomfort. The interrelations among types of construction, varying casual gains and heating systems are not well understood at the moment and this is reflected in the variety of systems being tried. It may turn out that the differences in energy consumption among the various solutions are not great but more research will be required to establish this. We may also see a range of optimum solutions, even within one country, depending on the weather conditions, notably incident solar radiation and temperature.

Theoretical work, based on the admittance method (see Appendix 2) by British Gas,[19] has found that, for an insulated terrace house, a heavy structure requires a larger peak heat input than a light structure of the same level of insulation and that the slower overnight cooling of a heavy building means that its average internal temperature is higher than that of a light building. The consequence of the second point is that the heat losses over a 24 h period will be higher and to balance these the heavy-weight building will require a larger total heat input.

3.3 The internal environment

3.3.1 Comfort

The philosopher Karl Popper has suggested that it is much easier to work towards a society that minimizes unhappiness rather than one that maximizes happiness and much the same might be said of comfort. We have a fairly good idea of what makes most people uncomfortable but the provision of an ideal environment for all of the people all of the time has proven to be an illusive, and perhaps futile, goal.

Comfort is a subjective experience, dependent on individuals and such factors as people's past experience of heating and whether or not they themselves are to pay the forthcoming heating bill. In practice a designer tends to provide conditions that are acceptable to a majority of the persons occupying a space or building and not so unacceptable to the others that complaints are too vociferous.

Comfort conditions depend mainly on the air temperature, the mean radiant temperature of the surroundings, the air speed and the relative humidity. It is interesting to keep in mind the similarity of the body's temperature control system to mechanical control systems for heating and air-conditioning installations. The important temperature-sensing system in the hypothalamus at the base of the brain operates like a thermostat, monitoring changes in blood temperature caused by the body's internal metabolic changes and by temperature gradients across the skin. The system has a set point of about 37.0 °C which it tries to maintain. If the body temperature is sensed at less than this value, physiological responses such as shivering occur to increase the metabolic rate so that more heat is generated. If the contrary occurs, the body starts to sweat and the subsequent evaporation of this moisture from the skin provides cooling. Between the extremes of shivering and sweating the body controls its temperature by altering the blood flow to the surface. The goal of these responses is the maintenance of thermal equilibrium.

Seated at rest, adult males produce approximately 115 W (Table 3.2) of which about 80% is lost from the body by convection and radiation (and to a much lesser degree conduction) and some 20% by evaporation (mainly from the lungs but also from the skin). Thermal comfort depends upon achieving a balance between the heat being produced by the body and the loss of heat to the surroundings. Since the body is at a higher temperature than the surroundings and since we sense skin temperature rather than room temperature, the function of heating and air-conditioning systems is in fact to control personal cooling.

The principal relationships between physical factors of the environment and heat loss from the body are the following:

(1) The air temperature affects heat loss by convection and evaporation.
(2) The mean radiant temperature affects heat loss by radiation.
(3) The air speed affects heat loss by convection and evaporation.
(4) The vapour pressure of the air affects heat loss by evaporation.

In the 1950s and 1960s, in the midst of the 'building boom', interest in comfort was related to office environments, the effect of large glazed surfaces, overheating and air conditioning. There was a search for uniformity. Now it might be said that, with increasing concern about energy consumption and the recognition that passive solar gain can usefully reduce the space heating demand of buildings, the ethos of comfort research must also change. The importance of this is readily apparent if one considers that a 1 K lowering of the internal temperature reduces the energy required for heating by approximately 5–10%.[20] One encouraging sign in comfort research is Humphreys' suggestion, which admittedly is not shared by all, that since the indoor comfort temperature depends on the outdoor mean temperature the internal temperature be allowed to follow the external and that continuous change at a rate of 2 K per week would be virtually unnoticeable.[21] Another is work on comfort in passive solar heated buildings where the thermal environment is likely to be less uniform.[22] We may also expect to see studies of the effects on comfort of the lower air-change rates so often used to conserve energy.

3.3.2 Temperature

The temperature index presently used by the CIBS[23] is the dry resultant temperature (there is also a wet resultant temperature which takes into account humidity) given by the formula

$$t_{res} = \frac{t_r + t_{ai}\sqrt{(10\,v)}}{1 + \sqrt{(10\,v)}},$$

where

t_{ai} = inside air temperature;
t_r = mean radiant temperature;
t_{res} = dry resultant temperature; and
v = indoor air speed.

The internal environment

Table 3.5. *UK design conditions*[23]

Season	Occupancy/ category	Resultant temperature (°C)	Relative humidity (%)
Summer	Continuous	20–22	50
	Transient	23	50
Winter	Continuous	19–20	50
	Transient	16–18	50

At 'still' indoor air speeds of $v = 0.1$ m/s the formula is simply

$$t_{res} = \tfrac{1}{2}(t_r + t_{ai}).$$

The dry resultant temperature, more commonly abbreviated to the resultant temperature, is that temperature recorded by a thermometer at the centre of a blackened globe 100 mm in diameter.

The formula shows that both the air temperature and the radiant conditions are important for comfort and this must be kept in mind when selecting a heating system. Ideally, all surfaces should be at the same temperature but, of course, glazed areas will tend to have a lower temperature during the winter and a higher one in summer, especially if solar absorbing glass is used. (One of the advantages of double glazing is to improve comfort conditions by raising the mean radiant temperature of the internal surfaces.) In the past it was at times suggested that a higher mean radiant temperature than air temperature was preferred but research has shown that in fact this is not so.[24]

The temperature gradient, which depends on both the heating and ventilating system and the thermal insulation of the space, is also of importance for comfort. A guideline is that in a room at 22 °C the temperature at ankle level should not be less than 19–20 °C and the head-level temperature should not be more than 3 °C higher than the ankle-level temperature.[24, 25] Floor-heating systems tend to satisfy this comfort criterion well by minimizing the vertical temperature gradient but care should be taken to ensure that the surface temperature is not greater than about 27 °C.[25] Similarly, but on a higher plane, discomfort may be experienced if the head is subjected to radiation from a heated ceiling. The CIBS gives no specific temperature but cites the dependence of the maximum permissible temperature on room height and area and reports that temperatures of up to 40 °C have been found acceptable in rooms of up to 5 m square and of usual height.[25]

Table 3.5 gives the CIBS recommendations for design conditions and Table 3.6 for resultant temperatures in a variety of spaces.

Humphreys[20] provides an excellent discussion of desirable temperatures in dwellings. Fig. 3.2, from his work, shows approximate temperatures for thermal neutrality, a state in which the occupants experience no heat stress or thermal strain. In constructing the chart it was assumed for simplicity that the combined effect of the temperature of the air and the

Table 3.6. *Recommended design values for dry resultant temperature*[23]

Type of building	t_{res} (°C)	Type of building	t_{res} (°C)
Art galleries and museums	20	Hotels:	
		Bedrooms (standard)	22
Assembly halls, lecture halls	18	Bedrooms (luxury)	24
		Public rooms	21
Banking halls:		Staircases and corridors	18
Large (height > 4 m)	20	Entrance halls and foyers	18
Small (height < 4 m)	20		
		Laboratories	20
Bars	18		
		Law courts	20
Canteens and dining rooms	20		
		Libraries:	
Churches and chapels:		Reading rooms (height > 4 m)	20
Up to 7000 m³	18	(height < 4 m)	20
> 7000 m³	18	Stack rooms	18
Vestries	20	Store rooms	15
Dining and banqueting halls	21	Offices:	
		General	20
Exhibition halls:		Private	20
Large (height > 4 m)	18	Stores	15
Small (height < 4 m)	18		
		Police stations:	
Factories:		Cells	18
Sedentary work	19		
Light work	16	Restaurants and tea shops	18
Heavy work	13		
		Schools and colleges:	
Fire stations; ambulance stations:		Classrooms	18
Appliance rooms	15	Lecture rooms	18
Watch rooms	20	Studios	18
Recreation rooms	18		
		Shops and showrooms:	
Flats, residences, and hostels:		Small	18
Living rooms	21	Large	18
Bedrooms	18	Department store	18
Bed-sitting rooms	21	Fitting rooms	21
Bathrooms	22	Store rooms	15
Lavatories and cloakrooms	18		
Service rooms	16	Sports pavilions:	
Staircases and corridors	16	Dressing rooms	21
Entrance halls and foyers	16		
Public rooms	21	Swimming baths:	
		Changing rooms	22
Gymnasia	16	Bath hall	26
Hospitals:		Warehouses:	
Corridors	16	Working and packing spaces	16
Offices	20	Storage space	13
Operating-theatre suite	18–21		
Stores	15		
Wards and patient areas	18		
Waiting rooms	18		

(As, for example, Table A2.2.) (Reproduced from Section A1 of *The CIBS Guide*, by permission of the Chartered Institution of Building Services.)

The internal environment

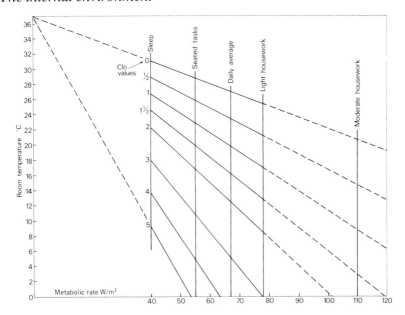

Fig. 3.2. Approximate temperatures for thermal neutrality for various weights of clothing (air velocity 0.1 m/s).[20] (BRE, Crown Copyright, HMSO.)

room surfaces is sufficiently accurately given by the room temperature of a globe thermometer, that the air movement is generally slight and that the humidity of the air has only a small effect on perceived warmth. (The last assumption is supported by McIntyre who, in investigating whether an increase in humidity would permit lowering air temperatures, concluded that the effect of humidity on warmth was so small that it might safely be neglected and thus that the air temperature could not be safely lowered.)[26] The metabolic rate in W/m² of body surface is a reflection of the energy expenditure of the occupants. 67 W/m² is an average daily value for men and women. Durnin & Passmore[27] give a complete range of energy expenditure in daily life.

The 'Clo values' refer to the amount of clothing worn. One Clo (0.155 m² °C/W) is the insulation afforded by a suit as normally worn.[28] Half a Clo corresponds to a skirt and blouse or trousers and shirt and one and a half Clo to a suit plus thicker underwear, a waistcoat and/or cardigan and thicker socks. A change to 1.5 Clo permits the acceptable room temperature to be decreased by almost 4 K at the average metabolic rate!

Lightweight indoor clothing could be developed so that higher insulation could be achieved without reverting to the bulky apparel of 50 years ago and Humphreys, although noting this, goes on to base his conclusions on the more common value of 1.0 Clo. He suggests that in winter the normal daytime temperature for a dwelling might be 19 °C with provision for raising or lowering the temperature by perhaps 4 K in a particular room for a particular occasion. To allow for different probable usages, normal settings of 21 °C for living rooms, 17 °C for kitchens and, say, 15 °C for passages and stairs are suggested. (In well-insulated houses these temperatures will tend to be closer to each other.)

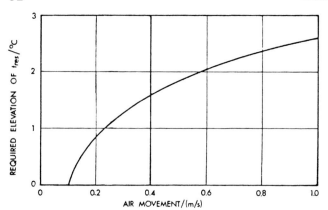

Fig. 3.3. Corrections to the resultant temperature to take account of air movement.[23] (Reproduced from Section A1 of the CIBS Guide, by permission of the Chartered Institution of Building Services.)

Bedrooms present a special case. With sufficient insulation a sleeping person can be warm enough in a bedroom at below 0 °C and it is common for many people in the UK to sleep at a temperature which is not far from ambient. However, Humphreys recommends 15 °C to allow in part for people reading in bed, for example with their shoulders exposed, and to permit moisture produced by sweating during sleep to evaporate from the bedding. Should the bedroom be used for study or as a sitting room it is suggested that provision be made for raising the temperature to 22 °C.

These results serve as an excellent point of departure for an actual design and indicate the subjective nature of some of the assumptions that must be made when recommending comfort temperatures.

It must also be remembered that specific local conditions can be of great importance. Sunshine streaming in a window, an open solid-fuel fire and an electric-bar fire are examples of radiant sources of heat which by heating the person permit lower air temperatures. The psychological effect of seeing the source of heat may also have a bearing on comfort. Finally, the needs of special groups must be considered and so, for example for elderly people, care must be taken to provide higher temperatures to compensate for the reduction in physical activity.

3.3.3 Room air movement, ventilation and relative humidity

A certain amount of air movement is desirable to avoid feelings of stuffiness and areas of stagnation. However, because of the cooling effect of air, its excessive movement should be avoided so that room occupants do not complain of draughts. One CIBS report[25] indicates that sedentary persons prefer air speeds of 0.1–0.2 m/s if the air is at ordinary room temperature — at lower temperatures only the lower part of the range is acceptable and, correspondingly, at higher temperatures (for example, discharge from warm-air grilles) the upper range or somewhat higher is satisfactory.

The CIBS guide recommends raising the resultant temperature when room air velocities are greater than 0.1 m/s. Fig. 3.3 shows the suggested corrections.

Local sources of discomfort may be dominant in determining whether

The internal environment

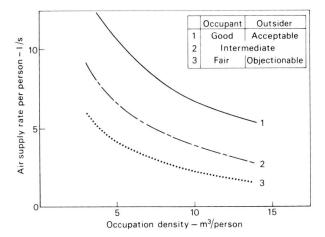

Fig. 3.4. Air supply required for odour removal – adults.[32] (BRE, Crown Copyright, HMSO.)

individuals consider themselves to be uncomfortable. Low-level draughts caused by infiltration and cold convective currents can be particularly troublesome[29] and due care should be taken to avoid them; if this is not done, higher internal temperatures (with correspondingly greater energy consumption) may be set by the occupants to achieve a similar overall standard of comfort.

Ventilation in buildings is required to provide a continuous supply of oxygen for breathing, to remove the products of respiration and occupation and to remove artificial contaminants produced by cooking, process work and so forth; in the latter category water vapour and heat may be included. Additionally, ventilation may be needed to provide air for a heating system.

The oxygen supply needed for breathing is about 0.009 g/(s p) (at an energy expenditure rate of 2500 kcal/d),[30] giving an air requirement of 0.03 l/(s p), but since this is significantly less than that required for other purposes it never serves as a basis for design. Similarly, to prevent the CO_2 produced by respiration from exceeding the maximum permissible level (assumed to be 0.5%), the fresh-air supply need only be 1.25 l/(s p).[31]

To keep the atmosphere fresh and reasonably free from odour, however, requires considerably more fresh air, as can be seen in Fig. 3.4. If we consider a closed living area of, say, 50 m³ with four people in it, the 'intermediate' situation, curve 2 of Fig. 3.4, calls for about 2 l/(s p) or roughly one-half an air change per hour.

The subjective impression of fresh air has important implications for energy conservation — the main reasons for opening windows in winter are to freshen the air, to avoid condensation and to avoid stuffiness.[33] Odours from heating appliances, particularly those burning paraffin, can also lead to a sensation of unpleasantness. (Additionally, large amounts of water vapour may be produced by such devices.)

Detailed air-supply rates are given by the CIBS[23] according to the type of space.

Less-commonly appreciated contaminants than those such as CO_2,

Table 3.7. *Range of open areas measured in six UK dwellings*[38]

Adventitious opening	Effective open area (cm²)	
	Before weatherstripping	After weatherstripping
Door	38–210	3–45
Window	6–110	3–14
Background (room with suspended floor)	52–150	52–150
Background (room with solid floor)	25–62	25–62

combustion products, tobacco fumes and body odours have been the subject of recent research. For example, formaldehyde (HCHO) is found in many products used by the building industry and studies have shown that in houses with low ventilation rates contaminant levels can be too high.[34] A second case is radon which is one of the products in the decay chain of uranium. Uranium is found naturally in most rocks, soils and common building materials. Radon and its daughter products emit alpha particles which, if inhaled, can damage lung tissue. Long-term low-level exposure levels have not been established but work is underway on the physiological effects of radon and on the modelling of alpha-particle activity levels at various air-change rates.[34] Brundrett has stated that 'present evidence suggests that to avoid danger the minimum ventilation rate in Britain should be 0.2 air changes an hour'.[35] Fuller, in the US, believes that 0.5 air changes per hour can keep indoor pollutants below critical levels.[34]

Natural ventilation is due to the pressure differences resulting from the action of the wind on open areas (wind effect) and the variation in the buoyancy of the air inside and outside (stack effect). Recent research indicates that the stack effect is important in two-storey buildings as well as taller ones. (Monitoring of a full-scale test structure[36] 4.9 m long by 2.4 m wide and 2.4 m high has shown that the stack-effect component of the infiltration is a function of the effective leakage area and its distribution, the degree to which the house is shielded from the wind, several geometric parameters and the square of the indoor–outdoor temperature difference.)

The open areas include purpose-provided openings such as vents, flues, chimneys and openable windows, cracks in and around room components (that is, doors and windows) and 'background leakage areas' which are left when the first two groups of openings are sealed – examples of these include joints between ceilings and walls and the porosity of room surfaces.[37] Table 3.7 shows the effective open areas in some UK houses and the importance of the background openings.

To make the ventilation rate less dependent on wind speed and external temperature it is necessary to reduce the total effective open area to a small value. It then becomes possible to use mechanical ventilation systems

The internal environment   55

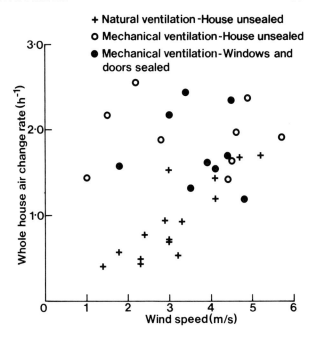

Fig. 3.5. Natural and mechanical ventilation rates in British Gas Corporation test house.[38]

(see Chapter 6) to control the air-change rate in the house because they are thus able to generate a relatively large pressure. If the total effective open area is not reduced a mechanical ventilation system simply results in a higher air-change rate as can be seen in Fig. 3.5.

To effectively reduce infiltration rates it is necessary not only to weatherstrip windows and doors but also to ensure that care is paid in the construction process to eliminate openings such as cracks between window frames and masonry walls, around light fittings and where pipes pass through the structure (services in general are a major source of infiltration), in floors and around skirting boards. This would reduce the total effective open area by attacking the problem of the background areas. Greater control on-site would be required and inspection procedures would have to be developed. By employing these measures and others such as external lobbies, higher glazing levels and the closing of chimneys when not in use, it should be possible to reduce the hourly air-change rate to a range of about one-half to one. A great deal of research is presently being undertaken to assess the cost-effectiveness of such measures and to determine how best to ventilate homes. The two principal options are mechanical ventilation in a very 'tight' house or natural ventilation with fewer adventitious openings but with purpose-provided openings which give some control over ventilation to deal with, for example, problems of condensation (see below) and stuffiness. At present, ventilation rates are not specified by the Building Regulations and ventilation is one of the least well-understood aspects of design. The situation may be similar to the lightweight v. heavyweight argument and research may prove both solutions to be suitable.

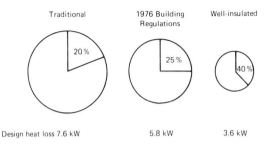

Fig. 3.6. Ventilation heat loss as a percentage of the total design heat loss (for a typical semidetached house of 90 m² floor area for three insulation levels; the specific fabric heat losses are: traditional, 1.4 W/(m³ K); Building Regulations 1.0 W/(m³ K); well-insulated house 0.5 W/(m³ K) (after Ref. [38]). (BRE, Crown Copyright, HMSO.)

In the past, little attention was paid to the heat loss that accompanies natural ventilation because it was a minor component of the total loss. However, as insulation standards increase it becomes more important, as can be seen in Fig. 3.6.

In traditional English houses, hourly air-change rates were approximately in the range of one-half to three (see Fig. 3.5).[38, 39] European construction, with its different techniques and sometimes higher standards, was able to reduce this considerably.

Fig. 3.7 shows some of the results from a study of ventilation rates in a group of modern Belgian houses conducted by the BRE and its Belgian equivalent (the Centre Scientifique et Technique de la Construction).[40] The Belgian houses were single storey with three bedrooms and were constructed of large prefabricated concrete panels with a central core of polystyrene. Joints between the panels were sealed. Windows were wooden framed and single glazed. The method of construction and the higher quality of the windows were the principal reasons for the relatively low air-change rates found.

Without a doubt, the initial effect of reducing infiltration is advantageous because less energy is wasted and because lowering the air speed in the room permits lower temperatures while maintaining the same degree of comfort.

Secondary effects may, however, be harmful. Care must be taken to ensure that the quality of the room air does not suffer, that sufficient air is supplied for heating appliances that are not room sealed and, as mentioned before, that local sources of discomfort are avoided. Nevrala & Etheridge[37] have discussed complaints of discomfort due to low-level draughts arising in certain situations such as a well-insulated room which retains a single-glazed window. They also point out that, with increasing insulation, heat emitters have become smaller and, in addition, the economic pressure to save capital cost has resulted in a curtailment of the former practice of oversizing heating-system components. The result is increasing sensitivity to the ventilation load. Care must be taken to allow for the 'background leakage areas' which become correspondingly more important as draught stripping is employed, and to take into account the magnitude and variability of the air-change rate which can arise from changes in wind speed and direction.

The greatest secondary effect is the danger of condensation and mould growth which even now are common problems in dwellings. As is well

The internal environment 57

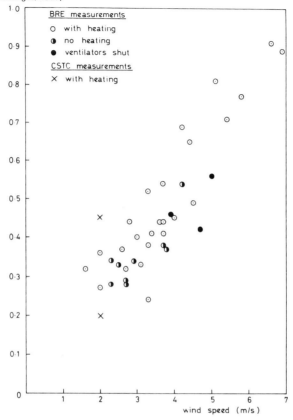

Fig. 3.7. Measurement of whole house natural ventilation rates.[40]

known, the amount of water vapour that air can hold is limited and at this limiting point the air is said to be saturated. The higher the temperature of the air the greater the amount of water vapour it can contain. The water vapour contributes to the vapour pressure of the air and the ratio of the vapour pressure of a given mixture of water vapour and air to the vapour pressure of a saturated mixture at the same temperature is called the relative humidity. Fig. A 3.2 shows how these variables are related. At a given temperature, as water vapour is added to the air the relative humidity will rise to the saturation point (100% rh) and any further vapour will be deposited as condensation. Alternatively, if the temperature of a given amount of air is lowered and the amount of water vapour remains constant the relative humidity rises because the cooler air can support less moisture. At a relative humidity of 100%, any further cooling causes water vapour to condense. For example, at point A in Fig. A 3.2 the dry-bulb temperature is 20 °C, the mixing ratio is 10.4 g/kg and the relative humidity is 70%. If this air is cooled to the dew-point temperature of about 14.7 °C where the relative humidity is 100%, point B, it can hold no more moisture.

The latter causes problems in buildings when, for example, warm, moist air reaches a cold single-glazed window or even the wall of an unheated bedroom. An important point to note is that condensation, particularly in dwellings, does not necessarily occur in the room where the water vapour is produced – any cooler surface in the house may provide a suitable site. Condensation can also occur in the building fabric since nearly all building materials are permeable to water vapour. The BRE has developed a method[41] of assessing the risk of this interstitial condensation which is outlined in Appendix 3, and the problem is further discussed in Section 3.8.

Mould growth can occur when the average relative humidity exceeds 70% for extended periods.[42] This situation may arise, depending on the standard of insulation, particularly in bedrooms which are normally unheated or only intermittently heated.

The figure of 70% is also an approximate upper limit for human comfort, with 40% being a lower bound. Low relative humidities can lead to electrostatic shocks, while physiological factors such as dry throats and cracking skin have been found to be related to low absolute water-vapour pressures.[23]

The effect of lower ventilation rates is to increase the moisture content of the air as can be seen in the equation

$$\bar{g}_i = \bar{g}_o + \frac{G}{\rho v}$$

where

G = mean rate of moisture emission;
ρ = density of air;
v = volumetric rate of air exchange;
$\bar{g}_i, \bar{g}_o$ = mean inside and outside moisture contents (mass of water per unit mass of dry air).

Typical moisture emission rates are shown in Table 3.8.

Cooking and drying clothes are the major moisture producers shown in the table. One way of combatting condensation and mould growth is to isolate moisture-producing activities by, for example, washing and drying clothes in a utility room separated from the rest of the house. Other suggestions include an extract fan in the kitchen (this may pose problems if the kitchen contains a boiler whose flue is not sealed from the room),[42] and increased insulation and additional heating in bedrooms.[43] In general, there is no ideal solution to the problem, short of moving to a more hospitable climate, and so care must be taken to allow some flexibility in the design by, for example, permitting higher ventilation rates than would be desirable from the point of view of energy conservation alone. Table 3.9 summarizes the advice to designers of one comprehensive guide to condensation in dwellings.

If there is no ideal solution there is at least a practical radical one which is a combination of the ideas discussed above. It starts from a reasonable position – that people are people – then moves to one less easily defensible – that people won't change and will therefore continue, for example, boiling vegetables until the kitchen is like a steam bath. The idea is that

Table 3.8. *Moisture emission rates*[42]

Sources where moisture emission is low	Moisture emission per day (kg)	Comments
5 persons asleep for 8 h	1.5	Typical moisture emission rates
2 persons active for 16 h	1.7	
Cooking	3.0	3 h cooking by gas cooker
Bathing, dish washing, etc.	1.0	Estimated
Daily total	7.2	
Additional sources where moisture emission is high		
Washing clothes	0.5	Estimated
Drying clothes	5.0	Measured, for 3 kg dry weight of clothes, spun dry
Paraffin heater during evening	1.7	1.7 l paraffin, e.g. 4 kW for 5 h
Daily total	14.4	

ventilation rates should be kept high, as in the past, for personal comfort, to avoid condensation and to ensure that the building fabric is free of danger. To avoid a profligate use of energy not only would a mechanical ventilation system be used, but heat recovery on the extract air would be incorporated (see Chapter 6). Such systems have been recommended by many, tried by some and are commercially available for the domestic scale. The only problems are the cost of sealing the house (reducing the total effective open area), as we have seen it is necessary to do, and the cost of the equipment.

3.4 Size and type

Reducing the size of a building to a minimum will lower the energy costs of construction, as well as those of heating. Given the human need for space, which is albeit culturally conditioned, this tends to be a fairly unpopular solution to energy conservation; however, where the density of occupation (volume per person) is naturally high as in, say, schools compared to homes, the scope for economy may be greater.

A change in building type may prove more acceptable. The appearance of grouped condominium housing in the US may hearken a development towards the terraced housing of the UK.

The relative advantages of terraced homes, which of course have fewer walls exposed to the outside air, are evident in Fig. 3.8 (the absolute seasonal heat requirements shown have decreased in all cases due to the introduction of higher insulation standards). Work at the BRE has given similar results.[45]

For the Basildon housing discussed previously, the effect of varying the terrace length was considered and it was concluded that very short terraces

Table 3.9. *Precautions for avoiding condensation*[44]

		Key aims
1		
	1.1	Assure reasonable bedroom temperatures. Target: maintain a temperature of at least 10°C at all times
	1.2	Provide fan extraction to kitchens
	1.3	Provide ventilated clothes-drying space
2		Make sure structural U-value is adequate in all places, that the insulation is positioned to the best advantage and is not broken by 'cold bridges'
3		If vapour barriers are necessary, ensure that they are properly positioned
4		If interstitial condensation is likely to occur, make certain that it can escape and that vulnerable materials are protected
5		Insulate cold-water pipes
6		Make certain that flat- or pitched-roof construction has been properly considered in the light of local climatic conditions
7		Consider carefully the method of paying for heating in relation to the occupiers' probable economic circumstances
8		Ensure that the heating system has adequate controls so that it will be used to the best advantage in combining efficiency and economy
9		Ensure that secondary air supply to warm-air heating is correctly positioned to avoid recirculation of moist air
10		Avoid dangerous 'dead-end' positions where heating or ventilation may be inadequate
11		In window design allow for easily reached and easily controlled ventilation without rain penetration
12		Ventilate larders to the outside
13		Ventilate other cupboards as necessary to the interior
14		Make arrangements for occupiers to be issued with simple advice on the avoidance of condensation, especially on the method of operating the heating

increase the overall energy demand by up to 10% (see Fig. 3.9). However, increasing the terrace length above eight or so houses gave little overall advantage.

3.5 Form and orientation

Dear to the heart of all designers is the question of form. Fig. 3.10 shows 11-year-old Judith Cornick's winning entry for a rotating solar house.

Fortunately or unfortunately, according to one's point of view, the building industry is unlikely to adapt to such novel structures in the near

Form and orientation

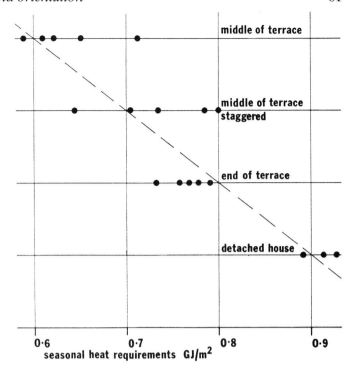

Fig. 3.8. Analysis of comparative space heating requirements for typical terraced and detached houses. (The dots indicate specific results; the dashed line shows the trend of energy demand.)[12]

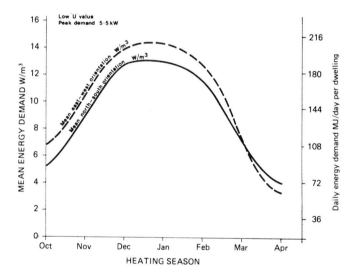

Fig. 3.9. Effect of varying terrace length.[46]

future (Buckminster Fuller's Dome and Alex Pike's Autarkic House encountered similar problems) and so we are forced to consider more-conventional solutions.

For large buildings, the constraints imposed by the site and intended use may be very important and both heating and cooling loads will have to be considered. The examples selected by Kasabov,[48] ranging from three-storey rectangles to sprawling structures with courtyards, indicate

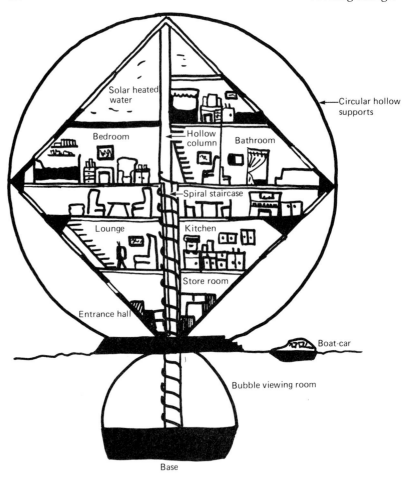

Fig. 3.10. A rotating solar house.[47] (Courtesy of J. Cornick.)

that a wide variety of forms are currently claiming, perhaps justifiably, to be energy efficient. There is a strong, easily understood hope on the part of architects that considerations of energy use will not greatly restrict the forms available to designers and hence the scope for creativity. It is a hope we share but, at present, insufficient data are available to decide whether it is one that is justified. Hawkes & MacCormac, in arguing for office buildings with interior courtyards, have presented theoretical calculations showing the possibility of significantly reduced energy demand.[49]

For fairly simple forms, computer programs are available, which will evaluate the heating and cooling requirements.[50, 51] Results tend to indicate that lower buildings of more compact shape are more efficient, that is, have lower running costs, but one must be wary of the assumptions made when studying such analyses. In particular, the results can be sensitive to the amount and distribution of the glazing.

Most studies have found that if passive solar gain is not accounted for, the aspect ratio (the ratio of the plan length to breadth of the rectangle that has the same area and perimeter of the proposed form of the building) has relatively little effect on energy consumption (see Fig. 3.11). Nonethe-

Form and orientation

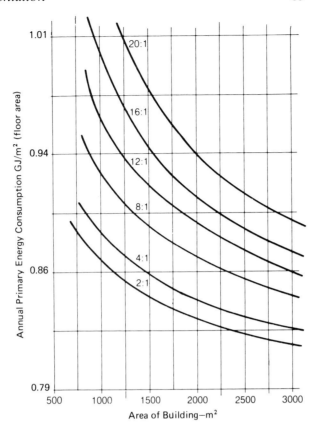

Fig. 3.11. Variation of primary energy consumption for heating and lighting, with aspect ratio and building area.[52] (No allowance has been made for solar gains or for orientation.) (Crown Copyright.)

less, for small buildings a first guideline should be to avoid the temptation to use complex forms since they will tend to have an increased surface area and may result in self-shadowing.

If solar gains are considered, as is preferable, we must first consider the energy balance at windows. Fig. 3.12 provides the necessary data.

With south-facing glazing and no curtains, over the heating season the energy balance is negative but, as Burberry has shown by curtaining such windows, there is neither a loss nor gain (Fig. 3.13). By going to south-facing double glazing, even without curtains, the energy balance over the heating season becomes positive and if an insulating shutter is drawn at night, at no time during the year is the energy balance negative. The advantage of south-facing glazing compared to other orientations is also evident from Fig. 3.12.

It might seem then that there is a clear-cut case for all glass south-facing walls but unfortunately this is not so. Designers agree that, whenever possible, glazing should be to the south rather than other orientations but the amount of glazing is a point of contention, largely because of disagreement as to how much of the incoming radiation is useful. This is related to the thermal performance of the structure, the heating system, the desired internal temperature and so forth and thus will require further study in monitored buildings. At Basildon (see Fig. 2.7) the (single-glazed) windows

Fig. 3.12. Energy balance for one square metre of glazing in the Bracknell area (assumed internal temperature: 18 °C).[53] (a) Daily average conduction loss. (b) Daily average solar heat gain through uncurtained single glazing. (c) Daily energy balance through south-facing glazing. (d) Daily energy balance through west-facing glazing. (e) Daily energy balance through north-facing glazing.

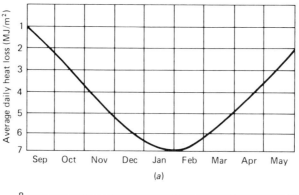

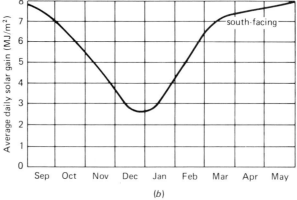

occupied 40% of the south wall area,[14] and at Pennyland, Milton Keynes, the comparable figure was 35%.[54]

In the meantime, if we assume conservatively that a curtained south-facing double-glazed window experiences no net heat loss during the heating season, the question of the optimum length to breadth ratio changes and elongated south-facing structures show savings. Recent developments of terraced houses orientated to the south and with glazing concentrated on the south, when viewed as units are very energy-efficient forms.

As we have seen (Chapter 2) with an actual site, departures from an orientation of due south are often required by planning and aesthetic considerations. For terraced houses, computer studies for Basildon indicated that deviations from the south of up to 22.5° had no significant effect on the energy demand;[46] similar work for Pennyland found that deviations from the south of up to 30° reduced the potential energy savings due to passive solar gain by only 2–5%.[54]

For semidetached housing with double glazing, a study of the effect of orientation has shown that groups with the long axis east–west and glazing concentrated on the south are likely to receive approximately 6.1 GJ of solar gain (4.7 GJ from the glazing; 1.4 GJ through the structure) during the period from October to March, compared to about 4.7 GJ (3.2 GJ through the glazing; 1.5 GJ through the structure) for

Form and orientation

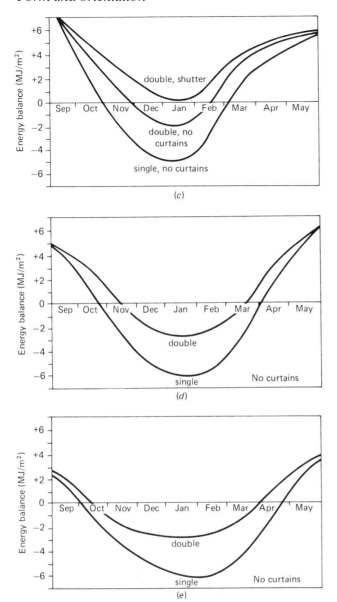

groups with the long axis running north–south and glazing concentrated on the east and west walls.[5]

For individual houses, a slight variation in form, which incorporates large south-facing glazing, is the trapezoidal plan of the Ouroborus House[55] in Minnesota (Fig. 3.14). To reduce the heat loss the north wall is shorter than the south and is bermed.

As to height, compact two- and three-storey buildings tend to be efficient because of their reduced surface area in relation to volume. Bungalows (one-storey detached dwellings) are particularly wasteful of energy.

Fig. 3.13. Net heat balance through one square metre of south-facing single glazing (assumed internal temperature: 17.1 °C; windows are curtained at night).[12]

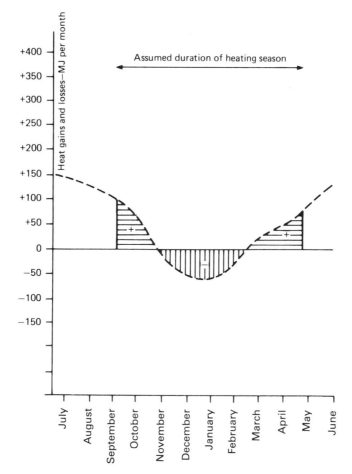

Fig. 3.14. Ouroborus House in Minnesota.

External and internal layout

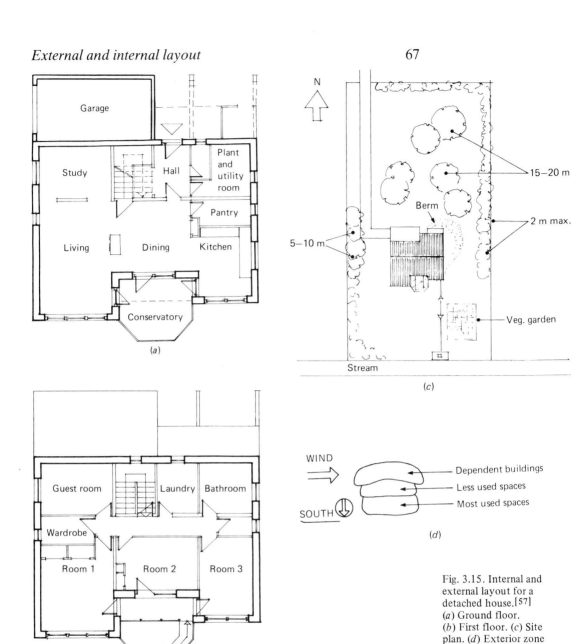

Fig. 3.15. Internal and external layout for a detached house.[57] (a) Ground floor. (b) First floor. (c) Site plan. (d) Exterior zone concept.

For larger buildings Markus & Morris[56] provide a very useful discussion and summary of the effect of form on energy consumption.

3.6 External and internal layout

Wherever possible, non-heated spaces should be used to shelter heated ones. Fig. 3.15 shows how a garage and berm form part of an exterior zone which is organized on a north–south axis.

If possible, the structure should be arranged to avoid self-shadowing

Fig. 3.16. Solar heated flats at Lewisham, London.[15] (Courtesy of Max Fordham and Partners.)

where passive or active solar gain may be used. The advantage of increased architectural interest which results from more-complex forms must be weighed against increased heat loss and potential loss of solar energy. South-facing bay windows, for example, will have a less favourable energy balance than a plane window because of the increased surface area and because all of the glazing is not at an optimum angle for solar gain. Additionally, by creating pockets of air pressure, bay windows and similar features can cause greater infiltration losses.[58]

Of course, on-site a compromise is inevitable. Fig. 3.16 shows the arrangement of the Lewisham Flats where the site lay between a row of terraced houses and a park. The site sloped to the south but faced north over the park, thus the new buildings were bound to overlook the terraced houses and would potentially block off their view of the park. To overcome this difficulty (and to receive planning permission) the development consists of a series of blocks with gaps in between to maintain visual access to the park. The south-facing roofs of the flats are much higher than those of the terraced houses and so are free from this form of overshadowing, but some shading will be caused by adjacent blocks and by the extended walls required by building regulations to retard the spread of fire.

Fig. 3.15 illustrates many of the guidelines for internal layout. Obviously, these will vary with house type, client preferences, choice of heating system and so forth but it should be possible to accommodate some energy-conserving measures in any design. A principal objective, well understood by the designers and occupants of nineteenth-century terraced housing, is to divide the internal plan into areas which can be treated dif-

External and internal layout   69

Table 3.10. *Temperature distribution upstairs and downstairs in a highly insulated semidetached house*[60]

	Downstairs (lounge, diner/kitchen)		Upstairs (3 bedrooms)	
	Energy consumption (MJ/d)	Average temperature^a (°C)	Energy consumption (MJ/a)	Average temperature^a (°C)
Internal doors closed	59.4	19	10.8	15
Internal doors open	93.5	19	3.6	16.5

^{a} Control temperatures were 19 °C downstairs and 15 °C upstairs. The house was unoccupied and had an open staircase.

ferently according to need. The north part of the plan consists principally of task-orientated spaces which are used relatively infrequently, while to the south are the major habitable rooms where the sun provides essential warmth and light. In general, services are concentrated in the north-east corner of the house to reduce pipe runs and to facilitate future installation of, for example, a heat pump to recover heat from the bath water.

Adequate lighting in the house is provided by the mainly south-facing glazing. The need for electrical lighting should be minimized and, as a rule, a small heat loss should be accepted in preference to a requirement for electrical lighting because of the primary energy demand of electricity (see Chapter 6). For this reason, the stairwell is lighted during the day by a small north-facing window. Where a choice exists, windows in walls are preferable to skylights because the thermal coefficient is lower (single glazing in a roof has a U-value of about 6.6 W/(m^2 K) compared to 5.6 W/(m^2 K) for glazing in a wall (see Table A 2.7) and, additionally, the temperature will be higher at the roof than the wall); any skylight installed should be provided with a shutter to cover it at night. (In commercial situations, however, where a certain level of illumination is required continuously, the question of horizontal v. vertical daylighting is more complex. It will depend on the illumination required and tables of frequency of daylight levels must be consulted.)[59]

The staircase is sealed to separate the bedroom temperature zone from the ground floor and to reduce heat losses due to the 'stack effect'. Sealing the staircase may also help to prevent warm, moist air rising from the kitchen–dining areas below to the cooler areas upstairs and causing condensation. Provision of a mechanical extract fan in the kitchen, perhaps operated by a relative-humidity sensor, is a useful precaution and will also remove objectionable odours. (Mechanical ventilation of bathrooms and toilets should also be considered.)

An alternative approach which is sometimes adopted is to leave the staircase open and use the ground floor to heat the first floor. If this is done, experimental evidence presented in Table 3.10 has shown that a significant saving in energy consumption (70.2 MJ/day compared to

97.1 MJ/day) may be achieved by closing internal doors and accepting a somewhat lower temperature in the upstairs bedrooms.

We would recommend that thermal zoning be used vertically and horizontally when possible.

Free heat from the sun can be used more effectively than that from the kitchen. The essence of passive solar design is to use the winter heat gain and avoid overheating which is perhaps most likely to occur in the spring and autumn but, in some instances, may also occur in winter[61] (see Chapter 4). Utilization of passive solar gain is increased when it can be redistributed among a number of spaces from areas where there is an excess. Cross-currents should be avoided except when specifically needed to prevent overheating. If for a particular design it seems likely that overheating may occur, it is best to plan the interior spaces so that warm air can reach, and then exit at, the top of the building.

The entrance hall acts as a draught lobby, a feature which should always be included. In designs where the entrance is from the south, the lobby can be a small glazed conservatory and thus increase the passive solar gain.

A woodburning stove for amenity purposes and to make use of any wood produced on site is provided and located centrally to take advantage of heat losses. (The flue must be equipped with a well-fitting damper to prevent high heat losses when the fireplace is not in use.) Linen cupboards with heaters and hot-water storage devices should also be located away from external walls to make the best use of their heat.

Fig. 3.17 shows the application of many of these ideas to a small terraced house, Ambient Energy Design House 1, which was designed by the authors and Lucy Krall (see also Chapters 6 and 12).

A point of note is the kitchen position. In a small house, options are restricted and in the design of Fig. 3.17, the north-east corner seemed the best place for the kitchen even though it meant that the window looked out on the street. It has been argued that kitchens should not be positioned on the south side in houses designed for passive solar gain as there is a risk of overheating caused by the combination of solar gains and heat generated by cooking, but this would seem to vary too greatly with specific designs to be accepted as a rule. As to whether the kitchen and dining room should be separated, at Basildon it was thought that if the kitchen had a mechanical extract there was little difference whether they were separated or combined, but if there were no such device there was probably a slight advantage in combining the two since the greater amount of air would aid in absorbing any moisture produced.[46]

3.7 Construction – general

For a number of years it had been recognized that UK insulation standards were dreadfully low compared, in particular, to northern European standards.[63] Fortunately, the recent trend has been towards more and more insulation, as can be seen in Table 3.11 which compares the Building Regulations requirements with a number of other recommendations.

GARDENS are longer than wide in the traditional manner which allows the family to enjoy a lawn and flowers close to the house while still having space to grow vegetables at the far end.

GARDEN CONSERVATORY/LOBBY. To reduce ventilation heat loss while still providing garden access and space to keep boots and tools.

WALL CONSTRUCTION. 250 brick-block cavity construction with cavity fill plus 100 mm polyurethane insulation fixed internally.
$U = 0.2$ W/m² °C

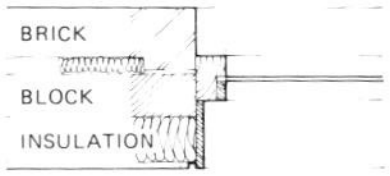

BRICK
BLOCK
INSULATION

LIVING AREA. Closing these doors on certain days when heating is required, the system permits the main living area to be heated exclusively by solar gain.

THE BOILER is housed in a separate space with access from the hall and with a grill to the living area to allow the recovery of any 'waste' running heat.

STORE e.g. FOR FUEL. Sealed off thermally by floor to ceiling partition – the lower part consists of insulating blocks which can be removed for access. Where solid fuel boilers are to be used, the floor slopes at 30° to facilitate movement of coal. (Future solid fuel deliveries from trucks through chutes could make possible completely automatic boiler feeding).

UNATAP to reduce water and energy consumption.

THERMAL ROCK STORE. This is 2 m³ constructed of blockwork and filled with fist sized rocks.

ENTRANCE LOBBY. To remove heat loss through ventilation. Both doors are ⅓ single glazed.

PARKING. One space per house is allowed.

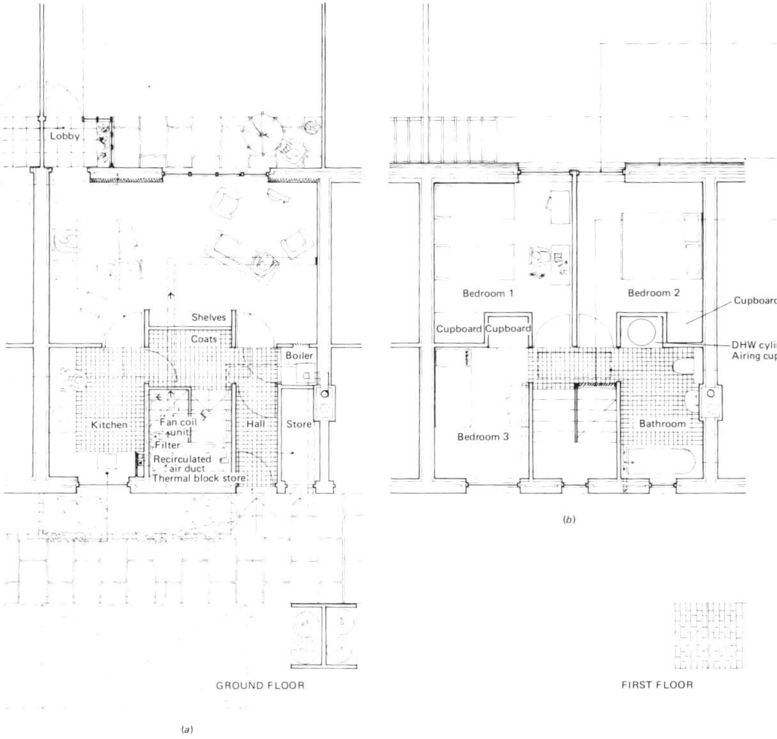

WINDOWS South facing windows are larger than those on the north side. The north windows are double glazed and have heat reflecting glass internally to reduce heat loss. All windows have timber frames to reduce heat transmission. The two south bedroom windows are close together to reduce the construction cost of the solar air heater on this part of the wall.

SOLAR COLLECTOR is fixed by timber posts onto the southern upper part of the facade as well as on the south facing roof. (See section).

HEATING. Upstairs all rooms are warmed from one air outlet at the top of the stairs. All doors upstairs have transfer grills to permit the warm air to circulate. The warm air is forced through a duct in the kitchen then runs in a duct between the floor joists and under the top step to emerge in the centre of the house from where the air can circulate freely and is extracted from the bathroom.

ACCESS to roof space for maintenance.

SHOWER fitted to permit reduction of water and energy demands.

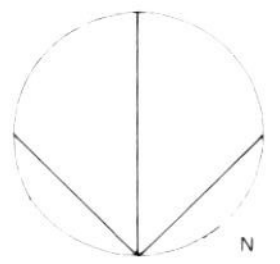

Fig. 3.17. Ambient Energy Design House.[62] (a) Ground-floor plan. (b) First-floor plan.

Table 3.11. U-values ($W/(m^2 K)$) for dwellings

Source	Ground floor	External wall[a]	Roof	Glazing
(1) 1976 Building Regulations[64]	–	1.0	0.6	single[b]
(2) 1981 Building Regulations[2]	–[c]	0.6	0.35	single[d]
(3) DOE better insulated housing[65]	–	0.5	0.3	single–double
(4) British Gas well-insulated house[9]	0.45	0.4	0.3	double
(5) BRS low-energy houses (terrace)[66]	–	0.35	0.35	double optional
(6) Electricity Council Research Centre target[67]	0.3	0.3	0.3	double

[a] Excluding glazing.
[b] Walls plus windows were required to give an average U-value not exceeding 1.8 $W/(m^2 K)$.
[c] If the floor is between a dwelling and the external air or between a dwelling and a ventilated space the maximum U-value is 0.6 $W/(m^2 K)$.
[d] Single-glazed window openings can be up to 12% of the perimeter wall area. If double or triple glazing is used the 12% may be similarly increased.

There is thus an encouraging trend towards substantially reducing heat losses. New buildings, other than dwellings, at present have a maximum U-value of 0.6 $W/(m^2 K)$ for walls, floors and roofs.[68] In the relatively near future we are likely to see all buildings, including dwellings, with U-values limited to 0.2–0.3 $W/(m^2 K)$. Double and triple glazing will become common. In Scandinavia, which is admittedly considerably colder than Britain, such insulation levels are often already attained.

One reason why insulation will be increased in new houses is that it is a very easy thing to do and one which the occupants should not need to worry about for the life of the house. Other ways of saving energy, such as heat-recovery devices, are often mechanical and fraught with the usual problems of maintenance, replacement of parts, and so forth. The difficulties of increasing the insulation level are mainly constructional and the BRE has recently started a study of how the highly insulated houses of the future will be built.[69] Part of the intention is to compare design details shown on the drawings and those achieved on-site and to investigate the additional work required to install the insulation.

A detail of some concern in the design of low-energy buildings is the avoidance of cold bridges. The term refers to the rapid passage of heat from inside to out if certain parts of the construction have a much lower thermal resistance than adjacent parts, as can occur when a metal framework is exposed both inside and outside. An example we are familiar with was the steel windmill tower of the Autarkic House, which passed through a highly insulated roof. The solution adopted was to adequately insulate all interior parts of the tower with sprayed polyurethane foam.

Construction – general   73

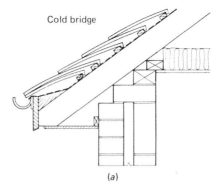

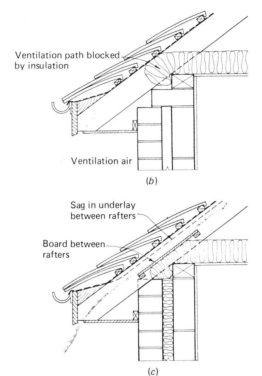

Fig. 3.18. Incorrect and correct roof construction techniques.[70] (*a*) Cold bridge. (*b*) Inadequate ventilation. (*c*) Acceptable: continuity of insulation at ceiling to wall junction while retaining ventilation path.

Another example is given in Fig. 3.18(*a*) which shows how a cold bridge may be formed in certain forms of roof construction. Fig. 3.18(*b*) indicates that other forms avoid cold bridges but instead block the ventilation path; and Fig. 3.18(*c*) shows one way of avoiding both problems.

A wide variety of insulating materials is now available to the designer.[71, 72] Plasterboard-backed polystyrene slabs (available with a vapour barrier), sprayed ureaformaldehyde foams, mineral-fibre mats and cellular glass blocks are but some of the materials on the market. The correct choice will depend principally on the application and cost but consideration should also be given to how the insulation level could be increased in the future since some materials lend themselves better to this than others.

Fig. 3.19. Penetration of formaldehyde vapour into a building.[73]

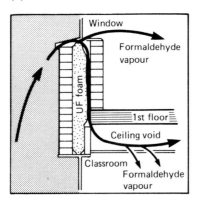

Safety of insulation materials is a raging issue at the moment and is likely to remain one for some time. Greater choice of insulants has not been without its drawbacks — manufacturers have rushed to put new products on the market before failsafe ways of incorporating them in buildings have been developed. To take but one notorious case, the use of ureaformaldehyde foam cavity insulation is being seriously questioned in the UK because it has proved difficult, if not impossible, to ensure that the cavity is completely sealed on top, bottom and sides to prevent toxic formaldehyde gas from entering the building. Fig. 3.19 shows a problem case in a school where gaps under window sills allowed air to be blown into the cavity pushing formaldehyde gas through openings in the inner leaf of the wall and into a classroom. Numerous complaints of headaches and respiratory problems resulted and the school has been closed pending remedial action.

Fire safety is also being scrutinized. The increasing number of plastic products in buildings and the large volumes of plastic insulants which are often used have given some cause for concern[74, 75, 76] and due caution should be exercised when using such insulants. To cite one example, the Agrément Board has insisted that where the cavity of a house is insulated with polystyrene beads the top of the cavity must be sealed with a non-combustible material to prevent the beads blowing out and into the eaves where they could constitute a fire hazard.[71]

Most insulation materials are, however, safe (government testing procedures are likely to become more stringent in the future) and the question the designer faces is how much of a material to use. In determining the insulation level it must be remembered that the law of diminishing returns applies since the thermal transmittance of a layer of material is given by the thermal conductivity of the material divided by its thickness (see Appendix 2). Thus to use an extreme example, if a building with 600 m^2 of surface area is relatively uninsulated (U-value of 4.0 W/(m^2 K)) and has a fabric heat loss of 48 kW, if 100 mm of expanded polystyrene is added to the walls the heat loss falls dramatically to about 3.8 kW. If a second 100 mm of insulation is added, the heat loss falls by only another 1.8 to 2.0 kW.

Increasing insulation also has a hidden cost at times. If, for example, one insulates an existing wall internally, living space is lost. On the other hand, with new houses especially (but in certain cases, such as loft insulation, in existing ones), increased insulation may affect only the material costs with the increase in labour charge being minimal and in such cases greater insulation can be very worthwhile.

A number of methods have been proposed to determine the economical thickness and cost-effectiveness of insulation and the interested reader is referred to the original work.[45, 77, 78, 79, 80] It is noteworthy that the intellectual acceptance of higher insulation levels (as evidenced by the figures in Table 3.11) is such that in many homes and buildings, more insulation is being used than is either required by law or calculated to be cost-effective. It would seem that many people feel that the wolf is at the door, that past recommendations have not been sufficiently stringent and so have squandered energy unnecessarily. If increased insulation is to be provided as insurance for the future, attention should be given first to the floor, walls and flat roofs since these are the building components which are not easily altered later.

A major design decision is the form of construction. Interest in timber-frame construction, for which techniques are well developed, is growing, in part because it lends itself to high levels of insulation. Other advantages claimed are facility in obtaining continuity of the insulating envelope and reduction of cold bridging.[81] Air infiltration can be lower and it may be easier to include an internal vapour barrier than in a masonry construction. It must be said, however, that the internal vapour barrier is also more necessary in timber-frame construction to guard against rotting of the wood.

Nonetheless, traditional masonry construction is likely to remain dominant in the UK for some time. With care it can incorporate relatively high levels of insulation and, additionally, is more suitable if thermal storage capacity is an important consideration. Traditional masonry construction, improved by insulation and vapour barriers, and timber-frame construction are thus the two most likely possibilities for houses in the near future and examples of both are given below.

Proper selection of materials and building type can also save energy.[82, 83] In the UK, typical two-storey houses have been found to require between 95 and 180 GJ of primary energy (for major building components only); for large blocks of flats, constructed in reinforced concrete the primary energy requirement for construction has been estimated as 230–265 GJ per flat.[84] Thus, in a house built to present standards, the primary energy required for space heating alone over a few years equals the primary energy required for construction. However, as the 'running' energy demands fall, we may expect more attention to be paid to construction energy costs. One sign in this direction is the use of recycled materials in building construction. In the US at least one house has been made almost entirely in this way with secondhand newspapers being transformed into insulation and wall coverings, old glass becoming bricks and insulation and recycled aluminium used for roof trusses and gutters.

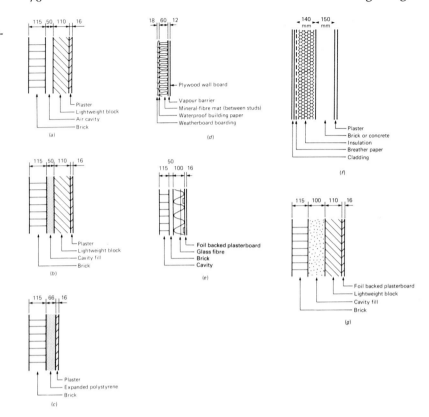

Fig. 3.20. Wall constructions.[2, 9, 78] Contemporary, $U = 0.89$ W/(m² K).
(b) Heavyweight insulated, $U = 0.40$ W/(m² K).
(c) Lightweight insulated, $U = 0.40$ W/(m² K).
(d) Timber-frame wall, $U = 0.44$ W/(m² K).
(e) Timber-frame wall – brick veneer, $U = 0.28$ W/(m² K).
(f) Highly insulated, $U \approx 0.3$ W/(m² K).
(g) Highly insulated, $U \approx 0.25$ W/(m² K).

3.8 Foundations and walls

In the US, in colder areas, consideration is being given to insulating foundations. In the UK, less concern has been voiced about such losses but at least one experiment with cellular glass insulation of the edge of the floor slab (basements are much less common in Europe than in the US) is underway.[65] In the Peterborough Houses discussed in Chapter 12, insulation is used under the whole floor. In the future such protection may become common.

Fig. 3.20 shows a number of wall constructions of low U-value and Fig. 3.21 gives two detailed sections used in actual houses (Appendix 2 explains how the U-values may be calculated).

Many of the designs incorporate a vapour check to reduce the possibility of interstitial condensation. During the winter the higher vapour pressure inside buildings causes moisture to move towards the outside (where the vapour pressure is lower). In traditional masonry construction it seems probable that the problem has been lessened by the ability of the masonry to harmlessly soak up a large amount of condensate until it can distil outwards or evaporate from the surface during favourable periods.[42] However, now, with more-heavily insulated buildings and new materials, it is not uncommon for the dew point of the moisture laden outgoing air to be reached within the building element. At this point inter-

Foundations and walls

stitial condensation occurs and this can lead to structural failure (in timber-framed homes the increase in moisture content could allow fungal decay)[85] and greatly reduced thermal performance (for example, if the insulation becomes waterlogged). (See Appendix 3 for the calculation procedure.)

The vapour check must be placed on the warm side of the construction. Commonly this is achieved by using aluminium foil-backed plasterboard or heavy-gauge polythene sheet just under the interior wall finish. Continuity of the protection is important but difficult to achieve in practice and because of this the term 'vapour check', rather than 'vapour barrier', is being used more and more to reflect what happens in actual construction — a vapour check, even if not completely effective, is still useful.[44] With aluminium foil-backed plasterboard it is desirable to seal the joints with a polythene strip (which will have to be put in place, for example, by nailing to battens, before the plasterboard is fixed). With polythene sheet, seams may be caulked or taped and then clamped between the various construction components. Achieving a continuous vapour seal around switch drops, socket outlets, flues, pipework and cables is also very important but is, needless to say, a meticulous and time-consuming task. Because of the likelihood of some moisture penetration into the building fabric, ventilation on the cold side of the insulation will be allowed for, in a good design, if possible, particularly behind areas where achieving a seal is arduous. One construction technique that bears consideration is to incorporate a 100 mm cavity in a brick–block wall but only insulate half of it, for example by attaching polystyrene slabs to the inner face of the block. While this raises the U-value (to about 0.45 W/(m^2 K)) it guards against condensation problems. In almost all forms of construction, given the difficulty of achieving an adequate vapour seal, it is important that the exterior surface be permeable to water vapour to avoid moisture accumulation within the construction.

Fig. 3.22 shows an example of a polythene ceiling vapour barrier stapled to prefabricated roof trusses. Joints were taped to effect a continuous seal and the polythene was taped to the trap door in the ceiling. Fig. 3.23 shows the opening for a switch drop in the same ceiling. With conventionally taped joints it proved very difficult to produce an effective seal at such points.

The correct position for insulation is related to the heating and occupancy patterns. Where intermittent heating is possible, insulation on the inner face of the wall will tend to warm up quickly when the room is heated and will thus give a potential reduction in energy use, provided the heating-system controls are suitably selected. For continuous heating, the position of the insulation is less significant.[12] One computer simulation study of internal and external insulation found that internally applied insulation was marginally more effective than external.[86] For example, for a semidetached house with 220 mm solid-brick external walls the internal use of 65 mm thick thermal board (12.7 mm plasterboard– 52mm polystyrene) to give a U-value of 0.51 W/(m^2 K) reduced the seasonal space heating requirement by 31%. Polystyrene insulation, 55 mm

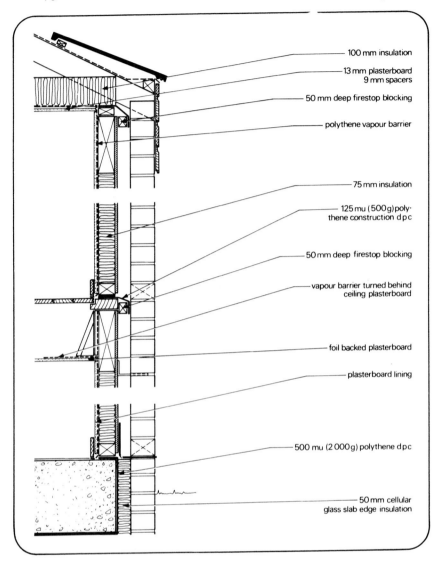

Fig. 3.21. Detailed sections through two external walls.[65, 14] (a) DOE HDD Better Insulated Housing (U-values in W/(m² K): wall 0.5, roof 0.3). (b) Basildon Housing Estate (U-values in W/(m² K): wall and roof, 0.55. (a) Crown Copyright, (b) Courtesy of Ahrends, Burton & Koralek.)

Foundations and walls

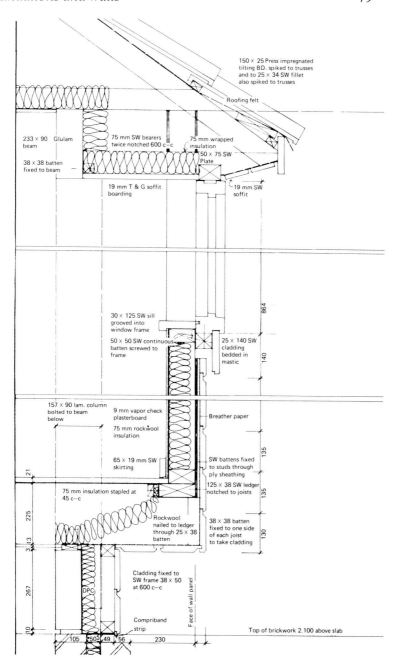

Fig. 3.22. Installation of a polythene ceiling vapour barrier.[65] (Courtesy of PSA.)

Fig. 3.23. Opening for a switch drop in a ceiling with a vapour barrier.[65] (Courtesy of PSA.)

thick with a 13 mm rendered finish, applied externally to give the same U-value gave a corresponding reduction of 30%. Our opinion is that the figures are too close to draw any conclusion as to whether internal or external insulation is preferable from an energy point of view. The same study confirmed that the response to heat inputs was much quicker with internally applied insulation than external.

Because heating systems and occupancy patterns will change during the lifetime of the building and because few data are available on actual energy consumption in varying building types, designers must make an informed

Foundations and walls

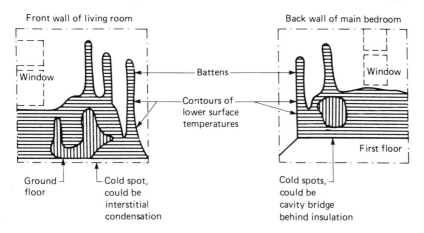

Fig. 3.24. Sketches from infra-red thermographs of Abertridwr Project Houses.[89]

guess as to the best solution. Some consolation for the lack of more-positive advice may be found in the growing awareness that, as insulation levels rise, variation in thermal mass, occupancy patterns, heating systems and position of insulation become, in a sense, less significant. Construction methods for new houses tend to favour insulation inside or in the cavity. If continuous heating is desired — and one argument for it has been the avoidance of serious condensation[87] — an inner cavity wall of high thermal capacity will give a structure that will warm up and cool down slowly. For existing homes it is possible to insulate either inside or outside. One approach to external insulation, which has the advantage of not reducing the living space of the house, is that used in the Granada House.[88] Semirigid slabs of mineral wool 100 mm thick were fixed to existing solid-brick walls and then covered with redwood weatherboarding. Another possibility (simulated in the insulation study described above) is fixing polystyrene boards with steel wire mesh reinforcement and waterproof rendering.

An interesting way of examining the effectiveness of construction techniques has recently come into more widespread use. Infra-red thermographs, which display temperature differences as varying colours (indicated by different patterns in Fig. 3.24), can be used to detect features such as cold bridges and the possibility of interstitial condensation. In practice, infra-red thermographs have often shown that insulation is not properly installed and so the prudent designer should make some allowance for his or her calculated U-values not being achieved on-site.

Colours and finishes will have an effect on the thermal behaviour of the building. The texture of the outer wall will have an effect on the convective heat transfer coefficient and the colour will control the reflection and absorption of shortwave radiation. These factors become less important thermally for opaque walls as insulation is increased, although the use of light-coloured walls to reflect radiation into the house through windows remains valid of course. Table 3.12 gives the radiation properties of selected materials.

Depending on the design of the house and factors such as the furniture,

Table 3.12. *Emissivity and absorptivity values for various surfaces*[90]

Surface	Emissivity at 10–38 °C	Absorptivity of solar radiation
Black non-metallic surfaces	0.90–0.98	0.85–0.98
Red brick, concrete and stone, dark paints	0.85–0.95	0.65–0.80
Yellow brick and stone	0.85–0.95	0.50–0.70
White brick, tile, paint, whitewash	0.85–0.95	0.30–0.50
Window glass	0.90–0.95	transparent[a]
Bright aluminium, gilt or bronze paints	0.40–0.60	0.30–0.50
Dull copper, aluminium and galvanized steel	0.20–0.30	0.40–0.65
Polished copper	0.02–0.05	0.30–0.50
Highly polished aluminium	0.02–0.40	0.10–0.40

(With permission of the American Society of Heating, Refrigerating and Air-Conditioning Engineers Inc., Atlanta, Georgia.)
[a] See Table 4.1 for more precise figures.

the use of lightweight finishes internally and lightweight partition walls may lead to a rapid rise in air temperature during periods of high passive solar gain. Materials with greater thermal capacity, on the other hand, will have a moderating effect on temperature changes. The use of dark colours internally will increase the absorption of solar energy at the expense of the reflected component. If the consequence of this is a need for greater artificial lighting such an approach should be avoided.

In the future we may expect to see some radical developments in wall construction. Currently, the trend is to increase the wall thickness to allow for greater insulation (the conservation house at the National Centre for Alternative Technology, Machynlleth (see Chapter 8) carried this to the point of a cavity wall incorporating 450 mm of glass fibre insulation),[91] but this increases the plan area and raises construction costs. The use of polished aluminium foil in (unfilled) cavities to reflect heat transferred by radiation (it has no effect on conduction heat transfer) has been appreciated and advocated for some time. The logical consequence of this has been the development of high-performance insulation[92] or 'super-insulation' composed of multiple radiation barriers in an evacuated slab. Evacuation can prevent the conduction heat transfer. Such a material was investigated for use in the Autarkic House[93] but no proprietary product of adequate thickness and sufficiently low cost could be found. Another possibility considered for the Autarkic House was foam-filled glass fibre reinforced plastic (GRP) modules which would be stacked on top of each other and attached to a frame. Research is also being carried out on solar 'skins' for solid walls. MacGregor[94] has suggested that a translucent layer of insulation be applied to the external surface of a solid wall so that the heat loss from the wall is reduced while the solar gain is increased by being transmitted and trapped by the translucent coating.

We may also expect to see insulation materials used structurally. The Danish Zero Energy House, for example, used modified 300 mm rock-

Floors

Table 3.13. *U-values for solid floors*[97]

Dimensions (m)	U-values (W/(m² K))	
	Four exposed edges	Two exposed edges at right-angles
30 × 15	0.36	0.21
30 × 7.5	0.55	0.32
15 × 15	0.45	0.26
15 × 7.5	0.62	0.36
7.5 × 7.5	0.76	0.45
3 × 3	1.47	1.07

Fig. 3.25. The Frusher Polyurethane Home.[96] (Courtesy of Building Design.)

wool batts as the load-bearing wall element.[95] And for those who favour 'organic' architecture, the Frusher House in Wisconsin, shown in Fig. 3.25, was constructed by foaming polyurethane around an inflated balloon used as a mould. The 180 mm thick walls are covered with a fireproof rubber skin.

3.9 Floors

For ground floors it is not possible to calculate U-values in the same way as for walls. In the case of solid floors in contact with the ground, the size and edge conditions of the slab must be accounted for and in the case of suspended floors allowance must be made for the higher temperatures (compared to the outside air) in the enclosed air space under the floor, which result from low ventilation rates in that space. Tables 3.13 and 3.14 give standard U-values.

Fig. 3.26 shows a typical construction for a solid floor and Table 3.15 gives the effect on the U-value of increasing the insulation.

Table 3.14. *U-values for suspended timber floors*[97]

Dimensions (m)	U-values (W/(m² K))		
	Bare, or with lino, plastics or rubber	With carpet or cork	With any surface finish and 25 mm quilt over joists
30 × 15	0.39	0.38	0.30
30 × 7.5	0.57	0.55	0.39
15 × 15	0.45	0.44	0.33
15 × 7.5	0.61	0.59	0.40
7.5 × 7.5	0.68	0.65	0.43
3 × 3	1.05	0.99	0.56

Table 3.15. *U-values for solid floors*[78]

Thickness of mineral fibre slab insulation (mm)	U-values (W/(m² K))	
	Detached house (8.9 m × 7.3 m)	Semidetached house (5.5 m × 8.2 m)
0	0.77	0.68
25	0.49	0.45
40	0.40	0.37
50	0.36	0.33
60	0.32	0.30
70	0.30	0.28

Similarly, Fig. 3.27 shows a typical construction for a suspended floor and Table 3.16 gives the effect on the U-value of increasing the insulation.

For suspended timber floors the BRE has stated that the safest course is not to include a vapour barrier because of the possibility of spillage or leakage from plumbing or heating installations.[97] If either of these occurred, the limited possibilities for the water to escape might lead to conditions conducive to fungal attack and eventual rotting.

Because of the difficulties of modifying the floor structure afterwards we recommend designing for a U-value of 0.2–0.3 W/(m² K).

3.10 Windows

Glass, as has been noted, can 'roast you, freeze you, cut you and blind you'. Nonetheless, its more attractive properties endear it to designers and occupants alike. The characteristics of glass and other translucent materials and their use in relation to passive solar gain are discussed in Chapter 4. Here we will briefly address certain questions concerning windows.

The first is whether or not double glazing should be installed. In a house built to 1981 Building Regulations Standards with single glazing, approximately 28% of the total heat loss is through the windows (see Fig. 3.1) – with double glazing this is halved. Double glazing all the nation's

Windows

Table 3.16. *U-values for suspended timber floors*[78]

Thickness of mineral-fibre mat (mm)	U-values (W/(m² K))	
	Detached house (8.9 m × 7.3 m)	Semidetached house (5.5 m × 8.2 m)
0	0.72	0.79
60	0.32	0.34
80	0.28	0.29
100	0.24	0.25
140	0.20	0.20
180	0.17	0.17

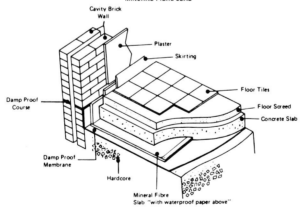

Fig. 3.26. Solid-floor construction detail.[78]

dwellings would save almost 1% of the national primary energy consumption, according to BRE calculations.[45] But opinion varies as to whether it is economical to do so. The BRE has found that only in the most favourable case of high future energy costs, combined with a new dwelling heated to a high standard with good controls, is double glazing 'just about cost-effective'.[45] It may be argued, however, that this is exactly what we

Fig. 3.27. Suspended-floor construction detail.[78]

GROUND FLOORS

(1) SUSPENDED TIMBER

INSULATION STANDARD MINERAL FIBRE MAT	DETACHED HOUSE 8.9m x 7.3m FLOOR AREA (65m²) U VALUE W/m² deg C	SEMI-DETACHED HOUSE 5.5m x 8.2m FLOOR AREA (45m²) U VALUE W/m² deg C
UNINSULATED	0.72	0.79
60 mm	0.32	0.34
80 mm	0.28	0.29
100 mm	0.24	0.25
140 mm	0.20	0.20
160 mm	0.19	0.18
180 mm	0.17	0.17
200 mm	0.16	0.15

Note 1) Insulation of Suspended Timber Floor assumed to use Plastic net technique and incorporating 25 mm airspace.

2) Semi-detached house assumed to be narrow frontage.

VENTILATED TIMBER GROUND FLOOR WITH MINERAL FIBRE MAT/PLASTIC NET

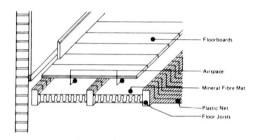

should be designing. Littler has found that double glazing is economical in new houses heated with electricity and marginally so if gas or solid fuel is used for heating.[98] At the Basildon homes described previously, double glazing was not judged to be cost-effective.[43]

Most calculations on the cost-effectiveness of double glazing are based only on the reduction in heat loss, but the numerous homeowners who have purchased double glazing in recent years have often done so for reasons of comfort, a less easily quantified aspect. Double glazing results in fewer down draughts and higher surface temperatures of the glass facing the room (approximately 4–7 K higher than that of single glazing for the same internal and external conditions),[3] thus increasing the comfort level and the space which may be fully used in a room. In buildings where windows are kept to a minimum area, comfort conditions are less affected than in spaces where large glazed areas are used to take advantage of passive solar gain. The choice is also related to how curtains are used. If heavy, well-fitting curtains are drawn regularly at night the relative advantage of double glazing is diminished. We recommend double glazing whenever circumstances permit and furthermore suggest that triple glazing be considered in certain applications. Some of the triple-glazed windows commercially available are of significantly higher quality without being much more expensive than a number of double-glazed units.

Window quality is especially important in reducing infiltration (see Table 4.5) but, unfortunately, it has often been sacrificed to cost. Additionally, workmanship around windows often suffers because the importance of weathertight construction is not appreciated. One solution to the problem could be the introduction of kits for wall windows which might include insulation, damp proofing, sills and lintels in order to bring the quality of the window and its surrounding construction up to that of some of the well-detailed roof windows now available. A first step towards energy conservation for all windows, and one that has been found unanimously to be cost-effective is draught sealing with synthetic rubber strips, felts, metal, plastics, wool or tufted polypropylene pile. It may also be possible in some cases, for example when a home has a mechanical ventilation system, to reduce infiltration losses by opting for fixed windows, but the possible energy savings must be carefully balanced against the loss of flexibility. Window frames are commonly of timber, metal or plastic. Table A 2.8 compares the overall U-value of glass and frames for several types (unfortunately, plastic frames were not available) and shows the advantage of wooden frames.

Wooden frames are also somewhat less likely to serve as condensing surfaces. In the past, designers often accepted that single glazing would act as a condenser and the higher-quality frames incorporated a means of draining out the condensate. Double glazing reduces the risk of condensation on glass because, as has been noted, the temperature of the surface exposed to the room is higher than with single glazing and is more likely to be above the dew-point temperature of the room air. In insulated, well-ventilated and heated buildings with proper consideration given to localizing moisture production it should be possible to avoid serious condensation problems with double glazing, but this is not to say that they will not occur, particularly if the ventilation rate is reduced to limit heat loss. The BRE has developed a condensation prediction chart which helps to assess the degree of risk.[3]

Curtains and shutters for windows are discussed in Chapter 4.

3.11 Doors

As with windows, doors should be draught proofed. The use of glass in external doors should generally be avoided because of the higher heat losses entailed. Where glass is used, for example in patio doors, thermal protection in the form of shutters or heavy curtains for night use should be provided. Self-closing doors, provided they are acceptable to the occupants, can help reduce ventilation losses and are becoming a feature of low-energy homes.

A British Standard is in preparation which will aid designers to select more-energy-efficient doors.[99] It suggests three grades of resistance to air penetration with the highest standard being a maximum air leakage of 12 m^3/(h m) of length of opening joint at a pressure differential of 150 Pa.

An additional concern with doors (particularly front doors) and to some extent windows, too, is that warping may set up new ventilation patterns which are unsatisfactory. In the Abertridwr Project referred to previously some warping resulted in stagnant pools of air favouring condensation and mould growth.

Fig. 3.28 Pitched-roof construction detail.[78]

ROOFS

PITCHED ROOF

INSULATION STANDARD MINERAL FIBRE MAT	U VALUE W/m² deg C	
	PITCH 30°	PITCH 22½°
UNINSULATED	2.01	1.97
60 mm	0.50	0.50
80 mm	0.40	0.40
100 mm	0.33	0.33
140 mm	0.25	0.25
160 mm	0.22	0.22
180 mm	0.20	0.20
200 mm	0.18	0.18

Note: Roof construction: tiles on battens on sarking felt with ventilated roof space and horizontal plasterboard ceiling.

MINERAL FIBRE MAT IN PITCHED ROOFS

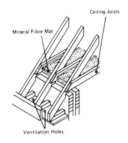

3.12 Ceilings and roofs

The roofs of both existing and new homes are often the easiest and most cost-effective sites for insulation. Materials used are usually preformed glass or mineral-fibre blankets and batts or loose-fill materials (often blown into the loft to settle between the ceiling joists) made of glass fibre or cellulose. Combustible products such as plastics and treated cellulose fibres are being studied by the BRE in order to develop tests which will ensure that such loft insulation materials meet fire safety performance requirements.[76]

U-values for insulated roofs are also being investigated by the BRE and one result of the work is the realization that more information is needed on air speeds in loft spaces.[100] Fig. 3.28 shows a schematic of a typical construction for a pitched roof (more care would be required in actual detailing to avoid a cold bridge — see Fig. 3.18) and Table 3.17 gives the effect on the U-value of increasing insulation. (If 200 mm seems high, it should be remembered that in Scandinavia it is not unusual to have 250 mm of mineral wool in the roof.)

Similarly, Fig. 3.29 shows a typical construction for a flat roof and Table 3.18 gives the effect on the U-value of increasing the insulation.

Vapour barriers are often used on the warm side of the insulation to prevent moisture entering the wall from the interior of the building. The risk of interstitial condensation may be assessed in a similar way to that for walls (see Appendix 3) and, unfortunately, the difficulty of achieving a satisfactory seal is also similar, so ventilating the roof space to the out-

Ceilings and roofs

Table 3.17. *U-values for pitched roofs*[a][78]

Thickness of mineral fibre mat (mm)	U-values (W/(m² K))	
	Pitch (30°)	Pitch (22.5°)
0	2.01	1.97
60	0.50	0.50
80	0.40	0.40
100	0.33	0.33
140	0.25	0.25
180	0.20	0.20
200	0.18	0.18

[a] Roof construction: tiles on battens on sarking felt with ventilated roof space and horizontal plasterboard ceiling.

Table 3.18. *U-values for flat roofs*[a][78]

Thickness of mineral fibre mat (mm)	U-values (W/(m² K))
0	1.54
60	0.47
80	0.38
100	0.32
140	0.24
180	0.19
200	0.18

[a] Roof construction: granite chips, three-layer felt, 25 mm timber boarding, 9.5 mm plasterboard with a vapour barrier.

side air, whenever possible, is wise. Ventilation through the soffit of projecting eaves on opposite sides of the building, as shown in Fig. 3.18, is often a simple and effective solution. Special care is required with shallow roof slopes, say of 25° or less since, here, any escaping vapour has a smaller volume of air to mix with and high humidity levels may thus result – purpose-made ventilation openings may be required.

A consequence of placing substantial insulation between the ceiling joists will be a lowering of the loft temperature. Fig. 3.30 shows representative loft temperatures as a function of the landing temperature.

In severe weather it is apparent that the loft temperature could fall below freezing, thus endangering cold-water tanks and pipes in the loft. If lofts are sealed from the house (to reduce the ventilation heat loss) and internal temperatures are lower (for example, when the occupants are away) the risk of damage is greater. British Gas thus suggests that where insulation levels are high the cold-water tank be located within the living area.[101]

The roof angle is particularly important if solar collectors are to be

Fig. 3.29. Flat-roof construction detail.[78]

FLAT ROOF : INSULATION BELOW DECKING

INSULATION STANDARD MINERAL FIBRE MAT	FLAT ROOF TIMBER JOIST		
	25 mm TIMBER BOARDING U VALUE W/m² deg C	25 mm CHIPBOARD U VALUE W/m² deg C	50 mm WOOD WOOL U VALUE W/m² deg C
UNINSULATED	1.54	1.60	0.93
60 mm	0.47	0.47	0.39
80 mm	0.38	0.38	0.32
100 mm	0.32	0.32	0.28
140 mm	0.24	0.24	0.22
160 mm	0.22	0.22	0.20
180 mm	0.19	0.20	0.18
200 mm	0.18	0.18	0.16

Note: Roof construction : Granite chips, 3 layer felt, timber boarding, chipboard or wood wool and 9.5 mm plasterboard with a vapour barrier.

MINERAL FIBRE MAT UNDER BUILT-UP ROOF

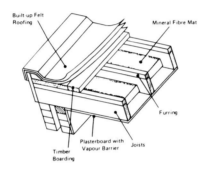

Fig. 3.30. The effect of ambient and landing temperatures on loft temperatures (with 200 mm glass fibre insulation in the ceiling).[101]

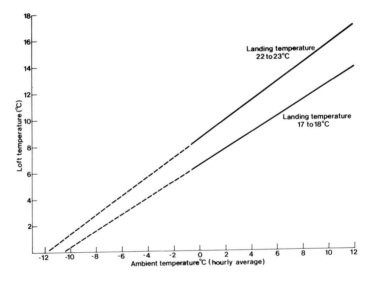

installed either at the time of construction or in the future. (We believe that homes should be designed with potential collector area and roof slope in mind. Normally, provision of a suitable south-facing area is not difficult and the cost of increasing a shallow sloping roof to an acceptable angle at the time of construction should not be very great — the increase in solar

References

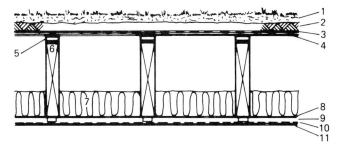

Fig. 3.31. A 'green' roof.[104] Key: 1. local turf; 2. peat; 3. plain faced bituminous felt; 4. 16 mm plywood decking; 5. firring piece; 6. joists; 7. 100 mm mineral-fibre insulation; 8. 50 mm mesh chicken wire; 9. 37 mm × 20 mm softwood battens; 10. 500 g polythene; 11. 9 mm medium density hardboard (class 1).

collector efficiency can be significant. A secondary advantage of a greater slope is that if appropriate rafters are used the loft space is more likely to be available for extra living space at a later date.) Or if the loft space is to be used as a source of preheated warm air a correctly sloped roof will provide somewhat higher temperatures. In this case the roof colour should be as dark as possible to increase the solar radiation absorbed. Experiments are presently underway to evaluate the use of a heat pump combined with very simple roof panels used as air collectors.[102] (See also Chapter 6.) Another technique is the use of panels which serve both as solar collectors and roofing elements. Active solar collection is discussed in detail in Chapter 5.

A second aspect of the roof angle, which may affect the design of some houses, is its effect on the pressure distribution around the house. For roof slopes of less than 30° the wind-produced pressures are negative on the windward surface, while for slopes exceeding 30° the pressures are positive.[58]

Finally, two roof constructions to keep in mind are insulating ceilings which require no roof cladding and 'green roofs'. The first are perhaps best suited to industrial buildings. One example is a cold store in Yorkshire which used structural insulation panels composed of polystyrene sandwiched between prestressed steel skins — a weather-resistant external surface eliminated the need for cladding.[103]

'Green roofs' are those covered with vegetation, usually grass. Their chief attraction is probably visual since present designs, as opposed to traditional ones, call for conventional insulation to be incorporated too to reduce heat losses. Fig. 3.31 shows one design for a grass roof over hardwood joists.

References

[1] Le Corbusier (1972). *Towards a New Architecture*. London: The Architectural Press.
[2] The Building (Second Amendment) Regulations 1981. *Statutory Instrument 1981 No. 1338.* London: HMSO.
[3] Anon. (1972). *Double Glazing and Double Windows*. BRS Digest 140. Garston, Watford: BRS.
[4] Siviour, J.B. (1976). *Designs for Low Energy Houses*. ECRC/M922. Chester: The Electricity Council.

[5] Siviour, J.B. (1977). *Houses as Passive Solar Collectors*. ECRC/M1070. Capenhurst, Chester: The Electricity Council Research Centre.
[6] (1977). *CIBS Guide, Section A7*. 'Casual Gains'. London: Chartered Institution of Building Services.
[7] Siviour, J.B. (1977). *Calculating Solar Heating and Free Heat and Their Contribution to Space Heating in Buildings.* Capenhurst, Chester: The Electricity Council Research Centre.
[8] (1970). *IHVE Guide Book A*. London: Institute of Heating and Ventilating Engineers.
[9] Nevrala, D.J. (1980). 'Heat services for housing – the insulated house design requirements'. *Studies in Energy Efficiency in Buildings*, No. 2. British Gas, Watson House Research Station.
[10] (1980). *CIBS Guide, Section A3*. 'Thermal Properties of Building Structures'. London: Chartered Institution of Building Services.
[11] Everett, A. (1970). *Materials*. London: Batsford.
[12] Burberry, P. (1974). 'The choices for designers'. *Architects Journal*, 160 (37), 615–28.
[13] Anon. (1979). *Glass and Offices*. St Helens, Merseyside: Pilkington Technical Advisory Service.
[14] Hermsen, J. (1979). 'Building study energy conscious design at Basildon, Essex'. *Architects Journal*, 170 (41), 758–72.
[15] Anon. (1979). 'Brownhill Road Flats, Lewisham'. In: *Buildings the Key to Energy Conservation*, Kasabov, G. (ed.). London: RIBA Energy Group.
[16] Fordham, M. (1982). Max Fordham and Partners. Private communication.
[17] Anon. (1979). 'Solar housing for the elderly'. In: *Buildings, the Key to Energy Conservation*, Kasabov, G. (ed.). London: RIBA Energy Group.
[18] Justin, B., Guy, A.G. & Shaw, G. (1980). 'Monitoring the thermal performance of houses with solar walls'. *Helios*, 10, 1–5.
[19] Hitchin, E.R. (1980). 'The sizing of heating systems for well-insulated houses'. *Studies in Energy Efficiency in Buildings No. 2.* British Gas, Watson House Research Station.
[20] Humphreys, M.A. (1976). 'Desirable temperatures in dwellings'. *The Building Services Engineer*, 44 (8), 176–80.
[21] Humphreys, M.A. (1979). 'The variation of comfortable temperatures'. *Energy Research*, 3, 13–18.
[22] Cooper, I. (1981). *University of Cambridge Martin Centre Conference on Comfort and Energy Conservation in Buildings.*
[23] (1978). *CIBS Guide, Section A1*. 'Environmental Criteria for Design'. London: Chartered Institution of Building Services.
[24] Brundrett, G.W. (1973). *Thermal Comfort in Buildings*. Electricity Council Research Centre, Capenhurst.
[25] Billington, N.S. (1974). *Domestic Engineering Services*. London: IHVE.
[26] McIntyre, D.A. (1977). 'Can humidification save on heating bills?' *Heating and Ventilating Engineer*, 51 (596), 10–11.
[27] Durnin, J.V.G.A. & Passmore, R. (1976). *Energy, Work and Leisure*. London: Heinemann.
[28] Gagge, A.P., Burton, A.C. & Bazett, H.C. (1941). 'A practical sys-

References

tem of units for the description of the heat exchange of man with his environment'. *Science*, 94 (2445), 428–30.

[29] Nevrala, D.J. (1979). 'Some energy implications of thermal comfort criteria'. In: *Studies in Energy Efficiency in Buildings*. London: British Gas.

[30] Mayo, A.M. & Nolan, J.P. (1964). 'Bioengineering and bioinstrumentation'. In: *Bioastronautics*, Schaefer, K.E. (ed.). New York: Macmillan.

[31] Billington, N. (1978). 'Comfort'. *Building Services*, October, pp. 37–8.

[32] Anon. (1977). *Ventilation Requirements*. BRE Digest 206. Garston, Watford: BRS.

[33] Brundrett, G.W. (1976). *Ventilation: A Behavioural Approach*. ECRC/M902. Chester: The Electricity Council.

[34] Fuller, W. (1981). 'What's in the air for tightly built houses'. *Solar Age*, June, pp. 30–2.

[35] Brundrett, G. (1981). 'Air pollution'. *Building Services*, March, pp. 38–40.

[36] Modera, M.P., Sherman, M.H. & Grimsrud, D.T. (1981). 'Long-term infiltration measurements in a full-scale test structure'. *AIC Conference on Building Design for Minimum Air Infiltration at the Royal Institute of Technology*, Stockholm, Sweden.

[37] Nevrala, D.J. & Etheridge, D.W. (1977). 'Natural ventilation in well-insulated houses'. *ICHMT Seminar. Heat Transfer in Buildings*. Dubrovnik.

[38] Etheridge, D.W. & Nevrala, D.J. (1979). 'Air infiltration and our thermal environment'. *Building Services and Environmental Engineer*, 1 (7), 10–13.

[39] Anon. (1956). *Domestic Heating – Estimation of Seasonal Heat Requirements and Fuel Consumption in Houses*. BRS Digest 94. London: HMSO.

[40] Guillaume, M., Ptacek, J., Warren, P.R. & Webb, B.C. (1978). *Measurement of Ventilation Rates in Houses with Natural and Mechanical Ventilation Systems*. Garston: Building Research Establishment.

[41] Anon. (1969). *Condensation*. BRE Digest 110. Watford: Building Research Station.

[42] Loudon, A.G. (1971). *The Effects of Ventilation and Building Design Factors on the Risk of Condensation and Mould Growth in Dwellings*. BRE CP 31/71. Watford: Building Research Station.

[43] Anon. (1979). 'Condensation and mould growth'. *Domestic Energy Note 4*. Department of Energy.

[44] Anon. (1970). *Condensation in Dwellings. Part 1. A Design Guide. Part 2. Remedial Measures*. Ministry of Public Buildings and Works. London: HMSO.

[45] Bakke, P. (1975). *Energy Conservation: A Study of Energy Consumption in Buildings and Possible Means of Saving Energy in Housing*. BRE CP 56/75. Garston, Watford: BRE.

[46] Berry, J., Emerson, R., Harrison, J.W. & Kasabov, G. (1977). 'Conservation of energy in housing'. *Building Services Engineer*, 45, 288–93.

[47] Anon. (1981). 'Children's designs for a solar house'. *What's New in Building*. March, 121.

[48] Kasabov, G. (ed.) (1979). *Buildings, the Key to Energy Conservation.* London: RIBA Energy Group.
[49] Hawkes, D. & MacCormac, R. (1978). 'Office, farm, energy and land use'. *RIBA Journal,* 85 (6), 246–8.
[50] Rodrigues, C.M., Cole, R.J. & O'Sullivan, P.E. (1978). 'Computerised thermal design models'. *The Building Services Engineer,* 2 (46), 19–27.
[51] Anon. (undated). *Building Better with BEEP.* The Electricity Council – EC 3819.
[52] Anon. (1981). 'Guidelines for environmental design and fuel conservation in educational buildings'. *Department of Education and Science Architects and Building Branch Design, Note 17.*
[53] Anon. (1979). *How Windows Save Energy – Technical Appendix.* St Helens, Merseyside: Pilkington.
[54] Turrent, D., Doggart, J. & Ferraro, R. (1980). *Passive Solar Housing in the UK.* London: Energy Conscious Design.
[55] University of Minnesota Project Ouroborus.
[56] Markus, T.A. & Morris, E.N. (1980). *Buildings, Climate and Energy.* London: Pitman.
[57] Penz, F. & Thomas, R.B. (1979). Entry for the Competition 'The Rational Use of Energy'. Unpublished.
[58] Mattingly, G.E. & Peters, E.F. (1977). 'Wind and trees. Air infiltration effects on energy in housing'. *Journal of Industrial Aerodynamics,* 2, 1–19.
[59] Hunt, D.R.G. (1979). *Availability of Daylight.* Garston, Watford: BRE.
[60] Siviour, J.B. (1980). 'The effect of increasing house insulation on electric heating equipment design'. *IEE Conference on Effective Use of Electricity in Buildings.*
[61] Pickup, G.A. & Miles, A.J. (1979). 'The performance of domestic wet heating systems in contemporary and future housing'. *Studies in Energy Efficiency in Buildings.* British Gas Watson House Research Station.
[62] Krall, L., Littler, J.G.F. & Thomas, R.B. (1979). 'Ambient Energy Design House 1'. In: *Ambient Energy Design.* Cambridge: privately published.
[63] Hayes, D.H. (1971). 'Higher standards of thermal and acoustic insulation – but when?' *IHVE Journal,* 39, A39–A41.
[64] The Building Regulations 1976. *Statutory Instrument 1970 No. 1681.* London: HMSO.
[65] Campbell, P. & Bratt, J. (undated). *Better Insulated Housing.* DOE Construction 21. Department of the Environment, Housing Development Directorate.
[66] Anon. (1979). 'BRS low energy house laboratories'. In: *Buildings, the Key to Energy Conservation.* Kasabov, G. (ed.). London: RIBA Energy Group.
[67] Siviour, J.B. (1976). 'A new approach to space heating requirements in houses'. Electricity Council Research Centre. *Energy Management in Buildings Conference,* Aston University.
[68] The Building (First Amendment) Regulations 1978. *Statutory Instrument 1978–723.* Department of the Environment. London: HMSO.

References

[69] Anon. (1980). 'Buildability of highly insulated housing'. *BRE News*, 52, 17.

[70] Anon. (1982). 'Building Regulations Changes – Second Amendment'. *Architects Journal*, 175 (13), 59–71.

[71] Barratt, P.V.L. (1979). 'Ways to meet the changed thermal insulation regulations'. *Building Services and Environmental Engineer*, 1 (12), 28–31.

[72] Martin, D. (ed.) (1980). *Specification 80*. London: The Architectural Press.

[73] Coates, J. & Sudjic, D. (1982). 'Toxic gas drives pupils out of classrooms'. *The Sunday Times*, 16 May, p. 4.

[74] Rogowski, B.F.W. (1976). *Plastics in Buildings – Fire Problems and Control.* BRE CP 39/76. Garston: Building Research Establishment.

[75] Anon. (1979). *Cellular Plastics for Building.* BRE Digest 224. Garston: Building Research Establishment.

[76] Rogowski, B. & Sutcliffe, R.J. (1980). *Fire Performance of Loft Insulating Materials.* BRE IP 25/80. Garston: Building Research Establishment.

[77] Venning, P. (1979). 'How to cost energy conservation in building'. *RIBA Journal*, 86 (10), 446–7.

[78] Boys, J. (1979). 'Insulation'. *The Architect.* November, pp. 41–9.

[79] Potter, J.G. (1980). 'Assessing the economic priorities for improving building insulation'. *Building Services and Environmental Engineer*, 3 (4), 28–30.

[80] Anon. (undated). 'Economic thickness of insulation for existing industrial buildings'. *Department of Energy Fuel Efficiency Booklet 16.*

[81] Ferraro, R. & Godoy, R. (1981). 'Thermal insulation'. *What's New in Buildings*, June, pp. 47–8.

[82] Stein, R.G. & Serber, D. (1979). 'Energy required for building construction'. In: *Energy Conservation Through Building Design.* Watson, D. (ed.). New York: McGraw Hill.

[83] Herman, P.R. (1975). 'The energy cost of the construction and habitation of timber frame housing'. *Building Science Special Supplement: Energy and Housing.*

[84] Gartner, E.M. & Smith, M.A. (1976). *Energy Costs of House Construction.* BRE CP 47/76. Garston: Building Research Establishment.

[85] Thorogood, R.P. (1981). *Vapour Diffusion Through Timber Framed Walls.* BRE IP 1/81. Garston: Building Research Establishment.

[86] Farbon, J.W. (1981). 'Inside or out?' *Building Services*, 3 (6), 50–1.

[87] Randell, J.E. & Boyle, J.M.A. (1979). 'Low energy housing – Salford's answer'. *Building Services and Environmental Engineer*, 1 (10), 18–21.

[88] Wilson, D.R. (1977). 'Existing buildings – experience from the "House for the Future" project of Granada TV'. *University of Nottingham Conference on Ambient Energy and Building Design.* Welwyn: Construction Industries Conference Centre.

[89] Anon. (1979). 'Abertridwr Project (new)'. *RIBA Journal*, 86 (10), 450.

[90] ASHRAE Handbook (1977). *Fundamentals.* New York: American Society of Heating, Refrigerating and Air Conditioning Engineers.

[91] Anon. (1978). 'Energy needs reduced to a minimum'. *RIBA Journal*, 85 (6), 235–6.

[92] Hnilicka, M.P. (1961). 'Engineering aspects of heat transfer in multilayer reflective insulation and performance of NRC insulation'. In: *Advances in Cryogenic Engineering.* Timmerhaus, K.D. (ed.). New York: Plenum Press.

[93] Littler, J.G.F. (1975). 'Multi-layer insulation evacuated or filled with air or krypton'. Department of Architecture, University of Cambridge. *Autonomous Housing Study Working Paper 26.*

[94] MacGregor, A.W.K. (1979). 'A solar skin for solid walls'. *The Passive Collection of Solar Energy in Buildings, UK–ISES Conference* C19.

[95] Esbensen, T.V. & Korsgaard, V. (1976). 'Dimensioning of the solar heating system in the Zero Energy House in Denmark'. *UK–ISES Conference* C12, pp. 64–77.

[96] Blythe, J. (1979). 'House of foam'. *Building Design*, February, pp. 26–7.

[97] Anon. (1972). *Heat Losses Through Ground Floors.* BRS Digest 145. Garston: Building Research Station.

[98] Littler, J.G.F. (1980). 'Low energy houses'. *Ambient Energy*, 1 (1), 57–62.

[99] Covington, S.A. (1981). *The Selection of Doors and Doorsets by Performance.* BRE IP 2/81. Garston: Building Research Establishment.

[100] Anderson, B.R. (1981). *The Assessment of U-values for Insulated Roofs.* BRE IP 3/81. Garston: Building Research Establishment.

[101] Pickup, G.A. & Miles, A.J. (1979). 'Energy conservation – field studies in domestic heating'. *Studies in Energy Efficiency in Buildings.* British Gas Watson House Research Station.

[102] Anon. (1981). 'Research in the UK – University of Aston in Birmingham'. *Helios*, 11, 12–13.

[103] Anon. (1980). 'Insulation panels need no cladding'. *What's New in Building*, October, p. 81.

[104] Ollis, J. (1977). 'Timber – available methods and systems'. *The Architect*, December, pp. 59–62.

4

Passive solar design

4.1 Introduction

Most buildings receive sunlight via windows, and in the design of passively heated buildings this warming effect is more fully exploited.

The five tasks to be accomplished in a successful design are as follows:
  to increase the amount of sunlight entering the building by enlarging the glazed area;
  to avoid causing excessive overheating or glare;
  to reduce the large heat loss through the glazing;
  to distribute the heat acquired;
  to store the surplus for use when the sun stops shining.

There has been some sterile argument about terminology. ' "Passive" and "natural" solar designs' are terms coined to differentiate the use of sunlight which provides warmth without the use of complicated controls, pumps and fans, from 'active solar designs' which employ solar collectors and fairly complex controls. The meaning of the terms will become more clear whilst reading the descriptions of the built examples later in this and the following chapters.

It is of interest to realize that active solar systems associated with very large thermal storage could be used to provide *all* the space heating and hot water in a building, even in the UK. On the other hand, passive designs can never be expected to do this in northern Europe, and there they must be regarded as fuel savers. In the US and other countries blessed with more-intense and regular winter sunshine, it is possible in cold but clear regions to rely entirely on passive solar space heating. Apart from the number of sunny days in a particular region, there are other constraints acting on the design team when dealing with passive solar buildings. For example, the density in housing layout in northern Europe is higher than one might find in many parts of the US and the elevation of the sun is lower, thus overshadowing is a severe problem. This point is enhanced because passive techniques usually rely on replacing *walls* rather than roofs with glazing and, of course, the walls are more subject to shading by houses nearby. Secondly, passive designs constrain the floor plan inside the building. This feature should be compared with the effect of using active solar collection, which can be regarded as an auxiliary boiler and which does not need to influence internal layout. It may be imagined therefore that, in favourable areas, a passive solar architecture will emerge, which sympathetically

Fig. 4.1. Annexe to St. George's School, Wallasey, near Liverpool.[2]

resolves the problems introduced by the designs described later in this chapter.

A building of great interest in this context is the Annexe to St George's School at Wallasey (1961; Fig. 4.1) in the north-west of England. One of the first in the world consciously to exploit passive gain, it illustrates several of the tasks confronting the design team, led in this case by the architect A.E. Morgan. The whole of the southerly wall is glazed (see Fig. 4.2 which shows the floor plan) – thus very large amounts of sunlight may enter the fabric. The windows contain two panes of glass, thus roughly halving the rate of heat loss through the enormous area of glazing (about 500 m^2.[1, 2] For convenience, the opening windows are single glazed and, of course, framing members take up roughly 10% of the facade. Overheating is reduced by the massive construction (ground and intermediate floors are of 250 mm solid brick). The exposed surfaces of the heavy fabric absorb energy and help to smooth out midday solar peaks. Overheating and glare are reduced by the deep format of the window wall which

Introduction

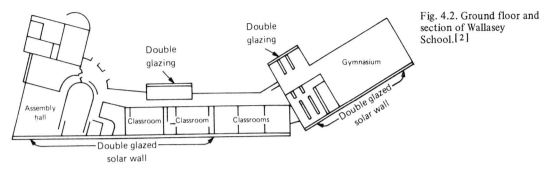

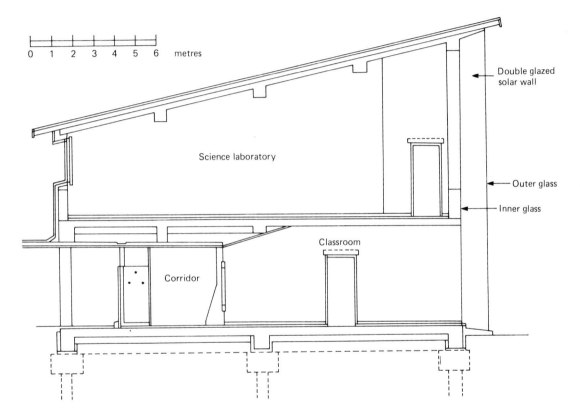

Fig. 4.2. Ground floor and section of Wallasey School.[2]

creates 1500 mm overhangs and, clea..y, these are more effective at the appropriate times, when the sun is high in the sky. Heat distribution occurs naturally since the glazing opens onto the heated spaces and a uniformly warm environment is achieved.[1] Teachers report that the students, who sit facing parallel with the window wall, do not find glare a problem.

In spite of the massive construction, the thermal mass contributes little to heat storage over a 24 h period, and not at all over a period as long as a month.

Auxiliary heat is supplied by the students and the lights, the latter being switched on well before the school opens. It is notable that a building occupied from 9 a.m. until 4 p.m. is well suited to passive solar heating, although one in use from, say, 11 a.m. to 7 p.m. might be even more successful.

The only complaints arising at Wallasey (where the back-up boiler system has never been used) concern odours, which are a result of the low air-change rate. The building is heavily insulated with 125 mm of polystyrene applied externally (giving a U-value of about 0.25 W/(m^2 K) for the walls and roof. Davies estimates the floor coefficient as 0.34 W/(m^2 K)). (U-values are defined in Appendix 2.)

It must be clear that the solar contribution expressed as a fraction of the heating load of a building increases as more measures are taken to reduce this load by insulation. It should also be clear that the absolute solar contribution will actually fall with rising insulation, since the passive energy (for a given glazed type and area) will more often exceed the demand, resulting in a rejection of the surplus heat.

Having introduced the concept and outlined some of the problems involved in buildings which deliberately use large amounts of passive solar gain, the rest of this chapter will concentrate on features of the components involved, on a description of the various building types which have evolved and on built examples of these basic designs.

4.2 Elements of passive solar systems

4.2.1 Glazing

Exacting demands are made on the glazing systems used in passive designs. Firstly, the skin must be weatherproof to driving rain and winds. Secondly, it must be highly transparent to solar radiation but opaque to long wavelengths characteristic of heat radiated from objects at 15–30 °C (about 3–20 μm). Thirdly, it must have a reasonable lifetime (say ten years) when warm and illuminated, and sufficient strength to withstand wind loads and small blows. Clearly, in situations where people might fall against the skin, the glazing must fail safely.

Finally the cost should be low, handling easy, and the appearance must be pleasing since, almost by definition, the area will be large and the impact great. Given these requirements it is often sensible to choose glass. Table 4.1 lists some of the relevant properties.

The Glass and Glazing Federation[3] has issued an excellent book concerning handling properties and installation, and Linsley[4] gives some very useful information concerning sealants and glazing compounds.

Passive-gain applications usually do not warrant the extra cost of the more highly transparent water-white glass unless it is non-tempered, especially in view of the higher dirt accumulation likely on vertical glazing compared with sloping active solar collectors. The loss of transmission from dust can be as high as 20%, but in areas which are not subject to large amounts of industrial pollution, a more reasonable figure is 2–5%. Clearly, it pays to wash down windows fairly frequently!

In the domestic situation, glass should not be continued down to ground level. The Building Regulations indicate that below 800 mm above

Elements of passive solar systems

Table 4.1. *Properties of glass*

Material	'Window glass'	Tempered[b]	Clear wired	Rough cast, often U-section	Water-white
Examples of manufacturers	Pilkington Brothers	Pilkington Brothers	Pilkington Brothers	e.g. Profilit by Pilkington Brothers	a,c
Thickness (mm)	3, 4, 6	3, 4, 6	6	6, 7	3, 4, 5
Thermal expansion coefficient (10^{-6}/K)	6–9	6–9 (Steel 11–13, aluminium 22, timber 3–6)	6–9	6–9	6–9
Effect of heat	Chipped edges may cause cracks	Resists thermal shock. Breaks into small harmless pieces	May crack but does not disintegrate	Cast edges less likely to be chipped	see tempered
Ease of installation	Little problem	More difficult[b]	Little problem	Fixed widths (262, 332 mm) special glazing bars	a
% solar transmittance at normal incidence	3 mm 86 4 mm 84 6 mm 80	Depends on iron content	≤ 80	80–85	89–92
% transmittance of heat radiation	About 3 (all types)				
Approximate price in £/m² 1981 (not installed)	9	21	15	14	ASG 5 mm, $20/m² in USA

[a] Water-white glass by ASG (Kingsport, Tennessee, USA) is tempered.
[b] Tempered and toughened are synonomous. Tempered glass must be ordered to the correct size and cannot subsequently be cut.
[c] No European supply with a UK outlet has been found.

Table 4.2. *Safe areas of low-level glazing*

Area of glass (m²)	Minimum thickness (mm)
0.2	3
0.5	4
0.8	5
2.5	6

the ground level (inside or outside the building) glazing becomes 'low-level'. In the domestic situation, if annealed glass is used, it should have the thickness indicated in Table 4.2. Toughened, wired or laminated panes could be less substantial.

Thus, for a typical situation in which the glazing is set to match 600 mm stud spacing, a window 2 m high would have to be made with 6 mm glass. Toughened glass or plastic would not of course suffer this restriction. Some passive collection designs benefit from a covering which provides neither sharp reflections nor perfect vision. The appearance of large areas of float glass over an opaque wall (see subsequent sections), or

Fig. 4.3. Use of Kalwall glazing in a roof. (Courtesy of Kalwall Corporation, Manchester, New Hampshire.)

even an attached sun space, may be softened by a mildly textured surface. Cast glass provides such distortions.

Some designs go further and use diffusing glass, which has the advantage of distributing the solar heat gain all over the thermally absorbing room surfaces. The problem of glare can be made better or worse in this situation since the whole of the window is now a source of high illumination. If the floor plan encourages occupants to sit and work so that *side* illumination is provided, then daylighting is enhanced and ceiling reflection avoided, but if the users *face* the window wall, glare will be a problem.

Glass requires glazing compounds for its installation. Generally, in passive applications the frame around the sheet will not rise above 50 °C. But the temperature excursions over a Trombe wall may reach this value, and putty will be unsatisfactory. Two-part rubberizing compounds, sealants or preformed compression gaskets are especially suitable.[4] Generally, cracking caused by thermal stress is not a problem in passive-gain designs. As a matter of good practice, the depth of glass hidden by the frame should be kept to the minimum for satisfactory mechanical performance, to avoid a high temperature gradient at the edges — where feathering cracks produced during cutting may develop into major cracks.

Table 4.3 illustrates some of the properties of plastics. Plastics have some strong advantages and penalties. The cost (1981) of inexpensive glazing bars for 3 mm glass in greenhouses is about £3/linear m. The maximum area of 3 mm glass recommended for vertical use (aspect ratio 1:3) is about 1 m^2, implying about £5/m^2 for the glazing bars. However, some plastics can be screwed down (Fig. 4.3) under a wooden batten placed over the lapped sheets and a strip of glazing compound sealant.

A second benefit is that the plastic will not break into dangerous shards either during handling or when installed. Additionally, its lightweight nature (for example, ≈ 1 kg/m^2 for Kalwall) may allow cheaper structures to be designed and certainly makes for easier installation. There are no regulations concerning the size of plastic glazing in the domestic situation, but some of the materials present a fire hazard.

There are other severe drawbacks to the use of plastics. In general, they age, and although one might expect the rigid glass fibre or polycarbonates to survive for 20 years, most of the other materials will either be totally

Elements of passive solar systems

Fig. 4.4. Kalwall glazing to the 'Ark' on Prince Edward Island. (Courtesy of Kalwall Corporation, Manchester, New Hampshire.)

disfigured by scratches, or will have blown away! They also lose transmission with age and thermal stress. The best product is probably rigid glass fibre which at 150 °C loses 9% of its transmission after 300 hours. Most plastics become yellow with age by ultra-violet action, although Filon (glass fibre protected with proprietary films, which also increase resistance to erosion) is believed not to suffer in this way. Those plastics which really are rigid (for example, polycarbonate) must be glazed in with great care – allowing a wide rebate (for example, 12 mm) to avoid panels slipping out under wind flexing, and yet to allow for thermal expansion. Even polycarbonate will bow in and out according to the humidity and temperature, and flexible covers such as Kalwall tend to present a wavy appearance which worsens as thermal cycling 'works' the anchoring points. In this case, the material can be stretched onto an aluminium frame which has roughly the same coefficient of expansion. Of course, this introduces another expansion problem! Plastic films, which often have large expansion coefficients, tend to sag when warm. Plastics have a different appearance from glass. This point is strongest with polythene and glass fibre. The latter contains both glass and resin and the different refractive indices lend a bluish cast to the material, rendering it translucent not transparent. For passive solar applications not involving areas of glazing which people look through, this can be an advantage. Finally it is important to notice that the plastics (and particularly the films) transmit more heat radiation than does glass. Fig. 4.4 illustrates a large roof glazed with glass fibre, which is also employed in the Trombe wall shown in Fig. 4.47.

4.2.2 Single and multiple glazing

Multiple glazing is used to reduce heat loss and noise transmission. For the latter, a wide space is needed (for example, 100 mm); but, for the former, gaps of 15–25 mm are suitable.

It is useful to consider why several layers of glazing can reduce heat loss. Heat passing through a material is usually quoted in terms of a U-value, which has the units $W/(m^2 \, K)$.

This, for example, means that a single pane of glass 1 m² with $U =$

Table 4.3. *Properties of plastics*

Material	Polycarbonate	Fibreglass	Ridged polyvinyl	Acrylic
Example of manufacturer or trade name	Rohme & Haas: Makrolon	Kalwall: Filon	British Industrial Plastics: Cobex	ICI: Perspex
Thickness (mm)	2–5	0.5–1.5	0.15–3.0	3, 5, 6
Surface spread of flame rating (glass = 0)	1–2	1	1	3
Thermal expansion (10^{-6}/K)	6.5	15	5–7	5–9
Effect of heat	Maximum temperature 130°C	No mechanical degradation up to 180°C	Softens at 80°C	Softens at 90°C, depolymerises at 200°C
Ease of installation	Requires flexible sealant to glazing bar	Maximum span 900 mm for 1.5 mm. Nail down	Can be solvent welded. Corrugated, can be nailed down	Can be solvent welded
% solar transmission	85 (3 mm)	85–90	77 (2 mm)	90 (2 mm)
% transmittance of heat radiation	3	5 (0.5 mm)	Very little. None beyond 6 μm	5
Approximate price in £/m² 1981 (not installed)	30 (3 mm)	8	18 (3 mm)	17 (3 mm)
Comments: all can be cut with scissors or a fine saw	Unbreakable. Appearance 'poor' after 6 years exposure	Standard width rolls 2', 3', 4' or 5' by up to 50'	Very brittle below 0°C	Can be molded in an oven. Appearance maintained over 7 years

5.6 W/(m² K), situated in a window with temperatures of 20 °C inside and 5 °C outside, loses heat at the rate of: (5.6 W/(m² K)) (20 − 5) = 84 W/m². If the glass has an area of 2 m² then the rate of heat loss is:

(84 W/m²) (2 m²) = 168 W.

Thus, over a period of 12 h, the amount of heat loss is:

(168 W) (12 h) = 2016 Wh or 2.016 kWh (= 3.60 MJ),

which is equivalent to a 'one-bar' electric fire burning for 2 hours!

The U-value is essentially a measure of conductance rather than resistance (R) and the two are related by:

$$U = \frac{1}{R};$$

clearly, the units of R are m² K/W.

The resistance to heat flow of the window is composed of several components illustrated in Fig. 4.5.

Film or flexible sheet					
Polythene, UV inhibited	PVC, UV inhibited	Polyester	Polyvinyl fluoride		Rigid polystyrene
–	–	Mylar	Tedlar	3M: Flexiguard	Amariplastics
0.075–0.2	–	0.025–0.2	0.08	–	2, 3, 4
–	–	–	–	–	–
16–18	4	18–25	–	–	–
	Collapses	–	–	Retains mechanical properties from −35°C to 150°C	–
				Does not shrink, thus easier than Tedlar	Cracks easily on sawing
Staple-or-nail under framing strip					
85–90	85–90	87 (0.13 mm)	91	–	–
up to 85	–	30 (0.13 mm)	50	> 90 when anti-reflected	–
≈ 0.1–0.5 (0.04 mm)	4 (0.04 mm)	≈ 1–3	≈ 5–12	–	10 (3 mm)
Ages rapidly	Softening temperatures low	Very strong	Extremely strong	–	Very clear and non-distorting

Typical values of these resistances are 0.045 m² K/W (r_1), 0.005 m² K/W (r_2 for 3 mm glass), 0.13 m² K/W (r_3) (see also Appendix 2). Thus the total resistance to heat flow ($r_1 + r_2 + r_3$) is about 0.18 m² K/W, and the U-value is about 5.6 W/(m² K).

When two panes of glass are used we have the situation shown in Fig. 4.6. We now have, instead of two film resistances, four plus the stationary trapped air, giving a U-value of 2.8 W/(m² K). Values of the resistances cannot be ascribed so simply in this case, since there is now radiation interchange between the glazing layers, which short circuits the conductive and convective resistances.[6] Similarly, three layers of glass provide a U-value of about 1.9 W/(m² K).

Clearly, the resistance to heat flow of a window is more or less independent of the thickness of the glazing material since the resistance of the glass or plastic is small compared with the film resistances. The other important observation concerns the wind-cooling effect. Roughly speaking, the film resistance R varies with the wind speed V and the tempera-

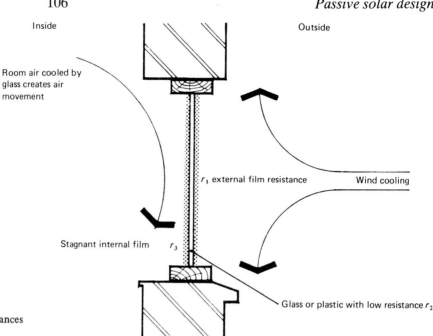

Fig. 4.5. Thermal resistances on a window.

ture T as[5]:

$$R = R_0 \left(\frac{V}{V_0}\right)^{-0.8} \left(\frac{T}{T_0}\right)^{-0.7}.$$

where T_0 and V_0 are the initial values. For example, if R_0 is 0.045 at a wind speed of 3 m/s, at $V = 6$ m/s it becomes:

$$R_{6 \text{ m/s}} = 0.045 \left(\frac{6}{3}\right)^{0.8} \left(\frac{T_0}{T_0}\right)^{0.7}$$

or

$$R_{6 \text{ m/s}} = 0.026 \text{ m}^2 \text{ K/W}.$$

Care must be taken with the relationship, which applies only to turbulent flow. Turbulence depends on the wind speed and window detailing and will be strongly affected by the shelter from neighbouring trees and so on. Roughly, one might expect laminar flow from 0–3 m/s and turbulent flow above 3 m/s. In the laminar regime the film resistance will be higher, decreasing rather suddenly at the transition to turbulent flow. Shelter is thus significant!

The gap between the glazing layers must be large enough to prevent short circuiting by conduction across a narrow air gap or through small convection cells, but not so wide that gross circulation is set up. Fig. 4.7 illustrates this effect and shows that a gap of 15–25 mm is optimal. Of course, any breaks in the seal between the layers, which allow air to pass in and out of the interpane space, will reduce the effectiveness of the multiple glazing.

Radiation was mentioned earlier. Two panes of glass or plastic at dif-

Elements of passive solar systems

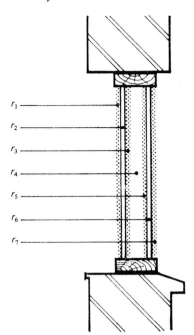

Fig. 4.6. Thermal resistances on a double-glazed window.

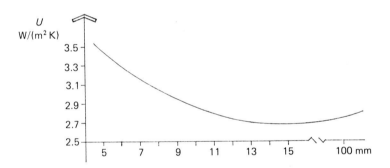

Fig. 4.7. Thermal resistance of a double-glazed window as a function of glazing separation.

ferent temperatures exchange heat by heat radiation (infra-red). Coatings can be applied which are more or less transparent to solar radiation (in other words they cannot be readily 'seen') but which look like metallic mirrors to infra-red. Thus these coatings reflect heat back into a room if a window is so coated on the inside. By the same token, when facing outwards they radiate heat much less well than uncoated glass, thus reducing heat loss. The optimal position on single glazing is facing inwards, and on double glazing on the outer surface of the inner pane (or on plastic between the panes). Since single glazing thus treated will be cooler, there is an added risk of condensation. This can be particularly serious in greenhouses.

Coated glass gives the U-values shown in Table 4.4.

Double glazing could be improved even further by filling the interpane space with krypton (an inert gas) when the U-value drops from 2.8 to 2.6 for the uncoated case, or from 1.8 to 1.0 in the coated case. (If double-

Table 4.4.[a] *U-values of windows*[b]

Glazing material	Coating	$U = W/(m^2 K)$
Single glass	None	5.6
Single plastic	None	6.6
Single glass	Coated	3.3
Single plastic	Coated	3.5
Double glass	No coatings	2.8
Double plastic	No coatings	4.3
Double glass	Coated[c]	1.8
Double plastic	Coated[c]	1.8
Double glass + interpane plastic	No coatings	2.0
Double glass + interpane plastic	Plastic coated	1.2
3M Quadpane[d]	Anti-reflected	1.5

[a] Taken from Ref. [6].
[b] NB: these values are approximate, depend on conditions and are open to dispute.
[c] Coated on surface 2 counting from the inside.
[d] Data from M. Rubin, Lawrence Berkeley Labs., 1983.

Table 4.5. *Infiltration rates through windows or door cracks in $m^3/(h\,m)$ of crack*

Element	Condition	Wind speed (km/h)			
		8	16	24	32
Double-hung-sash wood windows	Average, not weather stripped	0.7	2	3.6	5.5
	Average, weather stripped	0.4	1.2	2.2	3.3
	Badly fitting not weather stripped	2.5	6.4	10	14
	Badly fitting weather stripped	0.6	1.8	3.2	4.7
Between frame and wall	Not caulked	0.3	0.7	1.3	1.9
	Caulked	0.1	0.2	0.3	0.4
Metal casement	Unstripped	1.3	3	5	7

glazed coated windows could be *evacuated*, the *U*-value could drop to 0.5 W/(m² K).)

In 1982 Pilkington Brothers commenced production of sealed double-glazed units fitted with coated glass (called 'Kappafloat'). The unit has a *U*-value of 2.0 W/(m² K).

Air leaks around the window closure or frame can provide a serious heat loss. Typical values are shown in Table 4.5. Clearly, in passive solar gain applications it is essential to avoid ill-fitting windows. Most should be

Elements of passive solar systems 109

Table 4.6. *Conditions used in calculating the results shown in Table 4.7*

(a) *U*-value single-glazed unshuttered window 5.6 W/(m² K)
(b) *U*-value double-glazed unshuttered window 2.8 W/(m² K)
(c) *U*-value coated double-glazed unshuttered window 1.2 W/(m² K)
(d) *U*-value single-glazed shuttered window 0.53 W/(m² K)
(e) *U*-value double-glazed shuttered window 0.49 W/(m² K)
(f) *U*-value single-glazed curtained window 3.9 W/(m² K)
(g) *U*-value coated (heat mirror) double-glazed shuttered window 0.40 W/(m² K). Shutters consisting of 50 mm polystyrene 15 mm from the glass, and used at night (13 h)

Ambient temperature 5°C (day), 2°C (night)
Internal temperature 20°C (day), 16°C (night)
Glazed area 30 m²
Shutters in place at night only.

Table 4.7. *Heat loss with different window treatments*

Case	Heat loss over 24 h period (MJ)
(a)	210
(b)	105
(c)	47
(d)	112
(e)	61
(f)	175
(g)	29

fixed closed and, in general, the large area of glass is held in a frame which only encounters the masonry at the perimeter of the window and not around every pane of glass! However, a typical installation comprising 30 m² of glazing might have a perimeter length of 23 m and opening sections of 3 m² with a length round the frame of 7 m. Thus the total crack length could be 30 m and, where care has been taken, the infiltration at a wind speed of 16 km/h (typical of the UK) would be 13 m³/h, corresponding to a heat loss in the winter of about 70 W.

4.2.3 Insulating blinds and shutters

We have seen that the heat loss through glazing can be high – more than 5.6 W/(m² K) in exposed situations for single glazing. For example, 30 m² of single glass would typically (in the UK) lose about 200 MJ/d in winter if unprotected at night. Tightly fitting shutters 50 mm thick used at night could reduce this loss to about 100 MJ/d.

Table 4.7 summarizes what happens on a particular day with different window treatments. The conditions are listed in Table 4.6.

It is important to place these losses in context. A reasonably well-insulated house of 100 m² floor area might, on such a winter's day, endure a heat loss of 270 MJ. The solar gain through an unobstructed south-facing window of 30 m² might be 108 MJ on a January day in the southern UK. Reference to Table 4.7 indicates that large areas of glazing *could* be a source of heat loss rather than heat gain, unless defensive measures are taken.

Later sections of this chapter illustrate various designs adopted for using passive gain. Some of these involve attaching to the building glazed spaces which can be isolated under adverse conditions. In those cases, the windows may need no further treatment; however, when spaces are occupied under all weather conditions, it is essential that the glazing be protected at night. It is likely that in these cases curtains will be used. There is very little information concerning the insulating effect of curtains because they are ill defined and their manner of attachment is very important. If, indeed, a stationary layer of air were trapped between the drape and the glass, the extra resistance imposed would be about 0.2 m² K/W. Conventionally it is said that well-fitting curtains reduce the heat loss through single glazing by 20%, corresponding to a resistance of 0.08 m² K/W. The energy saved (compare cases (a) and (f) in Table 4.7), is small but the cost of curtains made to measure, plus the track, self-installed, is £10–20/m² (1980). The cost of effective shutters or blinds is £15–40/m².

Types of window insulation

Glazing can be protected on the outside, or between the panes of multiple units, or on the inside. External insulation implies systems which can withstand winds, whose sliding tracks, operating mechanisms, or hinges do not freeze up in winter, which can withstand ultra-violet radiation from the sun and which do not ruin the external appearance. Interpane systems can fail mechanically and inaccessibly, but they are protected from tampering, and can solve the 'storage problem'.

Internal shutters

The simplest form of insulation is a roller blind which when installed to leave a gap at the window head of less than 25 mm, no gap at the sill, and less than 5 mm at the sides, reduces the heat loss by 30%. Such a blind, when aluminized, conserves 45% of the normal heat loss. Fig. 4.8 illustrates an installation suggested by Cuchanan and Gresse.[7] The cost of roller blinds is less than £10/m² but minor modifications must be made to the window reveal to incorporate the vertical track which provides the edge seals. These are shown in Fig. 4.8. (Successful blinds can be made from cotton onto which is stuck aluminized Mylar. The cotton should be made more rigid with a fabric stiffener.)

The next step is to use quilted material for the blind.[8] For example, the 'Window Quilt'[9] (Fig. 4.9), which consists internally of three layers of polyesterfibre fill and externally of decorative colours, is said to have a U-value of 1.2 W/(m² K), and should reduce the heat loss through single glazing from 5.6 W/(m² K) to 1 W/(m² K). The reduction is thus 4.6 W/(m² K); if the Window Quilt were installed over double glazing, the drop would be from 2.8 W/(m² K) to 0.8 W/(m² K) or only 2 W/(m² K).

Elements of passive solar systems

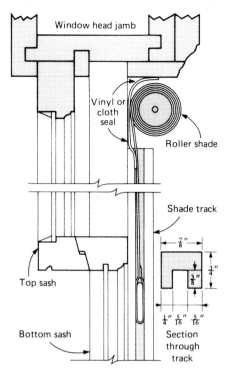

Fig. 4.8. Sealing the blind to the window frame. (Reprinted from *Popular Science*, with permission © 1979, Times Mirror Magazines Inc.)

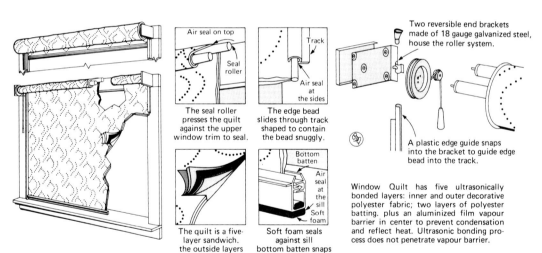

Fig. 4.9. Window quilt fitted into extruded plastic track.[8] (Reprinted from *Popular Science*, with permission © 1979, Times Mirror Magazines Inc.)

This observation is of course general: insulation over double glazing has less effect than over single glazing.

An even better roll shade is made from four layers of metallized fabric[10] (Fig. 4.10) and has a U-value of $\approx 0.5-0.6$ W/(m² K). This device shares with the 'High "R" shade' (Fig. 4.11) the ability to roll up flat (saving storage space) and to open up its layers by admitting air when covering the window. The High 'R' shade over a double-glazed window

Self-inflating sandwich

Curtain wall features automatic operation: Electric motor inside roller tube is controlled by thermostat and Freon sensor. Latter is enclosed in glass tube mounted outside house. Four layers of aluminized plastic film inflate and deflate via vents. Price of 10-by-20 ft. shade: about \$4.25 per ft^2.

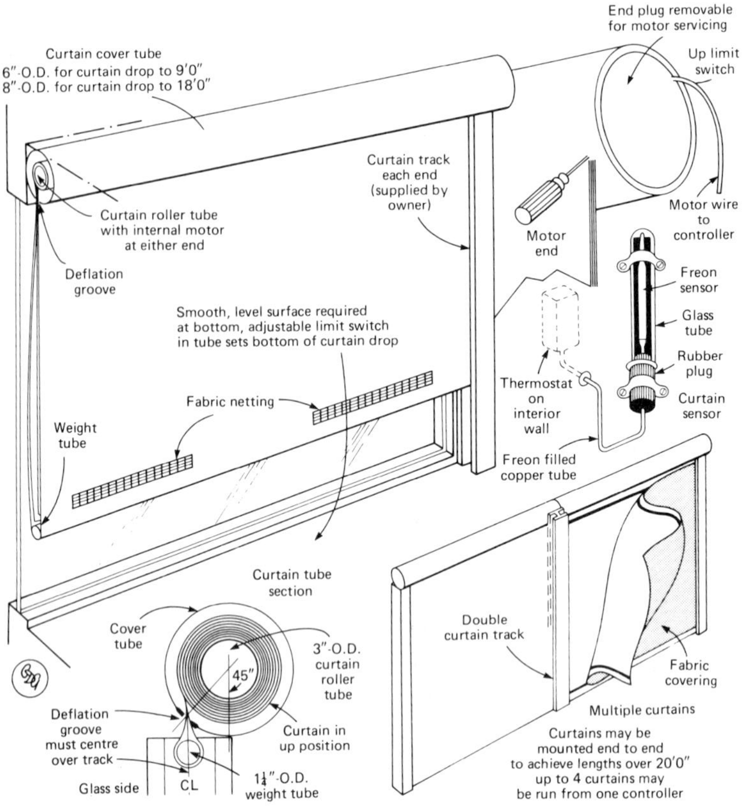

Fig. 4.10. Curtin wall.[10] (Reprinted from *Popular Science*, with permission © 1979, Times Mirror Magazine Inc.)

shows a measured thermal transmittance of 0.33 W/(m^2 K), indicating a transmittance for the shade of about 0.4 W/(m^2 K). A two-layer low emissivity blind (≈ £30/m^2), developed by the Energy Design Group, reduces the *U*-value of a single-glazed window to a measured 1.1 W/(m^2 K) and has very effective edge seals.[53]

Blinds such as these can readily be automated to respond either to sunlight or temperature differences. Obviously, they can also help to keep out sunshine if necessary.

The next type of internal insulation consists of a slab of expanded plastic foam (Fig. 4.12), which can be covered with fabric and either pushed into place in the window reveal, or stuck between fabric forming a rigid sheet. This is hinged like a casement, attached to the ceiling, or allowed to slide in horizontal track. To improve on the thermal performance of the blinds described above, the foam would need to be 70 mm thick (expanded polystyrene) or 50 mm (polyurethane) giving an overall thermal transmittance of about 0.3 W/(m^2 K) when installed over a single-glazed window. The success of the barrier will depend primarily on the seals to the reveal, (inwards) particularly at the sides and sill and, secondarily, on the seals (outwards) around the window itself.

Such large moveable panels (which should be sheathed in fire-resistant material), are awkward to stow away. When hinged, the swing negates a

Elements of passive solar systems

Fig. 4.11. High 'R' Shade.[10]

- 2-part extruded plastic head frame for easy access. Made with integral head seals.
- Compact single roll design with strong spring return
- 5 layers expand to form dead air spaces that slow convective and conductive energy transfer
- Radiant energy flow greatly reduced with low emittance materials
- Permanently shaped spacers conform tightly when rolled up yet separate the layers when pulled down
- Extruded plastic jamb frames with integral jamb seal
- Thermally effective summer through winter at windows and sliding glass doors

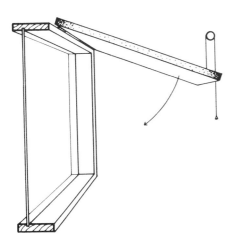

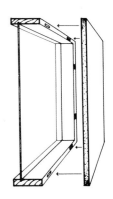

Fig. 4.12. Fold-up casement and push up internal shutters.[11] (Reprinted from *Home Energy for the 80's*, with permission, Garden Way Publishing.)

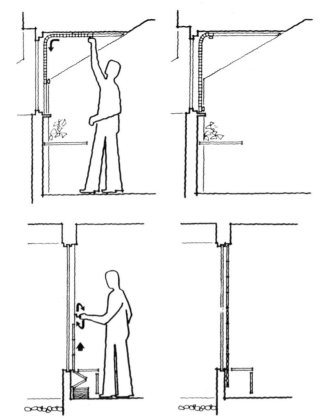

Fig. 4.13. Sliding shutters in Termorac House.[12]

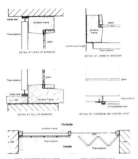

Fig. 4.14. Thermoblind.[13]

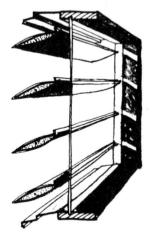

Fig. 4.15. Adjustable insulating louvres.[11] (Reprinted from *Home Energy for the 80's*, with permission, Garden Way Publishing.)

large amount of floor space. An alternative method (used in the Termoroc House in Malmö, Sweden) used many parallel strips of polyurethane measuring about 50 mm × 50 mm stuck between fabric, so forming a flexible sheet. This sheet slides in track mounted on either side of the window which extends in a curve onto the ceiling for 'parking' the shutter when not in use (Fig. 4.13).

Thermoblind[13] solve the stowage problem by a concertina action (see Fig. 4.14). The cost is about £30/m² and the U-value of the 20 mm version over single glazing is about 1 W/(m² K).

A still more convenient system is shown in Fig. 4.15 in which foam-filled louvres act like a very deep venetian blind. This type of system is also used in Guardia shutters[10] (Fig. 4.16), consisting of air-filled plastic extrusions which can be drawn back to the sides of the window. The U-value of a single-glazed window covered with such shutters is about 1.6 W/(m² K) and their cost is about £40/m². They also provide a deterrent to burglars!

Shurcliffe[14] gives a very useful review of innovative shutter designs.

External shutters

Anderson[15] has used huge panels of styrofoam hung from steel track outside patio windows. It is difficult to obtain a good seal to the

Elements of passive solar systems 115

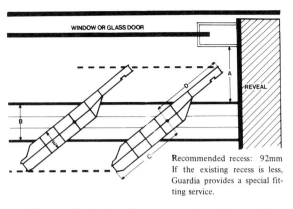

Fig. 4.16. Guardia Shutters.[13]

Recommended recess: 92mm If the existing recess is less, Guardia provides a special fitting service.

a) Minimum distance, edge of track to outmost projection on window or door frame - 58mm.
b) Width of track - 34mm.
c) Inner length of louvre - 68mm.) Total length of
d) Windowside length of louvre - 65mm.) louvre - 133mm.
e) Thickness of louvre - 18mm maximum.
f) Louvre overlap - minimum 35mm. maximum 50mm.
g) Number of louvres per installation - varies from 8 to 10 per metre run.
h) Angle of rotation of louvres - approx. 150 degrees.

Fig. 4.17. Zero Energy House 1975.

walls of the house or the window frames in this manner, and unless the lower edge of the shutter has a deep-pile draught seal or slides in a second track, the performance will not be very good.

The Zero Energy House[16] uses sliding shutters (Fig. 4.17) made of glass fibre boxes containing plastic foam.

The heat loss is reduced from 3.1 W/(m² K) (the value quoted by the Danish team for double glazing) to 0.4 W/(m² K). These shutters are

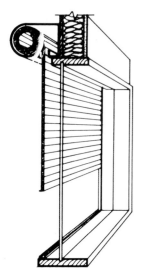

operated remotely from inside the house via a window mechanism which adds to their cost.

External shutters have not been widely used for insulation in solar houses; but 'rolladen' – rolling blinds (Fig. 4.18) consisting of narrow overlapping strips of hollow plastic or aluminium – are common in West Germany as a protection against forcible entry. In the Philips[17] Experimental Solar House at Aachen, such shutters are used. Unless tightly fitting they cannot be very effective and in order to achieve a reasonable storage volume when rolled up at the head of the window, the thickness is limited to 10–20 mm.

Interpane blinds

The third method for insulating glazing uses venetian blinds between panes of double glazing (for example, many Velux roof lights are so equipped).

The blind can be held with the slats at any angle but does not usually retract. The Massachusetts Institute of Technology (MIT) Solar V Home[18] (Fig. 4.30) uses narrow lamellae about 15 mm wide with a mirror finish on one surface. At night the louvres are closed to provide an interpane metallized sheet which reflects back the infra-red. Work by Berlad[19] indicates that similar windows 100 mm wide achieve U-values of 1.15 W/(m² K).

4.2.4 Shading

Especially in the case of direct solar gain to occupied spaces, it may be necessary to keep out the sun on occasions. Typical methods include permanent overhangs, or fixed, angled slats over the windows and mobile blinds or venetian shutters.

The south-facing Terry House (in New Mexico) illustrated later in this chapter (Fig. 4.31), uses wooden boards lined up east–west in echelon over the windows, as shown in Fig. 4.19.

In northern latitudes it is probably sensible to use a mobile device since admitting sunshine is usually desirable. The Zero Energy House employs external awnings which wind down from the window head.

If the aim is to exclude glare, but not heat, then an internal blind may be satisfactory – once the sunlight has been absorbed by the blind it will mostly appear in the room as heat. Only if the blind is white or shiny metallic will a large part of the solar energy be reflected out again. For this reason, external awnings are more successful in combatting overheating.

The metallized venetian blinds in the MIT V Home referred to earlier reflect the sunlight onto the ceiling of each room, thus avoiding glare.

When overhangs are preferred they tend to be rather large. If sun at an elevation of 50° is to be allowed to shine on only 25% of the window indicated in Fig. 4.20, then the overhang will be about 1 m.

An excellent survey of shading devices incorporated in aesthetically pleasing ways is given by Olgyay & Olgyay.[20]

Finally, an interesting device called 'cloud gel' has been invented by Chahroudi.[21] It consists of a gel sealed between double glazing and is

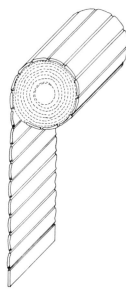

Fig. 4.18. Rolladen rolling shutters.[11]

Elements of passive solar systems 117

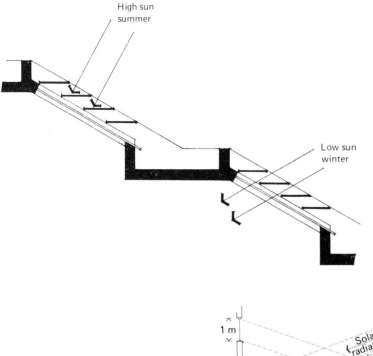

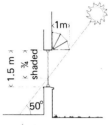

Fig. 4.19. Sun shading outside the Terry House, New Mexico, US.

Fig. 4.20. Overhangs in summer.

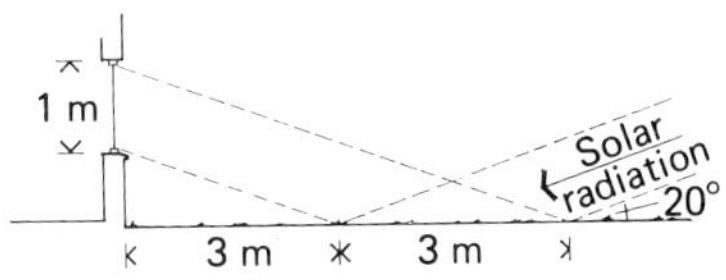

Fig. 4.21. External reflection enhancement in winter.

normally transparent. Above about 30 °C it becomes milky white and opaque, thus excluding unwanted radiation.

4.2.5 Radiation enhancement using reflectors

If the horizontal surface outside a window is made reflective, the amount of radiation entering the building is increased. However, for passive solar gain applications, one is primarily interested in the heating season when the sun is low in the sky. Fig. 4.21 indicates that rather large areas of reflector would be needed.

Since much of the radiation is diffuse, and the reflectors are not perfectly specular, the situation is much more complex than indicated in Fig. 4.21. Work by Baker, McDaniels & Kaehn[22] shows that, for a specular reflector with a reflectance of 0.8, the optimum tilt of the reflector is about 7° downwards (that is, sloping down away from the building) and that increasing the length beyond about twice the window height is of small benefit (Fig. 4.22). (This presupposes a configuration in which the reflector starts at the same level as the window.) The width required is shown in Fig. 4.23 and indicates little gain beyond a width of four times that of the window. The maximum enhancement is shown in Fig. 4.24 for various months and suggests that in the winter months about 50% more radiation would be directed at the window using a reflector of four times the width of the window with a length of twice the height of the window.

In the UK the enhancement will be reduced because of our predomi-

Fig. 4.22. Enhancement at 45° N with reflector length.[22] Enhancement in light gathered for the reflector-collector system at 45° N latitude as a function of reflector length. Calculations were performed for the month of January for three representative choices of reflector and collector orientation angles.

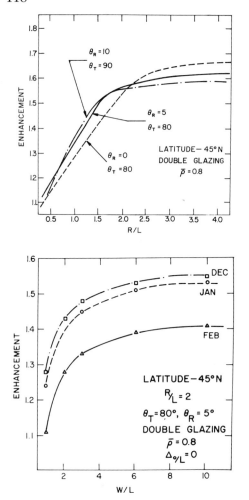

Fig. 4.23. Enhancement with reflector width.[22] Enhancement in light gathered plotted as a function of total reflector width for the months of December, January and February at 45° N latitude.

nantly diffuse radiation. McDaniels *et al.* quote a figure of $\approx 1.25 - 1.30$ for enhancement of diffuse radiation with the reflector dimensions above. Thus one would expect an overall gain in our climate, not of 50% but of $\approx 35-40\%$.

4.2.6 Thermal storage (see also Chapter 7)

The problem of thermal storage in passive solar design has bedevilled workers in this area from the beginning. It is plain that in clear climates such as the Pyrenees or New Mexico, frequent winter sun is available in quantities large enough to supply the major part of the space heating load of a building, and is regular enough to do this reliably. Thermal storage is essential to carry over heat acquired during the day until the sun again becomes powerful the following day, or, at worst, the next-but-one day. The situation is very different in the UK. For example, the ratio of solar radiation falling on south-facing vertical surfaces in the southern UK to north-eastern US in December is roughly 1:3. The percentage of possible sunny hours in the UK is about 35%, in New Mexico 80% and in Nice 60%.

Elements of passive solar systems 119

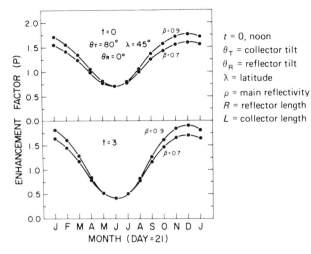

Fig. 4.24. Enhancement as a function of time of year.[22]

$t = 0$, noon
θ_T = collector tilt
θ_R = reflector tilt
λ = latitude
ρ = main reflectivity
R = reflector length
L = collector length

Enhancement factor P plotted as a function of time throughout the year. The abscissa labels time in terms of the twenty-first day of each month. A collector angle of $\theta_T = 80°$ and a reflector angle of $0°$ were assumed, with $\lambda = 45°$. The maximum value of R/L was 3·94 for 21 Dec. at $t = 3$ hr. Most of the winter months required R/L to be less than 2·0.

It is not yet clear how thermal mass should best be used in the UK in conjunction with large areas of glazing. It is, however, obvious that spaces which cool off during the night or on overcast days, and which contain large amounts of thermal mass, must be heated up again when the sun does shine. Thus heavyweight spaces opening directly to the sun may well be cold for a long time after the sun is up. For this reason it seems sensible in our climate to consider *detached* or controlled thermal storage.

Phase-change storage (see also Chapter 7)

Some materials, notably calcium chloride and a potassium salt of phosphoric acid, have hydrates which melt at low temperatures. Essentially, the salts dissolve in their own water of crystallization. For example, the material chliarolithe, developed by CNRS[23] in Nice, consisting of $CaCl_2$ water vermiculite and a stabilizer, melts at 28 °C with a latent heat of fusion of 55 kWh/m³. It has been tested over 6000 melting cycles. The material has a conductivity of 0.5 W/(m² K) compared with a value for water of 0.6 W/(m² K); the latter figure, however, neglects convection. Because the conductivity is low, large thin slabs are needed to absorb heat. To absorb 200 MJ in five hours would require a surface area of very roughly 100 m² if the temperature were allowed to rise 10 °C above the melting point on the outside of 100 mm thick slabs.

The French system places the chliarolithe in 50 mm pvc drainpipes; but these do not really provide the desired surface area to volume. Thermol 81 Rods also contain a mixture melting near 28 °C and are available commercially. Another American system[24a] uses hollow polymer concrete tiles filled with a eutectic salt, sodium sulphate decahydrate. The tiles weigh about 50 kg/m², the salt melts at about 32 °C and the tiles store

Fig. 4.25. MIT Solar V House: window detail.[18] (Reprinted from *Popular Science*, with permission © 1978, Times Mirror Magazine Inc.)

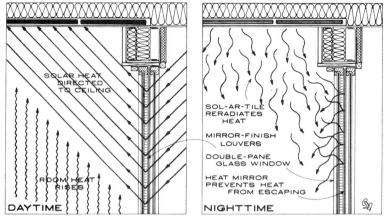

By day, solar radiation is reflected to ceiling by mirror-finish venetian-blind louvers. Ceiling tiles with chemical cores store heat. At night, tiles slowly reradiate heat as outside temperature drops.

Transparent Heat Mirror in window lets in solar radiation, but prevents room heat from escaping.

about 0.6 kWh over the range 30–35 °C. The salt is in two layers, each only 10 mm thick. Such a configuration is very suitable for absorbing heat at the correct rate. These tiles were used in the ceiling of the MIT Solar V Home mentioned in Section 4.3.1, and illustrated in Fig. 4.25 which indicates how the incoming sun is redirected by the reflecting blinds onto the ceiling tiles. For a room of 15 m^2, the storage provided using Sol-Ar[24a] tiles would be 36 MJ. Most phase-change systems available commercially change phase and absorb most heat at temperatures which are high for passive solar applications in the UK. A very good summary of products marketed in the US can be found in Ref. [24b]. Calor[24c] in the UK provides interesting materials, changing phase at 12.8, 31 and 35 °C.

Rock-bed storage

Occupied rooms warmed directly by sunlight may overheat; but the tolerable temperature swing is low – say 28–18 °C. Thus, at *best*, the temperature difference in the air available to charge the rock store is only 10 °C, and this is insufficient to drive much energy into the rocks. For example, with rock of density 2250 kg/m^3 and voids of 40% (that is, solid 60% and thus an effective density of 1350 kg/m^3), the heat capacity is about 1.15 MJ/(m^3 K). However, the rocks cannot all be raised to 28 °C, otherwise the room space becomes hot. A reasonable mean value is half the air temperature swing or 5 K. For a rock bed of 10 m^3 the heat capacity is thus (1.15 MJ/m^3) (10 m^3) (5 K) or 58 MJ. This situation then is totally different from that in which air-heating solar collectors charge a rock bed up to, say, 80 °C.

The process of charging the rock store with surplus heat from a passive-gain space, using air as the transfer medium, is more successful than reversing the procedure and warming the space in the evening with warm air from the store. This is because the air will exit from the store at a low

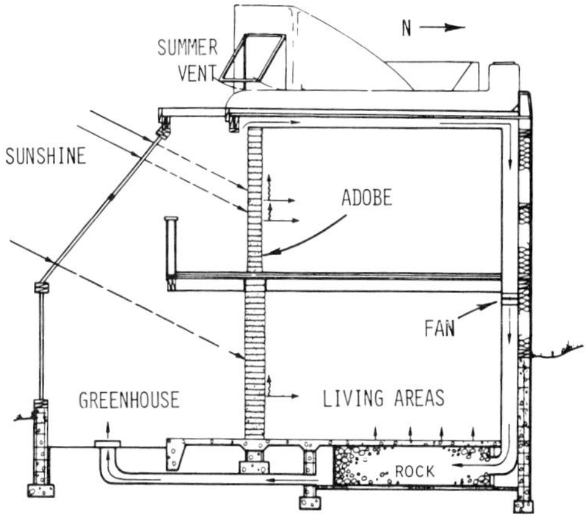

Elements of passive solar systems 121

Fig. 4.26. Rock-bed storage.[25]

supply temperature (that is, 25–18 °C in the cited case, depending on the state of discharge of the store).

To move appreciable quantities of energy at very low supply temperatures implies very high air volumes, high air speeds and chilling effects to the unfortunate occupants. The preferred method is to allow heat from a low-temperature rock store under the floor to rise by conduction, natural convection and radiation through the floor into the space requiring warmth. This process is also quite slow. One cannot expect a heat transfer rate from the floor of more than about 10 W/(m² K). If the store is of 10 m³ and measures 3 m × 4 m × 0.8 m, the upper surface area is 12 m². Assuming a room temperature of 18 °C and a store at 25 °C, the best that can be expected is a heat supply of (12 m²) (7 K) (10 W/(m² K)) or 0.84 kW.

Balcomb[25] describes how a rock bed should be sized, and what fan speeds and pressure drops should be expected. Kreith & Kreider[26] also discuss pressure drops across rock stores. In practice, it seems that a rock bed calculated to store Z kWh between certain temperature limits actually only stores and liberates about 0.75 Z kWh.

Following Balcomb one may, for example, design a system to store 72 MJ. Accepting that the temperature change in the rocks will be 10 K then the volume will be given by:

72 MJ = (volume m³) (5 K) (1.15 MJ/(m³ K)) or 12.5 m³.

Suppose the shape is 4 m × 4 m × 0.78 m, positioned under the floor, as shown in Fig. 4.26, then the face area is 4 m × 0.78 m or 3.12 m².

Next we calculate the air flow necessary to transfer the 72 MJ in a suitable time interval. This rate depends on assumptions about the length of time that the sun shines to produce the surplus 72 MJ. Suppose this period is five hours, then the rate of heat transfer is 4000 W and suppose that there is a temperature difference in the air stream of 10 K corresponding to an air flow rate of 1100 m³/h. Then the face velocity of the air at

Fig. 4.27. Rock-bed performance[25]

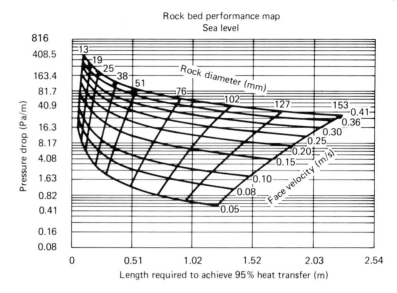

the entry to this store is

$$\frac{1100 \text{ m}^3/\text{h}}{3.12 \text{ m}^2}$$

or 350 m/h.

Reference to Fig. 4.27 shows that, with 125 mm diameter rocks, the distance the air travels into the bed before yielding up 95% of its heat is less than 2 m and therefore a 4 m long bed will satisfactorily absorb the heat. The pressure drop will be less than 2.5 Pa/m of bed or 9 Pa total. The pressure drop in the ductwork to and from the rock bed would probably be about 17 Pa giving a total of less than 30 Pa.

The amount of storage required will be reviewed when considering the solar energy available in passive solar designs.

Thermal storage in the building fabric

In sunny climates, large amounts of thermal storage are recommended in the building fabric. To quote Balcomb: 'It is not advisable to design for transfer of more than approximately 1/3 of the net heat out of a space to a rock bed. "Net heat" means the solar energy input minus the daytime losses. To exceed 1/3 will result in either excessive air flow rates or excessive (room and store) temperatures.'[25]

It is thus common in New Mexico to use 350 mm thick internal adobe brick walls. In the Unit 1 First Village House (Fig. 4.36), for example, there is about 40 m³ of adobe walling between the attached sun space and the rest of the house plus the mass of sun-space floor. The volumes of rock beds are 12.5 m³ and 10 m³. Thus the ratio of structural storage (45 m³) in the space heated directly by the sun, to the rock-bed storage: 12.5 m³ + 10 m³ with 4% voids + floor = 24 m³, is about 2:1. However, in the UK there are *few* days on which more solar energy is available during daylight hours than is required to heat the building during these same hours.

Passive solar heated buildings

In other words, apart from smoothing out possible short-term surplus towards early afternoon, thermal storage may be redundant. This point is illustrated more clearly in the subsequent discussion concerning the Unit 1 House.

There is, of course, a certain amount of thermal mass in any construction. For example, even an insulated timber-framed room lined on the inside with 15 mm plaster board has a heat capacity of about 24 kJ/(m² K), that is, a room with 60 m² of wall and ceiling has a heat capacity of 1.4 MJ/K in its plaster alone. Furnishings might add a further 0.4 MJ/K. For the whole house this adds up to about 10 MJ/K of thermal storage. For a house with an insulated brick and block cavity and solid floors downstairs, the value might be about 70 MJ/K. If we assume that the fabric of the house cools down overnight to 13 °C and requires warming up during the following day to 16 °C (thus allowing the walls to be cooler than the room air), then 215 MJ of heat will be needed. Very roughly, the solar energy available through a vertical surface in the UK in winter is 3.6 MJ/(m² d). Thus to supply this 215 MJ would require 60 m² of glazing. This calculation is, of course, illustrative only — it is open to great criticism, and is only intended to illustrate the pitfalls of drawing on experience from sunny countries where winter solar radiation is likely to be three times that experienced in the UK.

4.3 Passive solar heated buildings

For convenience in description, one can divide the designs used for passive solar space heating into six types:
- direct gain;
- attached sun space;
- convection loops;
- Trombe or water walls;
- roof ponds;
- roof space collectors.

Obviously, all the methods will depend upon exposing a large area of glazing to the sun and, apart from the roof ponds, the glazing will generally be vertical or steeply sloping.

4.3.1 Direct gain

Most buildings have some windows which admit the sun. In direct-gain designs, this area is greatly increased. The advantages over other passive forms lie in the conventional appearance and in the low cost which arises because the builder is not involved in unfamiliar techniques, and because no extra plan area is introduced to the building. Unfortunately, the design also carries some drawbacks. Firstly, unless diffusing glass is used the occupants may suffer from the glare of the direct sunlight, and furnishings can fade in the shorter wavelengths of light penetrating the glass. (Note that diffusing glass will distribute the light and heat to all room surfaces.) Secondly, even if a lot of energy is entering via the window, on a cold day the glass will be considerably below room temperature,

Table 4.8. *US passive buildings*[27]

Direct gain	23
Trombe walls	13
Water walls	12
Roof-space collector	4
Sun space	24

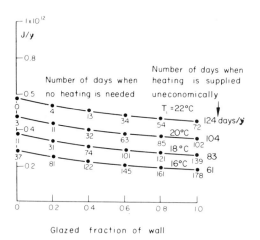

Fig. 4.28. Estimated yearly consumption of artificially introduced heat as a function of glazed area and design indoor temperature, 0.5 room air changes per hour.

thus providing a source of radiant cooling to the occupants. When the sun goes down the glazing causes a great deal of heat loss. Even multiple glazing and heat mirrors are unlikely to reduce this loss below 1 W/(m² K), which is three to five times greater than a well-insulated wall (per m²). Night insulation may thus be mandatory unless the space is not to be used until late the next morning. Finally, the temperature will swing widely unless controlled thermal storage is present, leading to overheating in summer.

The buildings described in the following section all try to moderate these drawbacks without adding greatly to the initial cost.

There are very few documented examples of most of the six basic passive designs in Europe, the most common being the use of Trombe walls or direct gain. Experience in the US suggests that this position will alter quickly, and Table 4.8 lists the number of passive designs which are very briefly described in a survey by the American Institute of Architects (1978).[27] The number in 1983 is probably about 100 000.

The Wallasey school building in the UK has already been described, and it remains only to add that lights are turned on from 6 a.m. to 8 p.m. to supplement the heat provided by the students and the sun. Davies[1] estimates the energy supplied by students and lights, and Fig. 4.28 indicates that, at an internal temperature of 20 °C, the number of days during a year when no heating from any non-solar sources is needed varies from three to 102 as the area for direct solar gain increases. The number of days when the loss through the glass exceeds the gain is estimated as 104. Of

Passive solar heated buildings 125

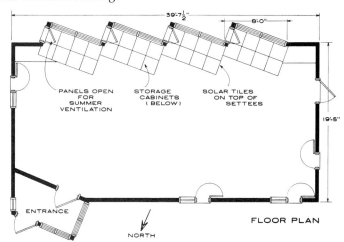

Fig. 4.29. MIT Solar V House floor plan.[18] (Reprinted from *Popular Science*, with permission © 1978, Times Mirror Magazine Inc.)

course, this figure is about half the days in the heating session and in many ways it is representative of a particular building — one which is not occupied at night and which has no insulating shutters on the glass wall.

The second UK example was built near London in 1956 and designed by E.J.W. Curtis.[28] One reason for including a large solar window was to add a feeling of spaciousness to a small ground plan of 11.2 m × 6.1 m (total floor area 132 m^2). The unobstructed south front is entirely made in ply-glass double-glazed units set in timber, with a total area of 61 m^2, and it was only on appeal that planning permission was granted — refusal having been on the grounds that 'it was too modern for the area'.

The brick walls consist of two 38 mm cavities separated by heavy aluminium foil, and the calculated U-values are: east and west walls 1.1 W/(m^2 K); roof 0.62 W/(m^2 K); north wall 0.79 W/(m^2 K) — these values reflect fuel prices and attitudes towards insulation in 1956 and are not satisfactory today. During one monitoring period from 23 March to 8 May 1957 the back-up heat pump was only 'required for short boosts as it was possible to heat the entire house from direct solar gains via the south and west window walls'.[28] From 8 May onwards, no back-up heating was required. Average temperatures were 21 °C (day-time) and 22 °C (evenings). Electricity consumption for the only back-up — the heat pump — was 24 GJ (day-time) and 5.4 GJ (night-time).

No condensation was found with the solar wall but, in summer, overheating was a problem — though not a serious one. It is notable that no special attempt was made to introduce thermal mass — the east and west walls were of cavity brickwork, the south wall of glass and the north wall of resin-bonded plypanels — the ceilings of fibreboard on timber joists — the ground floor was, however, in poured concrete.

One of the most exciting buildings using direct solar gain is the experimental MIT Solar V Classroom.[18, 29] This project avoids glare by reflecting incoming radiation to the ceiling with interpane venetian blinds, it uses phase-change tiles in the ceiling to moderate temperature swings and has a heat mirror film in the windows to reduce heat loss via the glazing. The floor plan is shown in Fig. 4.29 and the outside in Fig. 4.30.

Fig. 4.30. MIT Solar V House. (Reprinted from *Popular Science*, with permission © 1978, Times Mirror Magazine Inc.)

The south-angled windows are fixed closed with a total area of 18 m^2, and are triple glazed with two sheets of water-white glass and one coated plastic sheet half-way between them. The two gaps are each 20 mm and the solar transparency of the coated plastic is 70% implying a total window transparency of 57%. The heat loss through the treated window is said to be one-quarter to one-third of that through conventional double glazing over a 24 h period. The inner gap is occupied by a venetian blind whose horizontal slats are covered, on the upper slightly concave surface, with aluminized mylar whose reflectance is initially 85%.

The slats are very closely spaced (10 mm) and are 8 mm wide. The tilt can be varied through the season as the sun's elevation in the middle of the

Passive solar heated buildings

day alters, in order to redirect the direct solar radiation towards the ceiling tiles. The blind even in its 'open' position reflects out 5–10% of the solar radiation.

The half of the ceiling adjacent to the windows is covered with 100 tiles 30 mm thick and 580 mm square containing pouches of $Na_2SO_4.10H_2O$ and 9% NaCl to lower the melting point, borax to induce recrystallization and the thickening agent Cab–O–Sil. The ensemble melts at 23 °C. The total heat stored in the tiles is about 90 MJ for a 6 K temperature change. The tiles are darkly coloured to absorb radiation. Seats under the windows are also covered with tiles adding about 29 MJ/6 K of storage – they are charged because the blinds do not reach the bottom of the windows.

The floor consists of 100 mm of concrete resting on gravel and, in terms of a daily temperature cycle, its storage capacity is about 65 MJ for a 5 K temperature change. The floor is further stabilized by coupling to a large volume of earth which is partly enclosed by insulation. 100 mm styrofoam board extends 450 mm down below the walls and 900 mm beyond the corners of the building.

The walls contain 150 mm and the ceiling 280 mm of glass fibre batts giving U-values of 0.32 and 0.17 W/(m² K). The heat loss of the occupied building has been measured as 11.2 MJ/(K d) without a solar contribution.

Measurements indicate that the minimum night-time temperature is normally 17 °C and the maximum day-time temperature is 22 °C in winter.

This design offers features which may make it appealing in the UK climate. Glare is restricted by sacrificing some solar gain, much of the thermal storage is only 'present' at high temperatures, and thus the space is not kept cold unnecessarily and is of relatively fast response (it might be better for that reason to use a timber floor). However, the low transmittance of the windows means that if the building were in the UK, then typically, in winter, only about 29 MJ/d of solar heat would reach the interior space. Assuming a mean U-value of 0.9 W/(m² K) for the windows, the heat lost over 24 h would be about 29 MJ. It is thus not clear that insulating shutters used at night would not be a better solution in the UK.

Karen Terry's house is the last direct-gain design to be discussed. It is situated in the hilly country around Santa Fe, New Mexico, at 2200 m elevation and is located as far south as Gibraltar. The site experiences 3300 °C heating days per year. (1000 °C days means 100 days on which the ambient temperature was 10 K below a fixed temperature, or 200 days with a 5 K gap, etc. The number is usually assessed hourly and roughly represents the severity of the heating load in a particular site. The data here are expressed to a fixed temperature of 16.5 °C, and typical values for the UK are Thames Valley 2100, east Anglia 2400 and east Scotland 2700 (see also Appendix 2).)

Karen Terry supervised the building and made many of the internal fittings. David Wright was the architect, and this beautiful house (Fig. 4.31) cascades down a southerly hillside in a series of one-storey terraces. The walls are in adobe 350 mm thick, which are insulated on the outside with 50 mm of rigid urethane foam, rendered, and bermed on the north,

Fig. 4.31. Terry House, New Mexico, US.[52]

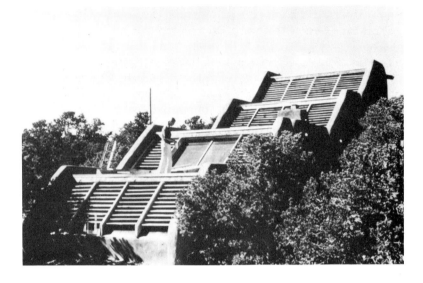

east and west sides. The floors are also made of adobe brick.[30]

Each of the four terraces has a south-facing roof of tempered double glazing 5 m × 2 m in four sections. These are insulated at night with internal shutters and protected against summer sun by fixed horizontal wooden slats on the outside (Fig. 4.19).

The incoming sunlight strikes the solid walls and floors and also hollow internal partitions filled with drums containing water. The total water volume is 0.35 m^3 with a heat capacity of 16 MJ/K. The 65 t of wall and 49 t of floor also add thermal mass whose heat capacity on a diurnal basis is roughly 70 MJ/K.

Back-up heating is provided by two woodburning stoves; but these are rarely needed. In the winter of 1975–76 only half a cord of wood – about 7.2 useful GJ – was burned.

Table 4.9 summarizes the performance of some of the houses discussed in the preceding section.

4.3.2 Attached sun spaces

The attached sun space in the Autarkic House shown in Fig. 4.32 is one of the very few examples of a solar space designed specifically to provide passive heating to the rest of the building in the UK climate. The performance was explored using hourly weather data as an input to the computer model which is described in Refs. [31] and [32].

Fig. 4.33(a) illustrates the floor plan and (b) shows insulating shutters between the conservatory and the house. In sunny weather these shutters would normally be open, allowing the transfer of heat acquired by the solar space to the rest of the house. In cold dull weather the shutters would be closed, thus isolating the glazed space (which would be cold) from the house. This facility to 'change the size of the inhabited volume' is an attractive feature of this type of design. When the living volume is diminished in this way, the amount of glazing to occupied areas is also

Passive solar heated buildings

Table 4.9. *Summary of direct-gain spaces for January and May (calculated)*[a]

Site and floor area (m^2)		Solar radiation penetrating appropriate glazing at correct tilt (MJ/(m^2 d))	South-facing area of glazing (m^2)	Approximate daily solar input (with no shading) (MJ/d)	Mean daily temperature (°C)	Approximate[b] heat loss through glazing without shutters (MJ/d)	Heat balance (MJ/d)
Terry House, Sante Fe (85)	Jan	14.4	36	518	−2	202	≈ +300
	May	16.9		608	+7	112	≈ +500
MIT V, Boston (88)	Jan	4.7	18	85	−1.6	27	≈ +50
	May	6.1		109	+0.5	25	≈ +100
Curtis, Rickmansworth (132)	Jan	3.2	61	202	+4.0	205	≈ 0
	May	5.8		360	11.0	108	≈ +250
Wallasey (2500)	Jan	3.2	520	1660	+4.0	1760	≈ −100
	May	5.8		3020	+11.0	900	≈ +2000

[a] By the authors.
[b] Mean internal temperature 18°C.

Fig. 4.32. The Autarkic House, Cambridge. (Photo by John Donat.)

reduced, the volume of the space is lowered, and the degree of wall insulation is increased. It should be noted that current Building Regulations require 10% glazing between a room which 'borrows' light and an adjacent conservatory. In turn, the conservatory then requires windows equal to 10% of its own, plus 10% of the room-next-door's floor area.

The sun space provides, in theory, extra and inexpensive space, and it can be fitted onto existing or new houses, thus giving one of the few credible ways of attacking the problem of existing buildings. The space is best considered as a general amenity rather than one for growing plants. Large areas of soil or foliage imply high levels of humidity with possible problems of odour and condensation. Further, there seems little basis in

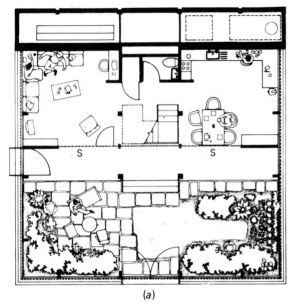
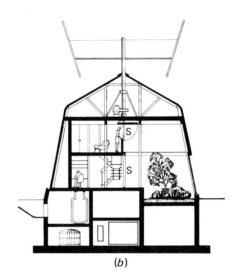

Fig. 4.33. The Autarkic House, Cambridge. Insulating shutters are marked 'S'. (a) Ground-floor plan. (b) Central north–south section.

fact for assuming that a few plants convert appreciable quantities of carbon dioxide into oxygen or, indeed, act in any way to clean the air.

The major advantages then of this design of passive heating are its potential low cost, the fact that it acts as a buffer between the outside and the inside (when the insulating shutters are closed) and the pleasant space created. Since the space is not *required* for occupation, its temperature can be allowed to swing more widely than that of a direct-gain room, and yet it moderates the temperature swing of adjacent rooms. Olgyay[33] illustrates the temperature excursions which are tolerable (Fig. 4.34). The conservatory can then be allowed to vary from about 12–30 °C before it becomes unusable in the UK, whereas an occupied space should preferably be kept between about 15 and 22 °C. (However, the growth of many types of plants would restrict the permissible temperature range in a conservatory.) Inherently then, this wider temperature range can lead to more energy storage and less rejection of surplus heat.

From the point of view of design, it is most difficult to predict the performance of the attached sun space. For one thing, the cost and thus the thermal integrity of the space can vary enormously and the performance will be very strongly influenced by the quality of construction. Furthermore, since the space will be occupied by people who may open windows or doors, the degree of human influence is high. Finally, if manual means are used for transferring heat to the rest of the house via switched fans or opening doors or moveable insulation, one cannot assume that these operations will be carried out at appropriate times – or even carried out at all!

During the design of the solar space in the Autarkic House, data were used for 1962–63 (a 'bad' winter) and 1964–65 (an 'average' winter). It

Passive solar heated buildings 131

Table 4.10. *Number of days (expressed as a fraction of the number of days of the indicated type) when the shutters between living area and sun space would be opened*[32]

Month	Overcast[a]	Intermittent	Sunny	Total
October 1962	9/10	11/11	10/10	30/31
November 1962	9/21	4/5	3/4	16/30
December 1962	0/16	0/3	4/12	4/31
January 1963	4/19	6/6	6/6	16/31
February 1963	2/15	5/6	6/7	13/28
March 1963	12/15	7/7	9/9	28/31
April 1963	10/11	8/8	11/11	29/30

[a] Overcast days are defined as those when the hours-of-bright-sunshine indicator is less than 0.1 (on a scale of ≈ 0–1 for every day-time hour). Sunny days are those which have an hours-of-bright-sunshine indicator greater than or equal to 0.5 for more than half of the day-time hours. Any other day is an intermittent one.

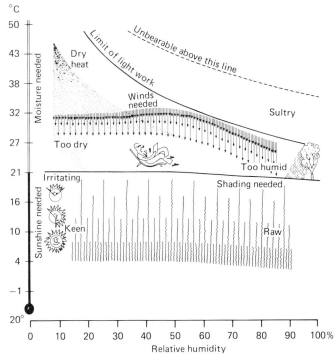

Fig. 4.34. Schematic bioclimatic index. (Reprinted with permission, Princeton University Press © 1963, from *Design with Climate: Bioclimatic Approach to Architectural Regionalism*, by Aladar Olgyay.)

was predicted that the sun space would acquire sufficient solar energy to raise its temperature to 18 °C on the number of days illustrated in Table 4.10. Thus, even in a bad year, the conservatory was sufficiently warm for use on 136 days out of 212 in the heating season. For an average year the corresponding number was 155 out of 212. Clearly, it was also warm enough in the summer.

During the calculation, allowance was made for the radiation incident

Table 4.11. *Statistics on energy collection in the attached sun space of the Cambridge Autarkic House*

Month	Number of days in the month when there is an excess	MJ energy saved by storing surplus day-time solar gain for heating in the:		
		Evening	Night	Next day
May 1962	27	0.0	0.0	23.0
June 1962	30	0.0	9.0	0.0
July 1962	30	0.0	16.6	4.0
August 1962	31	0.0	13.7	0.0
September 1962	30	0.0	5.8	0.0
October 1962	30	0.0	1.4	11.9
November 1962	9	1.1	0.0	37.8
December 1962	0	0.0	0.0	0.0
January 1963	1	1.4	0.0	0.0
February 1963	5	16.9	11.5	7.9
March 1963	19	10.8	6.8	46.1
April 1963	26	0.7	0.4	33.8
Totals	238	30.9	65.2	164.5

on the sloping glazing, reflection and transmission losses and thermal storage.

Surplus energy was stored for use later on the same or the following day. In fact, the number of days when more energy was supplied by the attached sun space than was needed to heat the house during the hours of sunlight was remarkably small (obviously these remarks refer only to the heating season). Table 4.11 illustrates the value or otherwise of thermal storage in this case. It seems clear, then, that thermal storage is required to smooth out the midday peak of radiation, but is not very gainfully employed over longer periods. The decision was thus made to limit storage to one day's excess solar input on a spring or autumn day, with due allowance for the inefficiency of charging a store.

When the radiation (on the horizontal — which is the only measured value available, and then not for all sites) has a value of 8 MJ/(m^2 d), typical of early March or early October, that incident on a south-facing vertical surface is about 9 MJ/(m^2 d). The radiation penetrating double glazing (suggested for the Autarkic House sun space) is thus about 6.3 MJ/(m^2 d). The space heating load of this particular low-energy house was about 72 MJ on a typical autumn day (including heat loss from the conservatory) and the glazed area was 30 m^2. Thus it can be seen that the amount of thermal storage chosen (36 MJ) was reasonable.

The principal choices to be made in the design of an attached sun space (apart from considerations of overshadowing, structure, access and so on) concern these points: Is the skin to be double or single glazed? How great should be the thermal mass? How is the heat to be transferred to the rest of the building and how is ventilation and/or shading to be effected for

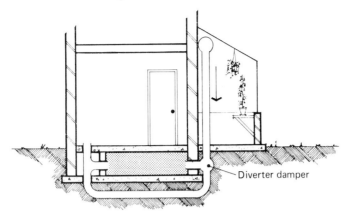

Fig. 4.35. Thermal storage for conservatories.

days when overheating might occur? To a great extent, the solutions to these problems depend on the future occupants and their demands on the space. If the space is to be used in the evening then single glazing with insulating shutters is probably the correct solution, whereas if the space is to be used predominantly as a solar collector it is probably better to use double glazing, although the cost is so high that it cannot be justified on energy grounds alone. In the view of the authors, *controlled thermal mass* in the form of a small rock-bed unit charged with warm air drawn from the top of the sun space and placed under an adjacent room in the house, or in the form of phase-change materials, is the sensible choice in the UK. In the latter case it is more difficult to use the stored heat for warming the rest of the building after sundown and this method is more appropriate to a conservatory which is used in the evenings. Warm air can be transferred from the space to the house by opening doors (which requires intervention) or by fans in a communicating wall, one at high level blowing from sun space to living room and possibly one at low level blowing in the reverse direction. A simple thermostat may control the fans, with an override if the living room rises above a preset temperature. If a rock-bed store is used, it may be possible to use the same fan system with a bypass either to the store or to the living room (Fig. 4.35).

Either system is expensive – one fan costs about £90 to install (1980), and simple high and low vents in the common wall, operated by mechanical greenhouse ventilator actuators may be more appropriate. The slot-area required can be calculated approximately:

Rate of heat transfer (kW) = $0.17 A (h)^{1/2} (Dt)^{3/2}$,

where

A = vent area in m^2;
h = height between vents in m;
Dt = temperature difference between upper and lower vents in K.

For example, a conservatory linked to an adjacent room by a high opening of 1 m^2 passing air at 25 °C in one direction, and a low opening of 1 m^2 passing air at 20 °C in the opposite direction should cause a heat-transfer rate of about 2½ kW. (In practice, the flow is likely to be less

Fig. 4.36. First Village Unit 1, Santa Fe, New Mexico, US (designed by Susan and Wayne Nichols).

because of vent shape and the restricted size of the roof in an attached conservatory, which tends to diminish the free flow of the column of air.)

Ventilation may also be required if the conservatory overheats and it is customary to provide an area of openings equal to 20% of the floor area of a greenhouse. However, this probably underestimates the requirement in an attached sun space whose surfaces, apart from the southerly one, are better insulated. In the height of summer 30–40% will be needed. It is more desirable, however, to limit the access of summer sunshine by shading, since windows and doors which can open are more expensive and will increase undesirable air infiltration in the winter. Sections 4.2.3 and 4.2.4 deal with some of the shading methods available.

One of the very few dwellings for which experimental information is available concerning the performance of an attached sun space is in New Mexico and is illustrated in Figs. 4.36 and 4.37.

This house consists of living accommodation folded round and sheltering a south-facing conservatory. In a terrace, a similar effect could be obtained by sinking the sun space in each house behind the south facade, rather than leaving it as an 'add-on' exposed piece. This concept is exploited in a proposal for a Steiner School residential building shown in Fig. 4.38. The glazed area need not be a conservatory and, as in the Steiner case, can serve as circulation and play space.

Returning to Fig. 4.37, it may be seen that the rooms are disposed suitably for morning and evening use.

The north wall of the conservatory, which leads to the other rooms in the house, is made of adobe, 350 mm thick on the ground floor and 250 mm thick on the first floor. The external walls are also about 300 mm thick and the floor is paved with adobe bricks. The space thus has a high heat capacity (see Section 4.2.6). In addition, hot air from the roof of the 6 m high sun space is extracted and blown down into two rock beds underneath the living room and dining room. The vertical (15 m^2) and

Passive solar heated buildings

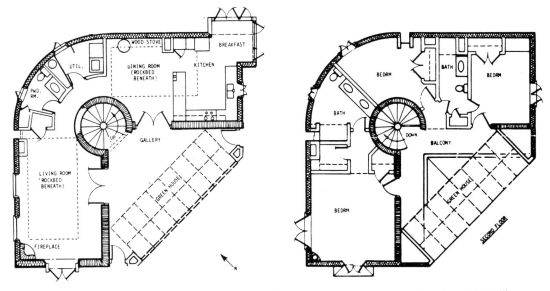

Fig. 4.37. First Village, Unit 1, floor plans.[39]

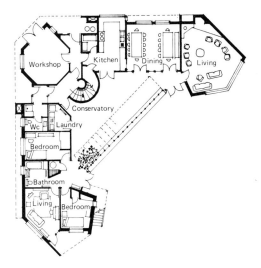

Fig. 4.38. Proposed new residential building for Steiner Schools. (Architects, Feilden Clegg Design.)

sloping south walls (25 m² at 50°) of the conservatory are double glazed. Fig. 4.39 illustrates the behaviour in winter. For example, on 28 December, as the insolation (on the horizontal) rose to 1.8 MJ/(m² h) at midday, the upper air in the conservatory climbed to about 42 °C, the air in the centre to 32 °C, yet the air outside was only 5 °C. The total insolation during the day was about 8.5 MJ/m² (horizontal) and, at that latitude and time of year, the ratio of radiation on a south-facing vertical surface to that on the horizontal[35] is about 1.2:1, and the ratio for a surface sloping at 50° is 1.3:1,[34] thus the total radiation incident is about 424 MJ/d. The transmitted value[34] is thus about 306 MJ/d. Now, heat loss via the

Fig. 4.39. Representative performance of First Village Unit 1 House, December–January 1978–79. Prepared by Los Alamos Laboratories.[39]

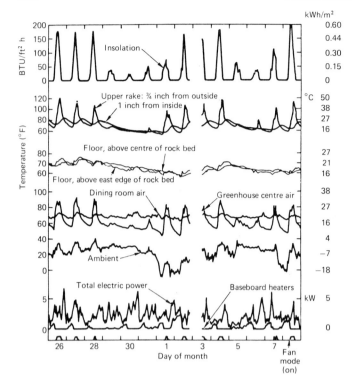

double glazing over a 24 h period with the temperature shown in Fig. 4.39 amounts to about 173 MJ (when the glazing is uninsulated at night), and the ventilation heat loss from the conservatory is about 94 MJ/d. Thus the conservatory gains a net 306 − (173 + 94), or 39 MJ/d. Some of this heat is put into the rock stores as indicated by the 'fan-on' marks. The fan was running for three to four hours on 28 December at a rate of 50 m^3/(h m^2) of glazing, and the temperature difference between the warm air pumped into the store and the store itself was about 6 °C. But only half of this temperature difference is useful for charging a store and, assuming a useful value of 3 K, one finds a heat transfer of 27 MJ. Of course, in reality the fans only came on because the heavy fabric of the greenhouse was not able to absorb heat fast enough in the middle of the day. The exposed surfaces of adobe brick in the sun space in the walls and floor would only absorb heat at a rate of about 7 MJ/h and at midday, this would be insufficient to prevent overheating − the rate of heat transfer via the fans would add a further 7 or more MJ/h.

The significant point to notice is the small back-up heat needed in the morning. In other words, in the climate of New Mexico, the house works beautifully. It is indeed a splendid example of sensitive design.

Our problem in the UK is not so simple.

Whereas the insolation in the New Mexico case was 8.5 MJ/(m^2 d), that on a good day in December or January in the average year, 1964–65 in the UK, was 3.2 MJ/(m^2 d) − by chance also on 28 December! (An average value would be 2.0 MJ/(m^2 d) for this time of year.) The ratio of

Passive solar heated buildings

radiation on the horizontal to south-vertical[34] is a healthy 1.6:1 and the same value is assumed for the 50° tilt. Thus the incident radiation in the UK would be 209 MJ/d. The transmitted value would be about 144 MJ/d. The heat loss via the glazing on that day would have been the same as in Santa Fe (173 MJ), and via ventilation probably greater, since in our humid climate an air change of 1/h might be inadequate — however, the same value, 94 MJ, is accepted in this calculation. Thus the heat balance is adverse (+ 144 − 173 − 94) or −123 MJ/d. The design would then require modification, and to be successful would probably move towards the isolable sun space of the Autarkic House. In this case, most of the heat loss would occur only during the daylight hours (being reduced to about 90 MJ/d) and the conservatory would then supply a surplus of 36 MJ to heat the house and the fabric of the greenhouse.

4.3.3 Thermal storage walls

Trombe and Michel built several houses, near the Solar Furnace at Odeillo, latitude 42°, in the Pyrenees, which use dark heavyweight walls covered with glazing in which solar radiation is absorbed during the day. Slowly the heat penetrates the walls to appear several hours later in the occupied rooms to the north. If heat is required during the day, vents can be opened to encourage air circulation — see Fig. 4.40. (In fact, as Balcomb has pointed out, this design was patented in the US in the last century but the description Trombe wall has remained.)

The advantages conferred by this method of passive solar heating are the lack of human intervention necessary, the fact that the wall can be structural thus off-setting some costs, and the straightforward nature of the system. However, it has several important drawbacks; possibly the worst of these in the UK context is the degree to which winter sunlight is prevented from reaching occupied spaces. (Thus a mixture of direct gain and Trombe wall becomes more desirable.) Secondly, the cost is high — the wall and foundations being far stronger than is needed for structural reasons. Thirdly, the system lacks the control desirable for a country with a solar regime which is intermittent. Finally, the appearance from the outside tends to be monolithic, and the nature of the blank wall influences internal layout.

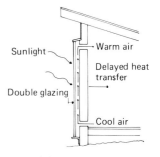

Fig. 4.40. Trombe wall. Double glazed Trombe wall showing convected heat rising from the warmed space. In practice there would be flaps on the slots to prevent cold air at night entering the room.

The system does however have possibly greater potential in the non-domestic sector, and the Benedictine monastery at Pecos in New Mexico is a good example of such an application. The office and warehouse attached to the monastery are 85% passively solar heated by a mixture of direct gain and a water wall. Table 4.12 indicates the heat storage capacities of various materials (see also Appendix 2).

It is clear that water has about twice the heat capacity of the other materials, and thus a hollow wall, or a series of steel drums, filled with water, will absorb more heat for the same volume. In addition, because of convection currents, the water will transfer the heat more rapidly, thus keeping the hot surface exposed to the sun at a lower temperature during the day. This reduces the heat loss outwards through the glazing.

The Pecos monastery shown in Fig. 4.41 contains a mixture of direct

Table 4.12. *Thermal properties of some storage materials*

	Specific heat (MJ/(m³ K))	Thermal conductivity (W/(m K))
Water	4.2	0.6[a]
Adobe brick	1.7	0.5
Common brick	1.7	0.7
Dense concrete	2.6	1.7

[a] In a configuration which encourages convection, this figure will be much greater, and a water wall will tend to have the same temperature throughout its thickness.

Fig. 4.41. Pecos Monastery, office warehouse.

gain (100 m²) and water wall (40 m²), all of which is double-glazed. The wall consists of 140, 200 l oil drums painted black, filled with water and treated with corrosion inhibitors. Beyond the glass, thin aluminium shutters provide some night insulation and serve to screen out the sun in summer. The drums are in a container which can be opened to admit more heat to the building and which is insulated with 100–150 mm of glass fibre. The building is well suited to passive gain since the warehouse section can tolerate wide temperature swings and low minima, although, in fact, whilst the temperature can go down to −13 °C, the normal daily swing is only ±3 K. The swing in the direct-gain space housing the offices is greater, at ±4–5 K.

The first of the Odeillo Houses (Fig. 4.42), designed by Felix Trombe and Jacques Michel and built by the Centre National des Recherches Scientifiques (CNRS), was established in 1967 and has a floor area of 80 m² with four main rooms. The walls have U-values of 1.3 W/(m² K) and the windows (20 m²) are single glazed with areas of 6 and 8 m² on the south and north sides respectively. The gap between the double glazing and the

Passive solar heated buildings 139

Fig. 4.42. Trombe wall house at Odeillo.

black painted Trombe wall (48 m²) is 120 mm. Some of the collected heat is passed into the house via air naturally circulating through slots at the top and bottom of the wall. The delay in heat penetrating the wall is 14–16 h for a thickness of 600 mm.

The house is 60–70% solar heated, and it is rare for the rooms to be too cold (occasionally this happens on winter mornings), or too warm (some autumn afternoons). According to CNRS,[36] such buildings in the north of France (Chauvency le Château, Meuse) achieve only 35–45% solar space heating.

It is significant that the average duration of sunshine during the heating season at Odeillo is 10 h/d.

A second series of (attached) houses was built as a terrace at Odeillo in 1974 (Fig. 4.43). Their superior insulation ($G = 1$ W/(m³ K) against 1.63 W/(m³ K) for the earlier houses) permits a reduction of collector area from 0.16 m²/m³ to 0.1 m²/m³ of inhabited volume. This reduction allows more ordinary windows to be used on the south facade. The wall U-values are about 0.4 W/(m² K). (G-values are global heat-loss coefficients from a space.) Table 4.13 indicates some of the parameters.

On a gloomy day (24 December 1974) when the radiation striking the glazing was only 1.15 MJ/(m² d) no temperature rise was observed on the Trombe wall which in fact consistently declined in temperature. Fig. 4.44, however, illustrates the behaviour of the 1974 houses on a better winter's day. The volume of air flowing between the slots cut in the lower and upper parts of the wall (see Table 4.13) is significant. On the day described in Fig. 4.44 the maximum flow for a temperature difference between the slots of 22 °C was 70 m³/h (see Section 4.3.2), representing a heat transfer of about 550 W for each panel of the collector, each of area 3.9 m². Many Trombe walls are fitted with dampers to control this movement since the air temperature may easily rise by about 30 °C in summer. Such dampers

Table 4.13.[a] *Data for the Odeillo Houses*

	Houses built in 1967	Houses built in 1974
Number of houses	1	3
Floor area	76 m²	180–250 m²
Collector area m²/house volume m³	0.16	0.1
Back-up heating	Electricity	Electricity
Trombe wall material	48 m² 600 mm concrete 2200 kg/m³ k = 1.75 C = 0.6	45–65 m² 370 mm concrete 2200 kg/m³ k = 1.75 W/(m K) C = 0.32 MJ/(m³ K)
Trombe wall surface	Black acrylic paint	Black acrylic paint
Trombe wall glazing	2 panes 3 mm glass	2 panes 4 mm glass
Air path	Glass to wall 120 mm	Glass to wall 120 mm
Circulation vents	3.5 m apart 565 mm × 110 mm	2.20 m apart (vertically) 840 mm × 95 mm
Each section of wall	4.4 m high × 1.3 m wide	2.5 m high × 1.6 m wide

[a] Data from Ref. [36].

Fig. 4.43. Trombe wall house at Odeillo.

Passive solar heated buildings   141

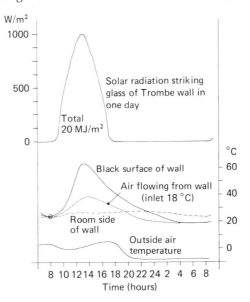

Fig. 4.44. Data concerning Trombe wall (1974) house for 23 December 1974.[36]

should be designed (for example, from light flaps of polythene) to prevent a reverse flow at night.

The time delay in heat transmission *through* the wall is large (14–16 h for the thick and 9–10 h for the thinner wall). In fact, this rate and the inner temperature depend on moisture content and the walls take at least a year to dry out.

The long-term efficiency of solar radiation capture is between 20 and 40% in the heating season, and at that southerly latitude it diminishes to less than 20% in summer. (These values refer to insolation on the horizontal.) The ratio of heat transferred by conduction through the wall to convection of the air is about 60:40 and the solar contribution to the space heating load is from 60 to 70%.

It is interesting to speculate on what would happen if these houses were built in the UK. When similar buildings are placed in Meuse, to the northeast of Paris, they perform less well, contributing 35–45% of space heating. *En passant*, remember that in New Mexico a good winter's day supplies 11 MJ/(m² d) to a vertical surface; in the UK the value is only 5.5 MJ/(m² d), which is very much less than the value of 20 found in Odeillo.

Now, data from Odeillo indicate some days with radiation at the UK level, and the data are reproduced in Fig. 4.45. The heat transferred on that day in the warm-air stream was less than 108 MJ/m² and that through the wall less than 215 MJ/m² of wall. The collection efficiency was between 5 and 10%. However, in winter, even days with radiation levels of 5.5 MJ/m² are rare in the UK.

One way to improve the performance of a Trombe or water wall is to provide night insulation. In the Baer House[38] the walls of water-filled steel drums are covered at night with large insulating panels which lie on the ground during the day, acting as reflectors (see Fig. 4.46). In the

Fig. 4.45. Data from Odeillo on Trombe wall house for 22 February 1975.

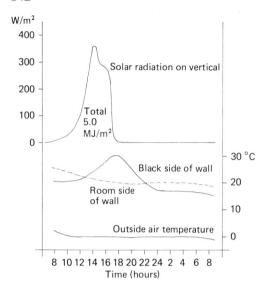

Fig. 4.46. Baer House.[52]

Tyrrell House (Fig. 4.47), polystyrene beads are blown upwards between the panes of the double glazing covering the Trombe wall, with a small fan, at night, in the so-called Beadwall system.

Fig. 4.48(a) illustrates a terrace of single-storey houses using Trombe walls and built near Liverpool. Fig. 4.48(b) shows their floor-plan and Table 4.14 summarizes some of their design parameters.

There is a large gap between the black wall and the glazing (about 600 mm), and the solar wall is pierced by the front door and lounge window. External canopy blinds provide summer shading.

The warm air from the solar wall rises to the roof space to be directed through the house from the north perimeter back to the base of the collector. Fans can be used to provide forced circulation.

Passive solar heated buildings

Table 4.14. *Acorn Close Houses*[37]

Trombe wall	Double glazed Black-brick density 2200 kg/m^3 Thickness 225 mm $U = 1.21$ W/(m^2 K) Width 6 m, height 3.8 m (approximately) Collector area m^2/house volume m^3 = 0.13
Building	Double glazed Cavity brick walls with 100 mm glass fibre, internal brick of density 2000 kg/m^3 $U = 0.29$ W/(m^2 K) Floor $U = 0.32$ Ceiling $U = 0.32$ Roof $U = 0.31$ Floor area 60 m^2
Back-up heating	Electricity

Fig. 4.47. Tyrrell House.

Table 4.15 illustrates the annual amounts of electricity used as auxiliary heating.

Some interesting conclusions can be drawn. Firstly, the *means* for space heating are 11.6 (Trombe) and 21.5 (conventional). The standard deviations are 3.7 and 5.3 respectively. From such a small sample it is thus not possible to indicate how much less fuel is likely to be consumed by such passive houses. Secondly, in spite of their occupants (elderly people, some on their own and with no children), heating hot water accounts for a large fraction of the total energy use. Thus, not all our attention should be focussed on reducing space heating loads.

A second terrace of low-cost maisonettes built with unvented mass walls and using Maxsorb selective surfaces is shown in Fig. 4.49.

144   Passive solar design

Table 4.15. *Annual fuel use in the conventional and passive solar houses at Acorn Close, between 8 September 1980 and 7 September 1981 (GJ)*

	Space heating	Water heating
Trombe houses		
3	10.3	5.7
4	8.2	6.6
5	18.9	7.0
6	10.4	4.9
7	9.5	4.6
Conventional houses		
8	15.8	8.6
9	24.6	5.7
10	15.2	6.8
11	23.2	7.8
12	28.8	3.9

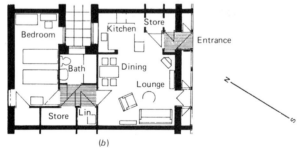

Fig. 4.48. (*a*) View of the solar terrace at Bebington looking west. (*b*) Floor plan of the sheltered housing at Bebington.

4.3.4 Roof ponds

Roof ponds are used where nocturnal cooling is useful in the summer and the sun is high enough in the sky in the winter to provide heat on a horizontal plane. They are not appropriate for homes in more-northern latitudes, but the concept may be useful for those commercial buildings which require cooling.

Harold Hay's House in Atascadero, California (outdoor temperatures

Passive solar heated buildings

Fig. 4.49. Low cost maisonettes in Bath using mass walls with selective surfaces (designed by Feilden Clegg Associates).

8–28 °C) is completely heated and cooled using the system illustrated in Fig. 4.50.

On the roof are 23 t of water in clear plastic bags with black polythene underneath. Insulating panels can be drawn over the water bags, and extra water can be sprayed on in summer. In hot weather, at night, the panels are opened allowing the bags to cool by radiation to the cold night sky, and on the following day the panels are closed permitting the cold bags to cool the ceiling (which is of steel decking). In winter the panels are opened during the day, allowing the bags to store solar energy.

The system has not been tried for cooling commercial buildings in the UK and the scheme might not work so well because our skies are much less clear than those over the desert areas of California.

Alternative methods of cooling buildings naturally are, of course, avail-

Fig. 4.50. Thermal storage roofs, summer and winter operation.[35]

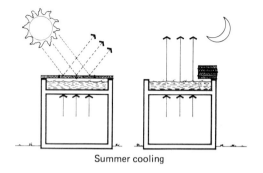

Summer cooling

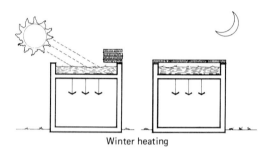

Winter heating

able and the wind-scoop towers of Persia and Egypt, which encourage draughts to pervade buildings, are well known.

4.3.5 Roof-space collectors

When a traditional pitched roof requires replacement, the question obviously arises 'is it worth glazing?'

Intuitively, a south-facing glass roof starts with several advantages over other forms of passive solar heating. No intrusion is made into living space, no extra space is added, the existing fabric of beams may be used and, perhaps most important of all, the roof is less likely to be overshadowed by neighbouring buildings.

The intuitive objection raised is that such shallow slopes (say 30°) *reflect* a great deal of the solar radiation striking the glazing. However, this is not true even for direct sunlight; loss in clear-sky radiation penetrating double glazing at a slope of 30° compared with a vertical window is less than 10% in the UK during a heating season.

In addition, much of our radiation, particularly during the heating season, is diffuse, coming partly from the direction of the sun but also from the whole sky-vault, which favours a sloping surface.

There are, of course, severe practical problems associated with trapping warm air in roofs. The space is notoriously leaky, the timbers may suffer from the high temperatures and the daily temperature cycling, and this problem may be particularly acute with the trussed-rafter construction common today.

A solar roof should be darkly coloured throughout on the inside and the unglazed surfaces must be insulated. Since the space is of light thermal

Passive solar heated buildings

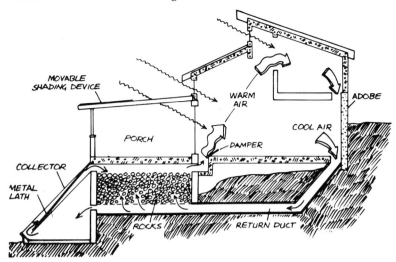

Fig. 4.51. Convective loop.

weight it will respond quickly to radiation, and either the warm air must be blown elsewhere, into the house or into a rock store, or thermal mass must be introduced into the loft. There is virtually no experience with this kind of system, but it shares with an attached sun space the advantage that the occupants can tolerate wide swings in its temperature.

At night, unless the glass is covered, nocturnal cooling may well reduce the temperature very drastically, thus the ceilings of the rooms below must be insulated. If the glazing is left uncovered at night, thermal storage should not be introduced into the roof space, otherwise it will become a 'coolth' storage.

An actual roof collector system is described in Chapter 12.

4.3.6 Convective loops

The final type of 'passive' solar energy system to be mentioned bears similarities to active thermosiphoning designs for heating domestic hot water. Fig. 4.51 illustrates a solar collector situated below the space to be heated, which allows warm air to rise either into thermal storage or into the space to be heated. Cool air may then return to the collector or be vented.

This type of system shares many of the problems inherent in active collectors — the possibility of very high temperatures in the unlikely event of stagnation of the air, high absorber-plate temperatures following from poor heat transfer to the air and thus a low collection efficiency and, from the design point of view, the collector is an 'add-on' piece of hardware. Since the air flow is not fan driven, it is below the desired rate for high performance efficiency, which affects the economics of the situation. Any rock storage system attached to a convective loop will introduce an extra pressure drop which may drastically reduce the performance of the collector.

Baer[38] recommends that the width of the flow path should be $\frac{1}{15}$ of the length from top to bottom of the collector — thus a 3 m length implies a flow path 200 mm wide; in addition to which must be added spaces between glazing and collector plate, and a thickness for insulation. The return-duct cross-section should be 5% of the collector area.

In very sunny conditions, the air flow in such a collector can be adequate; but in the UK weather it is doubtful whether such a system will work very well. Balcomb *et al.*,[39] for example, quote the very low value of 2 Pa for the pressure driving air around such a system on a good day (with an average collector temperature of 49 °C and a room (thus collector inlet) temperature of 21 °C). Of course, on the other hand, when there is no sunlight, the collector ceases to work forwards, but precautions must be taken to avoid it working backwards, when it produces a heat loss!

The principle of convective loops can of course be applied in many situations without formal acknowledgement. For example, when a greenhouse is placed below a rock bed or fusion-storage system, or an attached sun space transfers heat by convection into living rooms.

Ref. [39] presents an excellent review of the convective loop system; however, it is probably not the most suitable passive system for northern European conditions.

4.4 Design methods for passive solar buildings

4.4.1 Introduction

About 50 programs simulating energy use in buildings have been reviewed by one of the present authors[40] in order to identify those which could offer the possibility of use as comprehensive models for the evaluation of passive solar designs (see Table 4.16).

The major attributes necessary are discussed below and three models are identified as potentially the most suitable. An indication is provided of current deficiencies in the design tools presently available.

There are several ways in which the success, or likely success, of a particular passive solar design for a building, may be assessed — other than by intuition — and in terms of energy and cost.

The most expensive way, and the least flexible, involves constructing each design in a particular climate and observing its annual performance — for a variety of occupants and a variety of years with varying weather. The BRE method[41] of simulated occupancy removes some of these variables.

A much less costly method would use reduced scale models;[42] but so far, the models used have been of the Los Alamos Test Cell variety measuring about 2 m × 2 m × 3 m (high), which do however allow side-by-side comparison of various methods of passive solar capture in heated or free-running mode.

The third method of assessment would involve 'perfect' analogue or digital simulation models. Such a perfect model would be able to handle changes of climate, location, occupants and design. Ideally it would be interactive with the designer — it should be capable of quickly answering all the 'what if?' questions of the type 'should I sacrifice 2 m of rear northerly garden to reduce overshadowing by buildings on the other side of the street?'; 'is it cost-effective to install the hi-tech windows on the

Design methods for passive solar buildings

Table 4.16. *Possible passive models*

Program name	Location	Why the model is not a prime contender as a passive simulator, at time of printing
BLAST 3	CERL,[a] LBL[b] (US); Boeing CDC[c] Time Share (UK)	
DEROB 4	Univ. Austin Texas (US)	
DOE 2.1	LASL,[d] LBL[b]	
FREHEAT	Colo. State Univ. (US)	Limited documentation, only one zone, only mass wall or direct gain
SOLAR 5	UCLA,[e] Architecture (US)	Primarily an excellent building description model not an 'energy' model
SUNCODE	Ecotope Inc., Seattle (US)	
SYRSOL	Syracuse Univ. (US)	Uses only the sol-air method
TRNSYS	Univ. Wisconsin (US)	Heavily weighted towards active solar systems
2 ZONE	LBL (US)	
UWENSOL	Univ. Washington, Seattle (US)	
ESP	Abacus, Strathclyde (UK)	
SUNSPOT	LASL (US)	Excellent research tool for direct gain — but not organized for general users
PASOLE	LASL (US)	Excellent research tool for mass walls — but not organized for general users
THERM	Watson House, London	
TASS	Cranfield (UK)	
BEEP	CEGB[f] (UK)	Private code not accessible for changes or inspection of methods. Incomplete documentation, not user friendly.
BUILD	Univ. Nottingham (UK)	
HOUSE	ECRC[g] (UK)	
NBSLD	NBS[h] (US)	Superseded by BLAST
UWIST	UWIST[i]	Not documented.
BRISTOL	Univ. Bristol, Architecture (UK)	
UMIST hybrid	Building Dept., UMIST[j]	

[a] Construction Engineering Research Laboratories.
[b] Lawrence Berkeley Laboratories.
[c] Control Data Corporation.
[d] Los Alamos Laboratories.
[e] University of California at Los Angeles.
[f] Central Electricity Generating Board.
[g] Electricity Council Research Centre.
[h] National Bureau of Standards.
[i] University of Wales, Institute of Science and Technology.
[j] University of Manchester, Institute of Science and Technology.

Table 4.17. *Passive design strategies*

	Options					Surface treatment	
	Thermal storage						
	Storage walls						
Capture strategy	Isothermal	Non-isothermal	pcm[a]	Storage under floor	Insulated isolated storage	Painted	Selective coating[b]
Direct gain	✓	✓	✓	–	✓	✓	–
Attached sun space	✓	✓	✓	✓	✓	✓	–
Mass wall	✓	✓	✓	–	–	✓	✓
Trombe wall	✓	✓	✓	–	–	✓	✓
Thermosiphon	–	–	✓	✓	✓	✓	✓
Double envelope	–	–	✓	–	–	✓	–
Roof-space collection	–	–	✓	✓	✓	✓	✓

[a] Phase-change material, for example melting at 25 °C.
[b] Solar absorptive, but with low emissivity in the infra-red (for heating), or the converse (for cooling).

south side which only start to reject sunlight when the northerly rooms are above 22 °C?', and so on. Needless to say, such a perfect model does not exist.

4.4.2 Characteristics of a perfect design model

A perfect model would answer all possible questions and thus the problem of which criteria to use for the assessment of passive designs may be left aside. It needs to be said though that the criteria are not clear — does one use solar load ratios; annual back-up demand; annual solar energy usefully used per m² of floor area, per m² of window, etc? This problem has not been resolved.

The model would ideally be comprehensive, cheap to run and contain fast input and readily understood output; but these requirements conflict, and the only models which approach even one requirement — the desired degree of flexibility — are of medium size (say 5000–10 000 lines of FORTRAN code) running (at present) on main frames.

Table 4.17 indicates the range of strategies which a simulation model should practically be expected to attack. The great scope of physical problems which is implied by these strategies is indicated in Table 4.18.

4.4.3 Models approaching the required degree of flexibility

In 1979 The Design Office Consortium (DOC)[43] produced a report concerning 46 'Models for Energy in Buildings'. In 1980[44] Lebens assessed ESP, DEROB 4, HOUSE and THERM and James[45] examined ESP, DEROB 4 and PASOLE for use as passive simulators. Together with the results of a SERI[46] study, these reports provided a basis for drawing up the list of models shown in Table 4.16.

Some of the models can be discounted for this particular application on the grounds indicated in Table 4.16.

Design methods for passive solar buildings

Table 4.18. *Some of the physical properties to be modelled*

Radiation behaviour at glazing (multiple glazing, glass, plastic, infra-red and visible)

Radiation exchanges in spaces (on room and furniture surfaces)

Air movement causing heat movement (under the influence of temperature, gradients and infiltration) in zones and between zones and via doors, corridors, vents and stairs

Insolation and overshadowing

Building geometry

Control systems

Back-up heating systems

Infiltration

Heat loss to the ground

Comfort models specifically addressing passive buildings

Occupancy

Appliance usage

Moveable window insulation

In Table 4.19 the residual models are presented in more detail. This shortlist may then be reduced to Table 4.20. The simulation models, then, which seem potentially capable of handling passive systems are these: BLAST 3, DEROB 4, ESP (and whilst SUNCODE is specifically written for passive applications, it is designed to fit on a mini, and thus suffers restrictions). THERM is omitted because of its limited shading treatments. Table 4.21 indicates which characteristics can be handled by the four models.

It is probably fair to say that none of the models is yet in a state where it could be used fairly to evaluate passive solar designs. ESP and DEROB 4 currently offer the frameworks necessary within which (with changes and additions) most of the design questions could be answered. SUNCODE is a very convenient package which fits on medium-sized machines. For the smaller office it probably represents the optimal model available.

BLAST 3 has the framework which could accommodate all the routines necessary but, being a response-factor method with a 'time history', has problems in coping with rapid changes such as movable insulation. It seems likely that within a few years the outstanding problems associated with the models will have been solved, resulting in sophisticated design tools. The major problem areas seem to be: air movement (natural, buoyant and forced by infiltration, and fan forced); heat loss to the ground and via floors; model validation; better information concerning the use of passive buildings by occupants; and problems concerning comfort models specifically relating to passive designs.

Table 4.19. *Some details concerning prime contenders for passive models*

Program name	Research oriented	Architect oriented	Hourly time steps	Time period of calculation	Calculation method[a]	Reasons why a model is less suitable than others for passive simulation
BLAST 3	√	√	√	Any	rf	
DEROB 4	√	√	√	Any	fd	
DOE 2.1	√	√	√	Any	rf	Has been forced to handle passive and remains heavily weighted with HVAC plant
SUNCODE	√	√	√	Any	fd	
2 ZONE	√		√	Any	rf	Limited to 2 zones
UWENSOL	√		√	Any	fd	Not very well documented
ESP	√	√	√	Any	fd	
THERM	√		√	Any	rf[b]	
TASS	√	√	√	Any	rf[c]	Excellent input and output but less detailed in its handling of other problems
HOUSE	√		√	Four-day slots	fd	Only deals with four-day slots
BRISTOL Analogue	√		√	Any	Hard wired	Being analogue, presently difficult to alter[d]
UMIST Analogue	√		√	Any	Hard wired	Being analogue, presently difficult to alter[d]
UWIST	√		√	Any	fd	Essentially a research tool without the detailed effort necessary at input and output stages for ease of use

[a] finite difference = fd; response factor = rf.
[b] But the boundary temperatures are the wall surfaces not air temperatures.
[c] But an fd subset can be run for specific problems such as condensation.
[d] This position could alter with advances in analogue technique. The analogue method has great advantages of speed but disadvantages of linear treatments of radiation.

Design methods for passive solar buildings

Table 4.20. *Shortlist of models for passive use*[e]

Program name	Number of zones model can handle	Does it have shading routines by: Other	Self	Comparison of[a] computing units for superficially similar jobs	Machine space in thousands of words	Availability on a commercial basis	Machine type
BLAST 3	10	Yes	Yes	140	170	Public domain	CDC
DEROB 4	Maximum 9 in practice	Yes[b]	Yes	600	120	£250[c]	CDC
ESP[f]	Multiple	Yes	Yes	500	120	£5000	DEC 10 Prime 450 Honeywell 6060 HP 3000
SUNCODE[g]	10	No[d]	Yes	20	50	£750	DEC 10 and VAX
THERM	Multiple	Yes but crude	Yes	160	60	Time share only	

[a] These numbers predominantly reflect the difference between rf and fd treatments.
[b] Some users have experienced software crashes if the other buildings are completely detached.
[c] The *excellent* graphics are not included for this price.
[d] A shading 'horizon' is used.
[e] More models may be added in 1983 as documentation improves.
[f] Chosen for development by the European Commission in its Passive Solar Programme.
[g] Chosen for use by the UK Department of Energy and the US Department of Energy, in their Passive Solar Programmes.

Table 4.21. *'Best' codes*

	Topic handled	ESP[i]	BLAST 3	DEROB 4	SUNCODE
Designs	Direct gain	√	√	√	√
	Attached sun space	√	√	√	√
	Thermosiphon	–	√	√	√
	Roof-space collector	√	√	√	√
	Double envelope	–	–	Difficult	–
	Mass walls vented	√	√	√	√
	Mass walls unvented	√	√	√	√
	Under floor rock beds	√	√	√	√
Physical problems	Air–heat movement by convection – by fans	√	√	√	√
	Infiltration	*j*	*j*	*j*	*j*
	Solar radiation mapping round spaces[o]	√	√	√	*k*
	Variable glass emissivity	*m*	–	√	√
	Variable room colour	√	√	√	√
	Effect of furnishings	Difficult	Difficult	Difficult	–
	Air–temperature–stratification	*n*	–	√	–
	Moveable window insulation	–	Schedule	Schedule	Schedule
	Isothermal and non-isothermal storage	√	√	√	√
	Phase change walls	–	√	–	√
	Isolated storage	–	√ see Ref. [47]	√	√
	Adequate handling of beam radiation	–	√ Ref. [47]	√	√
	Adequate handling of diffuse fadiation	√	√	√	√
	Adequate treatment of sky temperature	–	–	–	–
Weather input	Complete 'set'	√	√	√	√
Daylighting		√	√	√	–
Surface temperature for providing comfort temperatures		√	Difficult	√	√
Validation		*a*	*d*	*b*	*c, d*
Documentation		v. good	poor	good	v. good
Graphics output		excellent	–	excellent	*p*
Building input via digitizing tablet		√	–	in principle	–
Building input format		Cartesian coordinates	Cartesian coordinates	Assemble standard shapes	*l*
Internal tables of construction parameters such as insulation values		√	√	–	√
Is it continuing to be improved?		√	√	√	√
Comments		*f*	*g*	*e*	*h*

Design methods for passive solar buildings

Notes to Table 4.21

^a Limited comparisons have been made between simulated (ESP) and observed data from conventional terraced houses in Scotland.[50]

^b DEROB 4 has been compared with results from Los Alamos Test Cells[51] with quite satisfactory results, and has been compared,[47] using runs on imaginary buildings with very large solar apertures, with BLAST 3, SUNCODE and DOE 2.1, with poor results. These poor results were due to convergence errors and subsequent runs are believed to be satisfactory.

^c SUNCODE has been validated against results from Los Alamos Test Cells and against a battery of analytic tests on simple enclosures, where the final results can be calculated exactly.

^d Recently (1982) SUNCODE and BLAST 3 have been extensively compared with experimental data on a conventional house at SERI with very good results.

^e (1) There may be problems with the convergence accelerators used and in some cases this may have led to simulation results at odds with identical runs on other large models (see Ref. [47]). There are four degrees of convergence. The first level corresponds to that achieved in DOE 2.1,[48] the third iterates on advection and the fourth on all heat transfers. Instabilities have also crept in when thermostat dead bands were at odds with the time stepping through the day. The CPU time can rise by a factor of eight at the highest convergence.
(2) DEROB 4 is exceptionally good at setting up the model equations for the user.[49]
(3) The overshadowing routines may be confused by large windows since the centre of the window is used as the viewing point for calculations.
(4) DEROB 4 is one of the longest established codes and in general ought for this reason to be bug free.

^f It is possible that fairly detailed air-movement routines will be available to run with ESP in the near future.

^g Work is in progress on adding a description of air movement.

^h SUNCODE does not have a ray-tracing facility.

ⁱ At present, ESP does not have routines called 'attached sun-space', 'Trombe wall', etc. but those ticked in the table can be simulated by use of the multiple-zone facility. Named routines are likely to be available in 1983–84.

^j The models can amend the infiltration value which is an input variable, according to wind speed and internal–external temperature differences.

^k The fraction of solar radiation absorbed by the walls or the floor in each zone is an input variable.

^l SUNCODE does not use a building plan. It accepts wall etc. dimensions.

^m A separate package called 'WINDOW' allows any window with multiple layers and coatings to be inserted into ESP.

ⁿ Under implementation for 1983–84.

^o Mapping could be handled by dividing surfaces into multiple zones. In all models this greatly increases computing time.

^p Under implementation 1983.

References

[1] Davies, M.G. (1975). 'Heating buildings by winter sunshine', *Building Science Special Supplement on Energy and Housing*, pp. 53–6.

[2] Pilkington Brothers (1979). *How Windows Save Energy Technical Appendix*. Pilkington Brothers R & D Laboratories, Ormskirk, Lancashire, UK.

[3] Pilkington Brothers (1978). *Glass and Glazing Federation Glazing Manual*. Pilkington Brothers R & D Laboratories, Ormskirk, Lancashire, UK.

[4] Linsley, G.F. (1977). 'Glazing flat plate solar collectors', *Practical Aspects of Domestic Solar Water Heaters.* London, UK Branch of the International Solar Energy Society, London.

[5] Berman, S.M. & Silverstein, S.D. (1975). *Energy Conservation and Window Systems.* National Technical Information Service, Springfield, Virginia, US.

[6] Littler, J.G.F. (1979). 'Thermal balance at windows'. *J. of Energy Research*, 2, 173.

[7] Balon, P. (1980). 'Save energy with ordinary roller shades'. *US Popular Science*, January, p. 54.

[8] Stepler, R. (1979). 'Insulating shades'. *US Popular Science*, October, p. 98.

[9] Information can be obtained from Appropriate Technology Corporation, Box 975, Brattleboro, Vermont, US.

[10] Information can be obtained from Thermal Technology Corporation, Box 130, Snowmass, Colorado, US; Guardia Shutters, Gallery House, Dunstable Street, Ampthill, Bedford, UK, and The Insulating Shade Company, Box 282, Branford, Connecticut, US.

[11] Clegg, P. & Watkins, D. (1979). *Home Energy for the Eighties.* Garden Way Publishing Co., Vermont, US.

[12] Rosengren, B. & Morawetz, E. (1976). 'The Termoroc house'. *European Solar Houses*, London UK. Branch of the International Solar Energy Society, London.

[13] Information can be obtained from Thermoblind Insulated Window Shutters, Waldorf Way, Dale Road, Wakefield, UK.

[14] Shurcliffe, W.A. (1977). *Thermal Shades and Shutters.* Shurcliffe Publishing, 19 Appleton Street, Cambridge, Massachusetts, US.

[15] Anderson, B. & Riordan, M. (1976). *Solar Home Book*. Cheshire Books, Harrisville, New Hampshire, US.

[16] Esbensen, T.V. & Korsgaard, V. (1976). 'Dimensioning the solar heating system in the Zero Energy house in Denmark'. *European Solar Houses*, London, UK Branch of the International Solar Energy Society, London.

[17] Bruno, R., Hermann, W., Horster, H., Kersten, R. & Mahdjuri, E. (1976). 'The Philips experimental house'. *European Solar Houses*. London, UK Branch of the International Solar Energy Society, London.

[18] Stepler, R. (1978). 'The solar laboratory'. *US Popular Science*, June, p. 96.

[19] Berlad, A., Jaung, R., Yoh, Y. & Tutu, N. (1978). 'Transport control in window systems'. *Mechanical Engineering Department*

Report. State University of New York, Stony Brook, New York, US.

[20] Olgyay, V. & Olgyay, J. (1957). *Solar Control and Shading Devices*. Princeton University Press, Princeton, US.

[21] Information can be obtained from SunTek Associates, 500 Tamal Vista Road, Corte Madera, California, US.

[22] McDaniels, D., Baker, S., Kaehn, E., Lowndes, D., Mathew, H., Reynolds, J. & Gray, R. (1975 and 1978). 'Time integrated calculation of the insolation collected by a reflector/collector system'. *Solar Energy*, 17.5, 277 and 20.5, 415.

[23] Schneider, M. & Berger, X. (1978). 'Utilisation de parois a chaleur latente'. *Groupe d'Ecothermique Solaire du Centre National de Recherches Scientifiques*, Nice, France.
and
Bordeau, L. & Jaffrin, A. (1978). 'Phase change collector wall versus water collector wall'. *Groupe d'Ecothermique Solaire du Centre National de Recherches Scientifiques*, Nice, France.

[24] (a) Information can be obtained from the Architectural Research Corporation, 40 Water Street, New York, US, and from the (c) Calor Solar Group, Calor House, Windsor Road, Slough, UK. Reference should also be made to (b) Eissenberg, D. & Wyman, C. (1980) 'What's in store for phase change?' *Solar Age*, May, p. 12.

[25] Balcomb, J.D. (1979). 'Designing fan forced rock beds'. *Solar Age*, November, p. 44.

[26] Kreith, F. & Kreider, J.F. (1975). *Solar Heating and Cooling*. McGraw Hill, Washington, US.

[27] American Institute of Architects (1978). *A Survey of Passive Solar Buildings*. American Institute of Architects Research Corporation, Washington, US.

[28] Curtis, E.J.W. (1974). 'Solar energy applications in architecture'. *Low Temperature Solar Collection of Solar Energy in the UK*. London, UK Branch of the International Solar Energy Society, London.

[29] Shurcliff, W.A. (1978). *Solar Heated Buildings of North America*. Brick House Publishing Company, Andover, Massachusetts, US, pp. 105–10.

[30] Shurcliff, W.A., ibid., pp. 186–7.

[31] Littler, J.G.F. & Thomas, R.B. (1984). *The Autarkic House*, to be published.

[32] Littler, J.G.F. & Thomas, R.B. (1979). 'Energy use in an autarkic house'. *Transactions of The Martin Centre*, Cambridge University, p. 2.

[33] Olgyay, V. (1963). *Design with Climate*. Princeton University Press, Princeton, US.

[34] Basnett, P. (1975). 'Estimation of solar radiation falling on a vertical surface from measurements on a horizontal plane'. *Electricity Council Research Centre Report M 846*, Capenhurst, UK.

[35] McFarland, R.D. & Stromberg, P. (1980). *Passive Solar Design Handbook*, editors Balcomb, J.D. & Anderson, B., Department of Energy, Washington, US.

[36] Trombe, F., Robert, J.F., Cabanat, M. & Sesolis, B. (1979). 'Some

performance characteristics of the CNRS house collectors'. *Passive Collection of Solar Energy in Buildings.* London, UK Branch of The International Solar Energy Society, London.

[37] Greenwood, P. & Ward, H. (1979). 'Solar homes for the elderly, Acorn Close, Wirral'. *Passive Collection of Solar Energy in Buildings.* London, UK Branch of The International Solar Energy Society, London.

[38] Baer, S. (1978). 'Solar design'. *New Mexico Solar Energy Association. Association Bulletin*, 3.2.

[39] Balcomb, J.D., McFarland, R.D., Jones, R.W. & Wray, W.O., in a series of papers, Los Alamos Laboratories, Los Alamos, New Mexico, US.

[40] Littler, J.G.F. (1982). 'Overview of some available models for passive solar design'. *Computer Aided Design*, 14, 15.

[41] Fisk, D.J. & Morrison, R.C. (1979). 'Energy conservation tests with simulated occupancy'. *Building Research and Practice*, p. 148.

[42] Littler, J.G.F. & Watson, M. (1981). 'Passive solar design and the use of reduced scale models and component tests'. *Report to the Department of Energy*, Energy Technology Support Unit, AERE, Harwell, Oxford, UK.

[43] Burgess, K.S. (1979). *Computer Programs for Energy in Buildings.* Design Office Consortium, Cambridge, UK.

[44] Lebens, R. (1981). Private communication. Architecture and Computer Aided Design, London.

[45] James, R. (1980). Private communication. CAP Scientific, London.

[46] SERI, 1980. *Analysis Methods for Solar Heating and Cooling Applications, Second Edition.* Solar Energy Research Institute, Golden, Colorado, US.

[47] Judkoff, R., Christensen, C., O'Doherty, B., Simms, D., Hannifar, M. & Wortman, D. (1980). 'A comparative study of four passive building energy simulations DOE-2.1, BLAST, SUNCAT-2, DEROB-3'. *Fifth National Passive Conference Proceedings.* US Branch of the International Energy Society, Boulder, Colorado, US. A valuable series of internal and published papers is available from Judkoff, A. and Wortman, D. at SERI, covering further code validation carried out between 1980 and 1983.

[48] Arumi-Noe, F. (1981). Private communication. Department of Architecture, University of Texas, Austin, Texas, US.

[49] Judkoff, R. (1981). Private communication. Solar Energy Research Institute, Golden, Colorado, US.

[50] Clarke, J.A. & Forrest, I. (1978). 'Validation of the ESP thermal simulation programme'. *ABACUS Paper 61*, Department of Architecture, University of Strathclyde, Scotland.

[51] Arumi-Noe, F. (1979). 'Field validation of the DEROB/PASOLE system'. *Third National Passive Conference Proceedings.* US Branch of the International Solar Energy Society, Boulder, Colorado, US.

[52] Further information may be obtained from the New Mexico Solar Energy Society, Santa Fe, New Mexico, who produce excellent colour slide sets concerning many aspects of solar design.

[53] Further information may be obtained from the Energy Design Group, 1 Canton Place, London Road, Bath.

5

Active solar heating

5.1 Introduction

5.1.1 Definitions

The somewhat arid discussion of the semantics of active hybrid and passive systems has been alluded to in the previous chapter.

Here, an active system is really taken to mean a bolt-on arrangement, which is usually not part of the building structure (for example, a solar hot-water heating system) and which often involves pumps or fans. Of course, some solar air heaters do form the roofs of buildings (see Section 5.2) and some hot-water systems flow naturally by a thermosiphon.

5.1.2 Popularity of solar heating

The UK Solar Trade Association[1] assesses how widely such systems are used, and the 1981 figures show that there are 60 firms manufacturing solar systems or components or installing them. From 1974 to 1981, 173 000 m^2 of collector were produced. About 21 000 m^2 of hot-water system were installed in 1981, comprising about 5000 systems. About 2400 swimming-pool systems have been installed from 1974 to 1981. The main complaints seem to involve misunderstandings either by clients or manufacturers of the likely output or benefit of a given system. There has been no 'Gallup Poll' of users' reactions in the UK, but in the US a study for the Solar Energy Research Institute[2] shows these conclusions: two-thirds of houseowners strongly wish to see solar energy developed over other sources; one-third of people feel solar is technically and economically practical today for homes; two-thirds of people have not considered investing in solar technology for their homes.

5.1.3 Outline of the active systems

A typical solar hot-water heating system is shown in Fig. 5.1. Cold water is heated in the normal hot-water cylinder by a coil through which hot water from the boiler circulates, or by an immersion heater. This coil or heater is at the top (hottest) part of the tank. Expansion in the tank is taken up through the vent. Hot water for taps is drawn off at the top.

Solar heating is introduced by adding a second coil at the bottom of the hot-water cylinder, through which hot water from solar collectors is circulated. This closed loop from the collectors is normally pumped, but if the collectors can be placed at least a metre below the tank, and large pipes

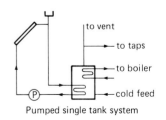

Fig. 5.1.[3] Pumped single-tank solar hot water system.

Fig. 5.2. Solar air-heating system.[4]

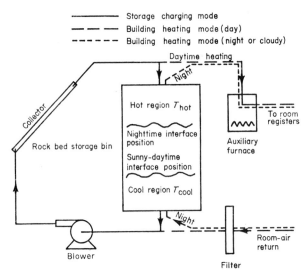

used, then the circulation will thermosiphon without a pump. Expansion of the solar loop occurs to the small tank shown, which could be closed with a flexible diaphragm.

Numerous variations are possible on this theme and will be briefly described in Section 5.3.

A typical solar air-heating system is shown in Fig. 5.2. This system is used to heat air for space heating but, as will be described in Section 5.2, domestic hot water can be added to such a design. Return air from the house is directed by a fan through the collector and back into the house via room grilles. Should the collectors be capturing more heat than is needed, the air is sent into a storage system which can be used for space heating at night by reversing the air stream. Many variations are possible around this theme, and thermosiphon air designs are quite often used when the system can be planned with a very low pressure drop.

5.1.4 Approximate energy available

At this point, some notion of the energy captured may be helpful. At the optimum angle, about 4 GJ of solar radiation falls per m^2 of collector surface per year in the UK. While collectors themselves might capture energy under good conditions with an efficiency of 50% the more useful figure is the ratio of solar energy delivered to taps or central-heating systems as a percentage of solar radiation striking the collector or radiation striking the ground. Very roughly, these two figures are both likely to lie between 10 and 20% on an annual basis. Thus a system of 5 m^2 of collectors will *deliver* the following amount of energy as hot water or hot air:

$$(0.15)\,(5\ m^2)\,(4\ GJ/y) = 3\ GJ/y.$$

The cost of this energy provided, using peak electricity, is £45 or £9/m^2 of collector (March 1982 price of 6 p/kWh including a typical proportion of the standing charge). The energy harvest is roughly 0.6 GJ/(m^2 y).

Solar air heating

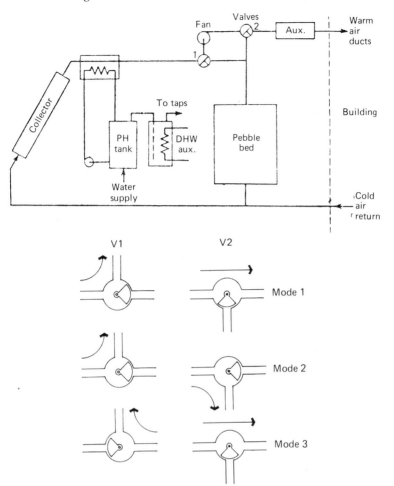

Fig. 5.3. Detailed solar air-heating system.[5] PH = preheat tank for domestic hot water. (Reprinted with permission, from *Solar Heating and Cooling*, Hemisphere Publishing Corporation.)

5.1.5 Hot water compared to hot water + space heating

Undoubtedly, most consumers are more likely to reap early economic benefit from a solar hot-water system than from a design which also supplies space heating. Families or institutions with a sustained and heavy use of hot water are likely to benefit most; whereas for someone living on their own, at the moment in the absence of tax incentives, the economics are not favourable. This is not so true in the US where incentives are available, or for those who carry out the installation themselves.

The Milton Keynes Polytechnic of Central London (PCL) House, which is described in Section 5.3, does use solar space heating and captured 15.3 GJ in the year April 1979–March 1980, for a collector area of 34.5 m² (0.44 GJ/(m² y)).

5.2 Solar air heating

Fig. 5.2 illustrates the basics of an air system. In Fig. 5.3 a more detailed system is presented.

In Mode 1, when dampers 1 and 2 are in the positions shown, cool air from the house is returned to the collector and passed back into the house after being warmed by the sun and topped up, if necessary, by the auxiliary heater.

In Mode 2, the house does not require heat and the solar energy is stored in the pebble bed which is charged by injecting heat from the top. This aids stratification.

In Mode 3, the house requires heat but the sun is not shining and energy is drawn from storage. Air flows through the pebble bed in the opposite direction from that in Mode 2, so that the hottest air is used to heat the house.

Other modes are possible and may be more useful in regions of low solar irradiance such as the UK. These will be discussed later but, before talking about the modes, it will be useful to present some of the design principles for the components of air systems.

5.2.1 Collectors

Commercial collectors are under system-testing at the Solar Pilot Test Facility[6] and some designers use site-built systems. *Solar Age* lists those commercially available in 1980 in Ref. [7].

There is again great disparity between the *collector performance* and the *system performance*. For example, the Denver Solar House[8] has a collector which transfers only 25% of the incident energy to storage or to hot water. Air collectors, on the other hand, can have instantaneous efficiencies comparable with any other collector of similar sophistication. Usually, however, the efficiency of air collectors is degraded by a fluid-mass-flow which is lower than that typically used in water collectors (the values being in the order of 20–70 $m^3/(m^2\ h)$ for air and 35–70 $kg/(m^2\ h)$ for water, where the data is expressed per m^2 of collector. Air has a density of roughly 1 kg/m^3.).

The principal reason for the lower flow rate used is the size of fan and the pumping energy. For space heating applications typical of an air system, 30 m^2 or more of collector would be used, implying air flowing at 600–2100 m^3/h (350–1200$^{'3}$/min). A fan to handle this flow at a pressure drop around the system of 150 Pa (0.6″ WG) implies a power consumption of about 150 W. The parasitic electrical load is thus not negligible.

The flow rate presents further problems. Velocity in duct work should not exceed 5 m/s giving a duct size of 350 mm × 350 mm at 2100 m^3/h, to which must be added at least 50 mm of insulation to keep duct heat losses below 10% of the collector output. Thus, the insulated ducts occupy a large space. Secondly, in Mode 1, when the house is being heated, there may often be a mismatch between the air volume desirable for efficient collector operation and that required for comfortable heating. Assuming a building heat demand of, say, 5 kW, then the air supplied at, say, 50 °C must flow at about 500 m^3/h, equivalent only to the lower limit for the collector. This limitation stems from one of the principal advantages of an air system, namely that the same fluid is used for cooling the collector and

Solar air heating

warming the home, thus avoiding losses across heat exchangers or radiators. A lower supply temperature (for example 35 °C) would raise the flow rate necessary in the home to 1000 m³/h but the air leaving a typical grille would then be travelling too fast and be too cool, leading to discomfort.

The second general comment about the collectors (and duct work) is that leaks degrade the system performance to a large extent. It is essential therefore that the collectors do not present too high a resistance to flow (< 100 Pa) and that they be of high-quality construction.

5.2.2 Air to water heat exchangers

In general, water for domestic use is required at a higher temperature than air for space heating and the heat exchanger for domestic use is placed shortly after the collector outlet. If there is danger of freezing, the loop should drain down or be filled with a fluid of low freezing point. The loop will probably be pumped and is thus very similar to a solar hot-water circuit. Adequate heat transfer is essential and, for example, to heat 100 l of water from 10 to 40 °C in an 8 h sunny period implies a heat transfer rate of 450 W. Since the air volume is high compared with the energy transferred, the duty of the exchanger is abnormal and care must be taken to check that the pressure drop is not too high.

5.2.3 Dampers

Three-way diverting dampers are generally used, and in the UK such units are relatively unusual in the domestic situation. Since it is important that the leakage past a closed damper should be low, it is probably advisable to use a packaged unit consisting of two dampers and associated duct work which is specifically designed for this purpose. Journals such as *Solar Age* regularly advertize such units.

5.2.4 Auxiliary Energy

The top-up system for the warm air ('aux' in Fig. 5.3) may obviously be of any kind, provided response is reasonably rapid in order to deal with bouts of cloudiness. The control for the auxiliary heater can be placed in the air stream to the house to achieve the desired delivery temperature.

5.2.5 Pebble beds

This is a difficult part of the system to design, both from the point of view of size and internal layout.

The store in a French house near Paris (48° 40′ north), which has 62 m² of air-heating collector, has a volume of 40 m³. The volume of the store adopted in houses with 32 m² of collector at Peterborough (52° north) is 4 m³, and that of the Wimpey Air House in London with 33 m² collector is 8.5 m³.

Roughly speaking, the specific heat of rocks (or bricks or pebbles) is 1 kJ/(kg K). The density is about 1400 kg/m³ and the void factor about 0.45, thus a bed of 40 m³ has a heat capacity of 25 MJ/K. For a temperature rise of 20 °C throughout the bed, 0.5 GJ would be stored. Typically, in northern latitudes the energy captured per day in spring or autumn would rarely exceed 8 MJ/m², and for a collector area of 62 m² the 40 m³ store would be very seldom fully charged. On the other hand, the daily heating demand of the house in spring or autumn is probably about 50 MJ and thus the store when fully charged could provide about ten days of heating if it started off at, say, 60 °C.

Clearly, the optimization to be made is fairly complex but unless the store is cheap it seems unlikely to pay off except in cases where it is quite small, satisfying perhaps one day's heating. This point is emphasized by the observation that, for a well-insulated house and its short heating season, the number of days when there is excess solar energy beyond that needed to heat the house and hot water during the day is quite small. It was for these reasons that the homes at Peterborough were fitted with small stores of 4 m³.

The second problem with pebble beds is their internal layout. Clearly, the overall pressure drop must be kept low (~ 50 Pa). Owing to the large surface area, there is not normally a problem with more or less total heat transfer from pebbles to air, or vice versa.

The heat transfer can be represented by:[10]

$$h = 650 \ (G/D)^{0.7},$$

where

h = volumetric heat transfer in W/(m³ K);
G = mass velocity of air (kg/(m² s) = air mass flow rate/bed frontal area;
D = pebble diameter (m).

For example, suppose 50 mm diameter pebbles are used in a bed 1 m × 2 m × 2 m (air flowing along a 2 m dimension); then with 30 m² of collector requiring air at 50 m³/(m² h):

$$h = 650 \left[\left(\frac{50 \times 30}{1.1 \times 3600 \times 2} \right) \frac{\text{kg/s}}{\text{m}^2} \bigg/ 0.05 \right]^{0.7}$$

$$= 1650 \ \text{W/(m}^3 \ \text{K)}.$$

Since the volume of the bed is 4 m³, for a temperature difference of 2 °C the rate of heat transfer is 13 kW.

The pressure drop Δp is given,[10] approximately, in Pascals by:

$$\Delta p \approx \frac{LG^2}{\rho D} \left(21 + 1750 \frac{\mu}{GD} \right),$$

where

L = bed length (m);
GD = defined above;
ρ = air density (roughly 1.1 kg/m³);
μ = absolute viscosity of air (roughly 1.8 × 10⁻⁵ N(s m) or kg/(m s)).

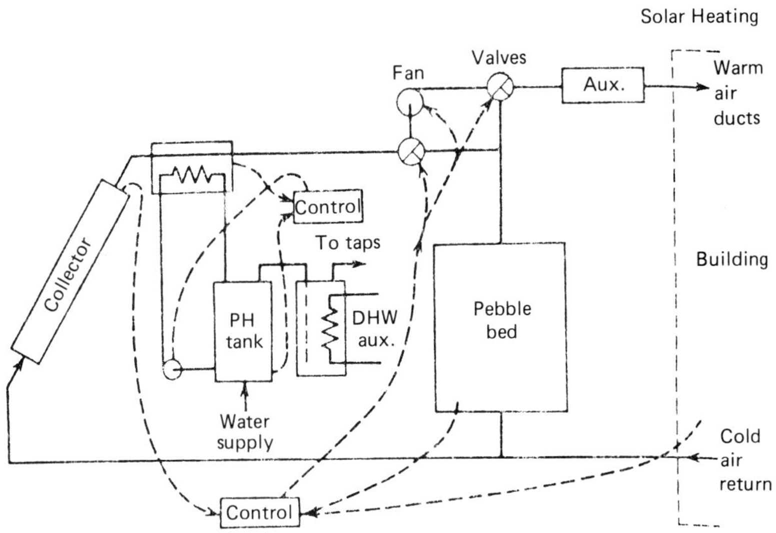

Solar air heating

Fig. 5.4. Control system for a solar air heating system.[5]

Thus

$$\Delta p \approx \frac{(2\ \text{m}) \left[\dfrac{50 \times 30}{1.1 \times 3600 \times 2}\ \text{kg/(m}^2\ \text{s)} \right]^2}{(1.1\ \text{kg/m}^3)\,(0.05\ \text{m})} \left[21 + 1750 \frac{(1.8 \times 10^{-5})}{GD} \right]$$

$$= 32\ \text{Pa}\ (\approx 0.13''\ \text{WG}),$$

thus achieving a low pressure drop is not difficult either; the main practical problems seem to be:

— arranging the inlet and outlet and rock pattern so that the bed (or stack of bricks, etc.) enjoys an even air flow which is well distributed;
— ensuring that the bed remains dry in the humid UK climate so that there is no build up of organic material (in practice this problem does not seem to arise);
— keeping out rodents, etc. during building operations;
— sealing the enclosure to retain the same integrity as the rest of the air system;
— installing the control sensors at appropriate positions;
— timing the building process so that the installation of the store does not interrupt the natural flow of site trades. For example, wet trades involved in blockwork may be undesirable in a timber-framed house.

5.2.6 Controls

Fig. 5.4 indicates the control lines for a typical installation. One controller runs the domestic hot-water (DHW) system by turning on a

Table 5.1. *Control modes in the solar air heated houses at Peterborough*

Mode	Mode characterization	YX	YS	$X\theta$	YZ	$Y\theta$	θ	Fan 1
1	Venting collector heats water (summer)	–	–	–	–	–	–	off
2	Collector heats water and circulates (summer)	–	–	–	–	–	–	off
3	Collector heats water and store	–	$Y>S$	–	–	–	satisfied or off	on if $X<80$
4	Store heats house and DHW, top up by gas if necessary	–	$Y<S$	$X>\theta$	–	$Y<\theta$	not satisfied	off
5A	Collector heats water, collector heats rooms, gas top up available	–	–	–	–	$Y>\theta$	not satisfied	on
5B	Collector heats rooms, collector heats water	–	$S>50$	–	–	$Y>\theta$	not satisfied	on
6A	Gas heats rooms	–	–	$X<\theta$	–	$Y<\theta$	not satisfied	off
6B	Collector preheats make up air, gas tops up	–	–	$X<\theta$	$Y>(Z+10)$	$Y<\theta$	not satisfied	off
Fail safe	Fail safe (power failure)	–	–	–	–	–	–	off

Note: Fig. 5.5 identifies the location of dampers D145 and D23. Other dampers, valves and temperature sensors are listed in the text.

pump to circulate water from the preheat tank, PH in the diagram, through the air to the water heat exchanger (AWE), when sensors indicate that the air at the exchanger is hotter than the water at the bottom of PH. Care is needed in choosing the temperature differential and, typically, a 5–10 °C difference might be suitable for switching on. This loop needs frost protection and boiling protection. Freezing can be avoided by antifreeze or a non-freezing fluid or a drain-down system. These are arrangements typical of a 'conventional' water-filled collector system. Boiling may occur in summer and, whilst a drain-down arrangement can cope with this problem, an antifreeze system cannot. In that case, provision must be made to ventilate the air collector to prevent the air stream reaching 100 °C.

The air circulation presents a more challenging problem of control and is discussed in terms of the Peterborough system,[11] illustrated in Fig. 5.5, for the house in Fig. 5.6. See also Table 5.1.

Dampers:
    DI   admits outside fresh air to collector system
    DS   safety vent with fusible link to prevent collector overheating

Solar air heating

Fan 2	CH switch	Summer–winter switch	DHW switch	Boiler pump	DHW solar pump	V3	D145	D23	DO	DS	DI
off	off	s	on off	on off	on if $A<Y$ and $A<60$	1	2	1	open if $Y>100$	open if $Y>120$	open if $Y>100$
off	off	s	on off	on off	on if $A<Y$ and $A<60$	1	2	2	closed	closed	closed
off	–	w	on off	on off	on if $A<Y$ and $A<60$	1	3	1	closed	closed	closed
on	on	w	on off	on off	on if $A<X$ and $A<60$	2	1	3	closed	closed	closed
off	on	w	on off	on off	on if $A<Y$ and $A<60$	1	2	3	closed	closed	closed
off	off	w	on off	on off	on if $A<Y$ and $A<60$	1	2	3	closed	closed	closed
on	on	w	on off	on off	off	–	3	1	closed	closed	closed
on	on	w	on off	on off	off	–	3	4	closed	closed	closed
off	–	–	–	off	off	–	–	–	open	–	open

DO summer vent to allow thermosiphoning to occur drawing air via the collector, without fans, over the DHW coil

Valves:
    V1 cuts out the heating coil from the gas boiler to the warm air stream
    V2 cuts out the heating coil from the gas boiler to the domestic hot-water storage tank
    V3 3-way from heat-exchanger in DHW tank to either air stream from collector position (1) or to heat-store exchanger (2)

Fans:
    Fan 1 is collector fan; Fan 2 is warm-air heating fan

Sensors:
    S heat-store temperature at the bottom
    X heat-store temperature at the top
    Y collector outlet temperature
    Z ambient temperature
    θ room set point
    A temperature at the bottom of DHW cylinder

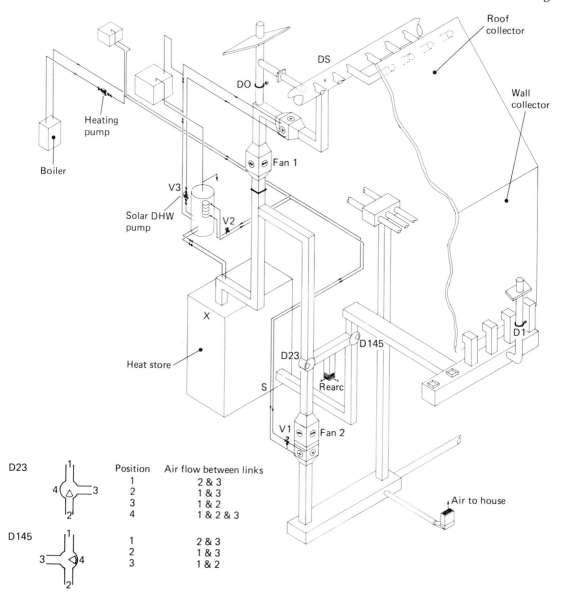

Fig. 5.5. The solar air-heating system for the Peterborough Houses.[11] (Courtesy of Peterborough Development Corporation, Chief Architect K. Maplestone.)

B   temperature in the middle of DHW cylinder

c   warm air heating supply temperature ($\approx 35\ °C$)

Commentary concerning the modes and their control:

Mode 1

In the summer, space heating is not required. In order to avoid unnecessary use of electricity, the fans are turned off by manually switching the summer–winter control. Air then circulates naturally by a thermosiphon. If, for example, air enters the collector at 20 °C and exits at 70 °C then the head of air is equivalent to about 4 Pa (0.01″ WG).

Solar air heating

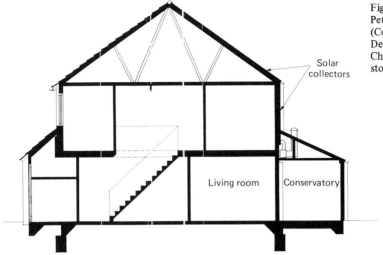

Fig. 5.6. Section of a Peterborough House.[11] (Courtesy of Peterborough Development Corporation, Chief Architect, K. Maplestone.)

The pressure created is given roughly by:[19]

$$P = 3400 \left(\frac{1}{T_a} - \frac{1}{T_m} \right),$$

where
P = pressure in Pa;
T_a = inlet air temperature in K;
T_m = mean air temperature in K;
and where T_a and T_m must be in degrees absolute, that is, °C + 273.

This low pressure is sufficient to drive roughly 20 m³/(h m²) of collector area through the ductwork from the outside entry point to the outside exit point.

The hot air passing over the DHW transfer coil heats the loop to the DHW tank through which water circulates, driven by the solar hot-water pump. The pump is on when the air temperature at the collector exit (Y) is above the temperature of the water at the bottom of the DHW storage tank (A). The pump is switched off if A rises above 60 °C to prevent scaling. The small pressure across the collector is not enough to drive air through the thermal store, and thus the control system ignores signals X and S in this mode. Similarly, space heating is switched off, so that signals from θ, Z are ignored.

Mode 2

This is also a summer mode, but the controller perceives that Y is < 100 °C and thus adjusts the dampers to close all connections to ambient, allowing circulation of air around the closed loop of the collectors and DHW transfer coil.

Mode 3

In mode 3 fan 1 is used to circulate air through the store (which measures 2 m × 1 m × 2 m high) and the DHW transfer coil. Space heating

Table 5.2. *Details of two solar air houses*

Project name	Ref.	Floor area (m²)	Design heat load (GJ/y)	Collector aperture (m²)	Collector type	Storage type	Storage volume (m³)	Solar fraction (%)	System collection efficiency (%)
Dourdain	9	197	126	62.5	Single glazed low ε glass black steel	Rocks	40	32	21
Wimpey	a	115	–	33.3	Matt black	No fines[b]	8.5	40[c]	16

[a] Report to be available from the UK Department of Energy late 1982.
[b] A Wimpey development consisting of coarse aggregate with no fine particles, bound by cement.
[c] Including small amounts (≈ 10%) of passive solar gain via windows.

is not needed. The store is charged both to provide space heating in the evening and to preheat the domestic hot water, since a closed circulation loop connects the heat store to the solar coil at the bottom of the DHW tank.

Other modes

The reader may follow the other controller operations by examining Table 5.1 and Fig. 5.5; but mode 6B is worth emphasizing in the context of northerly latitudes.

In this mode, the collector is used in conditions of low radiation intensity, merely to preheat makeup air. Normally, makeup air is not deliberately introduced in the Peterborough Houses, infiltration is assumed to provide fresh air, or occupants open windows; but infiltration in winter probably constitutes 40% of the load of poorly sealed but well-insulated houses. Pressurizing the house by drawing in fresh air decreases infiltration. The air volume required is small — perhaps 0.5 house air changes per hour or about 100 m³/h. This represents a flow of only 3 m³/(h m²) of collector which is greatly below that recommended for high collector efficiency. The mode is as yet experimental.

The controller

Clearly, to cope with such a complex series of modes a sophisticated controller is required, and the one used in the Peterborough Houses is built around a microprocessor.

5.2.7 Results of monitored systems

Data concerning one 'active air house' has been gathered by Energy Conscious Design within a Commission of the European Communities (CEC) programme monitoring solar performance.[13] Table 5.2 lists some of the parameters describing the house.

The Dourdain house is in an area with about 4.6 GJ/(m² y) of irradiation on the horizontal (compared with about 3.4 GJ/(m² y) in the southern UK, see PCL/MK house in Section 5.3). The performance is rather

Solar water heating

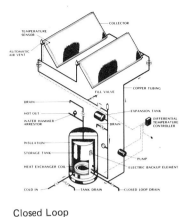

Fig. 5.7. Solar water-heating systems.[14]

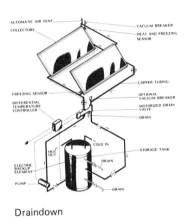

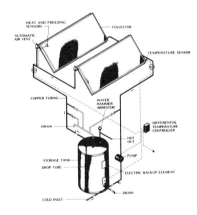

poor, and the report[13] indicates that the storage is too large, the system too leaky and the air distribution in the collection poor.

Wimpey has also built a test dwelling and some results are shown in Table 5.2.

5.3 Solar water heating

5.3.1 Outline of systems

There are four main variants of systems for heating water, and they are illustrated in Fig. 5.7.

In the drain-back system shown in Fig. 5.7(*a*), a closed loop carries water (with no antifreeze) from the pump through a heat-exchanger coil in the hot-water storage tank to the collectors and back to the pump. When the sun is not shining, the pump is switched off and water drains back to the water reservoir which is within the insulated volume of the building. On start-up, the pump must raise the water back to the collectors.

Provided the pipes all slope correctly, no water is left to freeze or to boil in the collectors under very cold or very hot conditions. Since the loop is closed, scale deposition and corrosion are reduced in the collectors.

In the closed-loop method shown in Fig. 5.7(*b*), the same loop is used but, being protected with antifreeze, it does not drain back. To take up expansion there is a small tank which may contain a diaphragm to reduce corrosion.

In Fig. 5.7(*c*) the heat exchanger is avoided, the water being heated is the water used at the taps. On drain-down at the end of the day, or when the sun goes in, a valve opens to discharge the water from the collectors either to waste or into a sink. Direct heating systems expose the collectors to mains pressure and to scale formation above about 65 °C; but the absence of a heat exchanger can increase system efficiency by 10%. The water circulation tends to destroy stratification in the storage tank, which may impair performance by a similar percentage.

In Fig. 5.7(*d*) the same water is heated, and fed to the taps, but freezing is avoided by circulating warm fluid from the storage tank, on demand by the freezing sensor. It is thus not suitable for cool climates.

The parasitic power associated with pumped systems is small, the pumps typically using less than 70 W (or about 0.4 GJ/y). If the vertical distance from the drain-*back* (Fig. 5.7(*a*)) reservoir to the top of the collectors is very large then a second pump may be needed to fill the loop in drained systems, but often one pump can be used both to charge and circulate.

It is important to note that mains pressure is used in drain-down systems to refill collectors, and can of course be reduced by a suitable valve to prevent damage. Such a system is fail safe on power failure if the drain-down valve is 'normally open', and the system has a vacuum breaker at the highest point.

In retrofit situations it may be difficult *either* to introduce a second heat exchanger into the DHW storage tank (below the boiler coil), for non-electric heating, *or* to build in a second preheat storage tank with a heat exchanger. In such cases the direct systems (drain-down and open-loop) have obvious advantages, and northern climates probably exclude the open-loop form.

Antifreeze can form sludges and its acidic break-down compounds can accelerate corrosion. Checks should be made every year on the freeze protection afforded, and at the PCL Milton Keynes house it appears that water containing antifreeze opens up small holes which otherwise might seal themselves, causing leaks.

A good review of other advantages and disadvantages of these systems may be found in Ref. [15].

Comparative figures for the various designs, (*a*) to (*d*), have not been found, but variations of more than about 10–20% seem very unlikely for strictly comparable set-ups designed according to Fig. 5.7. However, very large variations are probable stemming from more-mundane considerations. Table 5.3, for example, illustrates this point.

The cost per unit of energy (Table 5.3, published in 1981) varies by almost 100%. In addition to these variations for collectors operating under

Table 5.3. *From comparative performances estimated by the Tennessee Valley Authority for typical domestic installations*[14]

System	Estimated energy delivered by solar (GJ/y)	Cost of delivered energy (1981) (£/GJ)
Eagle Sun drain-back	15.7–18.9	3.8
Reynolds 1454	13.1–15.8	4.3
Exxon 3	11.7–14.0	5.0
Reynolds 1404	11.2–13.5	4.6
Rheem Ruud (insulated)	11.2–13.5	5.8
ITC	10.4–12.6	5.1
STC HWSG 72120	10.4–12.4	5.1
Sunmaster	10.1–12.2	5.6
Rheem Ruud (uninsulated)	9.2–11.0	7.0
Grumman	9.0–10.8	7.1
Exxon	9.0–10.8	5.9
Northrup	9.0–11.0	5.0
Solar one 5	8.8–10.6	5.6

Table 5.4. *Components of an active liquid system*

Collectors	Heat exchanger
Pump	Back-up heater
Controller	Drain-down valve
Sensors	Vacuum breakers
Thermal storage	
Overflow tank	Transfer liquid

identical conditions, the annual yields will depend on the draw-off patterns by the users (and the amount of hot water demanded), overshadowing, pipe and tank insulation, settings on the control system and so on.

It seems likely that 'good practice' at installation will do more to achieve high performance than will an exhaustive choice of the best 'laboratory' systems or the best 'laboratory' collectors.

However the components are assembled, they may include the items in Table 5.4. Several of these items are conventional and will not be discussed further. The special components are dealt with briefly below.

5.3.2 Collectors

A large number of manufacturers make collector panels which are very similar in principle. There are exotic versions such as evacuated tube collectors and xeolite heating and cooling panels, which may well sweep the market in the future. Traditionally, a solar hot-water panel operates by circulating water through a web of pipes joined by a plate. The plate is black, is insulated at the back and glazed at the front. Hutchins[16] presents a useful set of measurements concerning the transmittance of glass, low-iron glass and plastic cover sheets for collectors, plus measurements of

Fig. 5.8. Instantaneous efficiency of collectors on the PCL/MK House (measured by PCL).[9, 17]

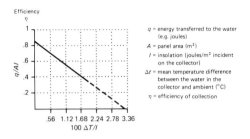

Fig. 5.9. Performance of a variety of collectors.[18] (Reprinted with permission, from *Solar Engineering and Thermal Processes*, J. Wiley, NY.)

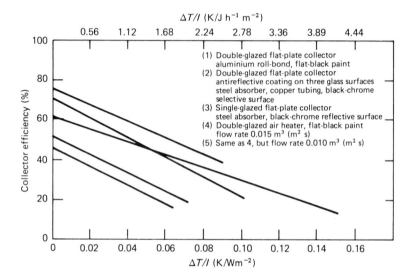

absorptance and emittance of coatings. The instantaneous performance of a collector is described by diagrams such as Fig. 5.8. The graph shows that, for example, when the solar radiation striking the collector is that prevalent on a clear summer day at noon (2500 J/(h m²) (≈ 700 W/m²)) and when the air temperature is 20 °C and the inlet and outlet water temperatures are 30 and 60 °C, then the efficiency of transferring energy from the solar radiation to the water is 55%, that is

$$\frac{\Delta T}{I} = \left(\frac{30+60}{2} - 20\right)/2500 = 0.01$$

which, from Fig. 5.8, gives q/AI as 0.55. At low insolation levels typical of a winter day (for example 0.3 MJ/(h m²)) the efficiency, even for heating water from 10 to 15 °C with an ambient of 8 °C, would be 0%. Typical efficiencies are shown in Fig. 5.9 for a variety of collectors.

The band of performance of the water heaters is not enormous, and it has been suggested already that system performance may not reflect collector performance. However, manufacturers ought to be able to provide this kind of graph and their product should not be markedly inferior to the ones shown.

Johanssen[19] has surveyed manufacturers. He finds that most give a

Solar water heating

five-year guarantee, many do not provide operating manuals, some do not supply design drawings and specifications, and some do not give performance data. A panel with a selective surface (which reduces heat loss by reducing radiation to the cover glazing) may reach 200 °C if the water stops flowing. Clearly then, the manufacturer should use components, including insulation, capable of withstanding such conditions. If the panel is installed between rafters there is some fire risk, and finally the anchor points must resist wind lift on the assembly.

Michelson[20] estimates the optimum size of a system and finds that for DHW systems, 5 m² of collectors and 250 l of hot-water storage is optimal. She assumed a demand of 250 l/d at 50 °C.

More information is available in Ref. [21], and there is now a British Standard, BS5919, covering solar collector installations.

The cost of such an installation would be in the range £1000–1500.

5.3.3 Pumps

Flow rates of 35–70 kg/(m² h) are recommended. Even at the higher value, a closed system with 5 m² of collector requires a flow of only 350 kg/h, and the system is likely to have a pressure drop of less than 40 kPa.

5.3.4 Heat exchangers

In a solar system, the heat exchanger transfers energy between the solar loop and the hot-water storage tank. Collector efficiency falls off very rapidly as the collector temperature rises, and thus it is desirable to keep the temperature drop between the collector loop and the storage tank as low as possible. This is done in two ways. Firstly, the heat exchanger is positioned at the coolest point in the storage tank, at the bottom. Secondly, the heat exchanger must have a large area. Unfortunately, the transfer rate from exchangers diminishes as the temperature difference between the metal and the water is reduced, at a rate faster than unity, that is, the heat-exchange rate per 1 K temperature difference is smaller the lower the value of the temperature difference. Thus values quoted for heat exchangers typically used between a boiler and a tank are higher than will be found if the same exchanger is driven from collectors.

The British Standard previously mentioned suggests that a 60% heat-exchange performance will only degrade system performance by 3% compared with a 100% efficient heat exchanger. Typically, such a calculation implies an exchanger with an area equal to about 0.15–0.2 of the collector area. Duffie & Beckman find[22] that, for every 1 K temperature drop across the heat exchanger, the collection efficiency falls by 1–2%.

5.3.5 Thermal storage

For a domestic hot-water system, approximately 50 l of storage per m² of collector is appropriate. Naturally, any optimization is difficult since it depends on the family size, etc.

For space heating, a much larger area of collector is often used. The Milton Keynes Solar Water-Heated House designed at the Polytechnic of

Table 5.5. *Solar heating in homes and flats (from Ref. [13])*

Place	Collector area (m^2)	Collector type	Storage (m^3)	l/m^2	System efficiency (%)	Solar fraction of load[a] (%)
South UK	34.5	Single-glazed, black paint	4.2	122	11	57
France (mid)	153	Single-glazed, black paint	10	65	27	43
France (mid)	340	Single-glazed, black paint	30	88	24	25
Italy (NE)	129	Single-glazed, black paint	3	23	14	18
Holland	52	Single-glazed, selective	4.1	79	35	26
Holland	35	Single-glazed, selective	2	57	15	32

[a] Fraction of space and hot-water load supplied by solar energy.

Central London has 122 l/m^2 of storage. Table 5.5 illustrates some other results.

Great play has been made in the past about encouraging stratification, so that cold water is returned to the collectors and hot water accumulates at the top of the storage tank. In direct systems the water is stirred and stratification is unlikely.

There is disagreement about the advantages of promoting a temperature gradient. For example, the Shell simulation model[23] suggests a 15% improvement in solar contribution with a stratified preheat tank. Duffie[35] has suggested a much smaller effect.

Chapter 7 deals in more detail with interseasonal thermal storage.

5.3.6 Drain-down valves

To optimize system performance, drain *back* (Fig. 5.7(a)) should occur at least once a day before nightfall, so that warm water is not left in the collector to cool down overnight. The implication is that the water reservoir should be insulated. In drain-*down* systems, where the water goes to waste, it is customary for the system to open the valve when the collector temperature drops to near freezing (see the freezing sensor in Fig. 5.7(c)). Drain-down valves are commercially available. Care should be taken with open-loop systems in hard-water areas, since the drain-down valve may not seal properly if scale builds up.

5.3.7 Transfer fluid

Water is most commonly used, with or without antifreeze. If a toxic antifreeze is used, some water authorities insist on double-walled heat exchangers. There are, however, non-aqueous fluids, for example 'Sun Temp', which freezes below −40 °C, boils at 355 °C, is non-toxic and, in principle, non-corrosive. Some problems have been experienced using non-

New types of system

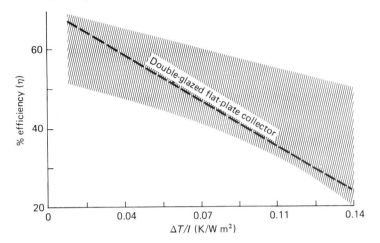

Fig. 5.10. The efficiencies, based on aperture area, of various evacuated collectors are shown hatched compared with the efficiency of a well-engineered, double-glazed, flat-plate collector with a selective coating on the absorber.[24]

aqueous liquids and thin aluminium roll bond collectors. When microorganisms growing in the cool part of a system are killed by heat and swept into the collectors, traces of metal in the organisms may set up galvanic cells causing corrosion.

5.3.8 Controllers

Controllers for all types of systems are available. Many contain digital read-outs for temperatures in various parts of the system and, clearly, if the owner is aware and sensitive to the behaviour of the system, he may be able to make improvements. Reports from the PCL/MK house[21] show that by 'tinkering' with the system the annual solar contribution to the space and water heating load was raised from about 15 to about 50%.

5.4 New types of system

5.4.1 Evacuated tubes

Fig. 5.10 indicates the superior performance of evacuated tube collectors particularly at low light levels or high exit water temperatures (that is when $\Delta T/I$ is high).

Many variants have been produced (see Ref. [24]), one of the simplest being shown in Fig. 5.11. At present, such collectors seem useful when high outlet temperatures are required, for industrial purposes or for driving absorption air conditioners.

5.4.2 Plastic collectors

High-temperature extrusions are available. The version illustrated in Fig. 5.12 is very suitable for the 'Do it Yourself' (DIY) market, since fixing the individual tubes to the heaters is simple but tedious. The web comes as a roll which makes access to roofs more convenient in some cases.

Plastic collectors are reviewed in Ref. [25], and their performance is

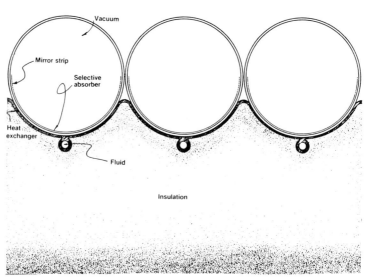

Fig. 5.11. Philips Mark Four solar collectors.[24]

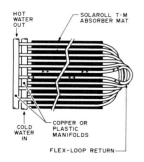

Fig. 5.12. Typical EPDM absorber layout[25] (EPDM = ethylene propylene diene monomer).

not markedly inferior to copper designs. Fig. 5.13 illustrates a glazed all-plastic collector.

It should be remembered that whilst polycarbonates have a low thermal ($\approx 2.5-50$ μm) transmittance of about 0.04,[16] their solar transmittance ($\approx 0.3-2.5$ μm) is also poor (0.76 for the twin-wall version). Some plastics have very high solar transmittance (for example, Teflon 95) but often high thermal transmittance too.

5.4.3 Zeolite heating and cooling collectors

Tchernev has produced a combined heat-pump collector, as shown in Fig. 5.14.[27] The zeolite filling is in a sealed box containing water under reduced pressure. When the zeolite (a yellow material) is hot it expels water vapour which on condensation transfers heat to the water loop. At night, when it cools off, the vapour pressure falls from its day-time value of around 7 kPa, to about 0.7 kPa, and water distills from the heat-exchanger area into the zeolite, thus withdrawing its latent heat of vapourization from the water loop, producing chilled water.

5.4.4 Refrigerant-charged loop

An extremely interesting system has been developed by R. French[28] and is shown in Fig. 5.15.

Freon R114 (freezing point $-94\,^{\circ}$C, boiling point $4\,^{\circ}$C at atmospheric pressure) is pressurized in the sealed loop to a maximum of 600 kPa. When the sun shines, the freon boils and vapour passes up into the heat exchanger (dashed in Fig. 5.15). Liquid freon runs back to the bottom of the collector. As the temperature in the store rises, the pressure goes up until, at 600 kPa ($68\,^{\circ}$C in the storage tank), the pressure limiter shown in the diagram closes. No liquid may now enter the collector, which contains only vapour. The liquid will all be in the heat exchanger.

Installed systems

- Glazing: transparent polycarbonate; UV inhibited
- Ultraviolet filter: TEDLAR® film—0.025
- Absorber panel: polysulfone, with black absorbtivity enhancer
- Insulation: 20 mm thick isocyanurate foam-vinyl and aluminum-foil clad
- Panel weight: 6 kg/m² dry; 11 kg/m² filled with water (approximately)
- Recommended operating temperature: 120 °C max.
- Recommended operating pressure: 205 kPa
- Maximum stagnation temperature: 150 °C
- Width: 0.025 m, 58.8 mm
- Length: 2940 mm standard
- Thickness: 50 mm

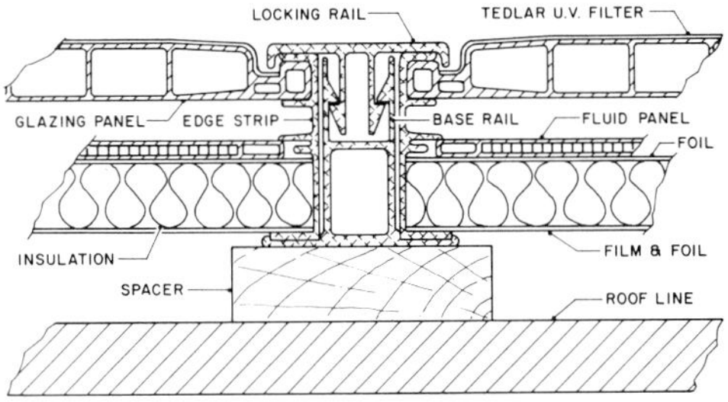

Fig. 5.13. Temtech collector.[26]

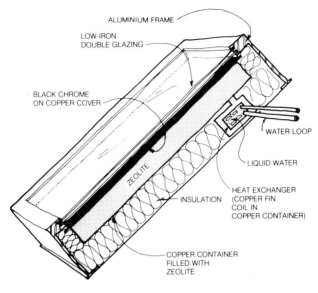

Fig. 5.14. Integrated zeolite panel with flooded-type evaporator–condenser.[27]

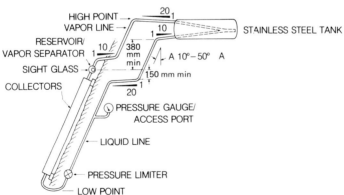

Fig. 5.15. Freon-charged collector.[28]

The system thus operates without pumps or controls. The heights indicated prevent reverse thermosiphoning at night.

5.5 Installed systems

The Performance Monitoring Group of the EEC has produced an interesting report concerning solar heating.[13]

Fig. 5.16. Polytechnic of Central London House at Milton Keynes (MK/PCL House) (from Ref. [9]).

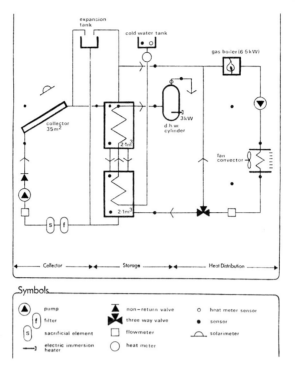

Fig. 5.17. Month by month performance of MK/PCL House.[9]

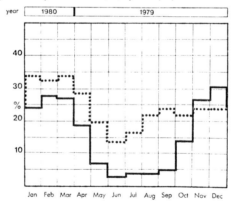

The scheme shown in Fig. 5.16 and discussed below is very typical of DHW and space heating plus DHW systems. The flow rate through the collectors is only 16 l/(m² h), which is considerably below that recommended, and the annual collector performance is 22%. Fig. 5.17 shows the performance month by month. Peak efficiency occurs with very low inlet temperatures and low insolation in the winter. The very large area of collector (32.5 m²) is partly unused in the summer, when the four-person family does not need all of the hot water available (which is why the mean storage temperature from May to September is 57 °C).

Installed systems

Table 5.6. *Energy summary for MK/PCL house*

	People and appliances (MJ)	Passive gain via windows (MJ)	Active solar contribution (MJ)	Gas and electricity for heating (MJ)	
January	1400	680	1206	3388	
February	1270	890	1664	1606	
March	1380	1470	2148	2042	
April	1310	1640	2329	688	
May	930	1250	1364	72	
June	*	*	659	7	
July	*	*	637	0	
August	*	*	651	0	
September	300	240	803	4	
October	1030	800	1325	162	
November	1330	880	1371	1242	
December	1400	610	1124	2506	
Sub totals	10350	8460	15281	11717	Total = 45 808 MJ/y
% of total	23	18	33	26	

*No space heating is needed in these months.

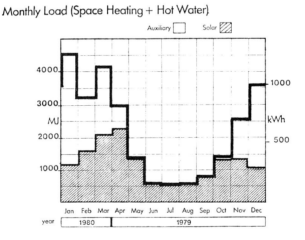

Fig. 5.18. Monthly loads for MK/PCL House.

The *system* efficiency is the sum of the energy delivered to the hot-water system plus that delivered to the space heating system, divided by the solar energy incident on the collectors. Energy losses from pipes and tanks are only counted as useful when the room temperatures are below the thermostat set point.

Fig. 5.18 illustrates the contributions to the heating load, the annual figures being roughly 50% solar and 50% gas plus electricity.

Table 5.6 suggests, however, that these figures do not tell the whole story. Firstly, it is important to realize that, even for this house, which is not extremely well insulated or sealed, the heating season is shortened by comparison with poorly insulated dwellings. Secondly, even though the house has most of its windows to the north (10 m^2 north, 8.5 m^2 south), quite a large fraction of the total load (18%) is met passively. But thirdly and most significantly, the occupants and appliances contribute 23% of the total load.

5.6 Multifamily installations

Sillman[29] finds that economies of scale occur above about 50 dwelling units with communal thermal storage. The cost of delivered energy to the consumer is not a smooth function of storage volume. Clearly, if 100% of the heat needed is available from stored solar energy, then capital may be saved on auxiliary heating plant. Thus a '100%' solar scheme may be more economical than a 50% solar scheme for large numbers of houses or flats.

The same is true at the one-house scale, but interseasonal storage is so expensive (partly because the heat losses are a large fraction of the energy stored) that such projects are unlikely to be economical until a different method of storage is developed.

5.7 Predictive methods

Properly formulated, computer simulations can be a powerful design tool. However, simulations require a significant amount of expertise and expense which often precludes their use for the design of small systems. In addition, the hourly meteorological data needed for the calculations are not available in many locations. Simulations can be used in another way which overcomes these difficulties. The results of modelling a particular type of system for a range of practical design parameters and a variety of climates can be used to generate a design method for this system type. A design method incorporates the information gained from the simulations in a correlation which relates the long-term performance of the system to the important system design parameters and the weather. A design method cannot provide as much or as accurate information as a simulation. However, the information from a design method is usually accurate enough for design purposes[35] and it can be obtained much more easily and with significantly less expense than by simulations or experiments.

Such correlation methods have been developed by Klein, Duffie and Beckman at Madison[30] in the US for active and passive systems and by Balcomb et al. at the Los Alamos Laboratories for passive systems. In the UK, Kenna[31] has produced a similar program.

Typically, such a design method requires the inputs shown in Table 5.7 which is the worksheet for input to f-chart, the 'Madison method'[36] (this correlation method uses the Hottel–Whillier equations). Tables 5.7(a) to 5.7(e) are included to give readers an idea of the detail required to use such a method. Complete explanations of the symbols are not given.

Table 5.8 illustrates the output from f-chart 4, using input data for the Peterborough Houses discussed in Section 5.2 and Chapter 12.

F-chart is probably the most common method in the US for sizing active

Table 5.7(a). *Collector input data to f-chart 4*

Flat plate	CPC concentrating parabolic collector	E–W-axis tracking	N–S-axis tracking	2-axis tracking		Collector parameters		
x	x	x	x	x	C1	Collector area	_____	m²
x	x	x	x	x	C2	*FR-UL*[c] product	_____	W/(m² K)
x	x	x	x	x	C3	*FR-TAU*-ALPHA (normal incidence)	_____	
		x	x	x	x	C4	Concentration ratio	_____	
	x				C5	*CPC* acceptance half-angle	_____	
x					C6	Number of covers	_____	
x					C7	Index of refraction	_____	
x					C8	Extinction coefficient·length (*KL*)	_____	
x					C9	Incidence angle codifier constant[a]	_____	
x	x	x	x	x	C10	Collector flowrate·specific heat/area	_____	W/(m² K)
		x	x	x	C11	Tracking axis (1 = t–W, 2 = N–S, 3 = 2-axis)	_____	
x	x		x		C12	Collector slope[b]	_____	degrees
x	x	x			C13	Collector azimuth[b]	_____	
x	x	x	x	x	C14	Ground reflectance[b]	_____	
	x	x	x		C15	Incidence angle codifiers (10, 20, 30, 40, 50, 60, 70, 80 degrees)		

[a] If parameter C9 is specified, then parameters C6, C7, and C8 are not necessary.
[b] Twelve values may be specified for parameters C12, C13, and C14. Only one value is necessary if these parameters do not vary by month.
[c] *FR* = Collector heat removal factor; *UL* = Overall collector energy loss coefficient.

systems. However, it is not sensitive to load profile, control strategy and freeze protection technique.[32] More-complex hour by hour simulation is necessary for an exploration of these more esoteric variables. TRNSYS in the US or programs from Cardiff University,[33] PCL[34] or ABACUS[37] in the UK are available for such an exercise.

Table 5.7(b). *Losses during heat transfer*

Heat exchanger	Duct or pipe heat loss	Air leaks	Transfer parameters	
×			T1 $EPS \cdot CMIN^b$ of collector store HX^c/collector area	_____ W/(m² K)
	×		T2 UA^a of collector inlet pipe or duct	_____ W/K
	×		T3 UA of collector outlet pipe or duct	_____ W/K
		×	T4 Collector duct leak rate (%)	_____
		×	T5 Duct leak location (1 = inlet, 2 = outlet, 3 = both)	_____

^{a} UA = U-value × area.
^{b} $EPS \cdot MIN$ = load heat exchanger effectiveness × smaller of the two capacitance rates in the heat exchanger.
^{c} HX = heat exchanger.

Table 5.7(c). *Thermal storage data*

Water	Rock bed	Phase change	Storage parameters	
×			S1 Tank capacity/collector area	_____ KJ/(m² K)
×			S2 Storage unit height/diameter ratio	_____
×			S3 Heat loss coefficient	_____ W/(m² K)
×			S4 Environmental temperature	_____ °C
×	×	×	S5 Hot-water auxiliary tank UA^a	_____ W/°C
×	×	×	S6 Hot-water auxiliary tank environmental temperature	_____ °C
	×		S7 Rock-bed capacity/collector area	_____ kJ/(m² K)
		×	S8 Phase-change volume/collector area	_____ m³/m²
		×	S9 Phase-change material density	_____ kg/m³
		×	S10 Void fraction	_____
		×	S11 Solid-phase specific heat	_____ KJ/(kJ/kg)
		×	S12 Liquid-phase specific heat	_____ KJ/(kJ/kg)
		×	S13 Heat of melting	_____ KJ/(kJ/kg)
		×	S14 Melting temperature	_____ °C

^{a} Parameters S5 and S6 are necessary only when domestic hot-water heating is considered.

Predictive methods 185

Table 5.7(*d*). *Data concerning the heat delivery*

Heat exchanger	Heat pump	Delivery device parameters		
×	×	D1 *EPS-CMIN* of load heat exchanger Hx	_____	W/K
×		D2 Minimum temperature for Hx operation	_____	°C
	×	D3 Delivery heat-pump number	_____	
	×	D4 Minimum heat-pump absorber temperature	_____	°C
	×	D5 Heat-pump bypass temperature	_____	°C

Table 5.7(*e*). *Input data concerned with loads*

Process heating	Space heating	Water heating	Combined	Load parameters		
	×		×	L1 Building *UA*[a]	_____	W/K
	×		×	L2 Room temperature	_____	°C
		×	×	L3 Hot-water use	_____	l/d
		×	×	L4 Hot-water set temperature[b]	_____	°C
		×	×	L5 Water-mains temperature[b]	_____	°C
×	×		×	L6 Total process or space heating load[b]	_____	MJ/d
×				L7 Hours per day	_____	
×				L8 Load-return temperature	_____	°C

[a] Note that the load for space heating is determined in a very approximate way using a global value for heat loss.

[b] Twelve values may be specified for parameters L3–L6. Only one value is necessary if these parameters do not vary by month. If L6 is specified for space heating systems, then L1 and L2 are not necessary.

Table 5.8. *Predicted performance of the Peterborough Houses*

* * * F Chart analysis (Version 4.0) * * *

Kew, London, UK   Latitude 51.5

Thermal performance

	HT (GJ)	TA (DEG-C)	SHLOAD (GJ)	HWLOAD (GJ)	FNP
JAN	4.82	4.3	4.64	0.97	0.05
FEB	5.86	4.9	3.88	0.87	0.19
MAR	9.50	7.5	4.24	0.97	0.39
APR	10.85	10.2	2.54	0.94	0.60
MAY	13.05	12.4	0.84	0.97	0.68
JUN	13.37	16.0	0.64	0.94	0.69
JUL	12.87	16.0	0.17	0.97	0.71
AUG	12.44	16.4	0.36	0.97	0.69
SEP	11.09	16.0	0.69	0.94	0.70
OCT	8.85	12.0	2.51	0.97	0.51
NOV	5.98	7.7	2.37	0.94	0.33
DEC	4.60	4.6	4.82	0.97	0.03
YR	113.28	10.7	27.70	11.42	0.35

HT   = total solar energy incident on the collector (for Kew UK)
TA   = air temperature
SHLOAD   = space heating load
HWLOAD = hot water load
FNP   = fraction of non-purchased energy (i.e. fraction of solar energy in this case)

References

[1] Information may be obtained from the Solar Trade Association, 28 Store St, London.
[2] Holland, E. (1981). 'Living with the sun'. *Solar Age*. April, p. 22.
[3] McCartney, K. (1978). *Practical Solar Heating*. Prism Press, Dorchester, UK.
[4] Kreider, J.F. & Kreith, F. (1975). *Solar Heating and Cooling*. Hemisphere Press, Washington, DC, US. Readers are also advised to consult the greatly enlarged, and invaluable 1982 edn.
[5] Duffie, J.A. & Beckman, W.A. (1980). *Solar Engineering of Thermal Processes*. Wiley, New York.
[6] Information may be obtained from the Building Services Research and Information Association, Bracknell, UK.
[7] Cuminsky, M. (1980). 'Directory of air collectors'. *Solar Age*. November, p. 39.
[8] See Ref. [5] above, p. 441.
[9] Information may be obtained from Energy Conscious Design, 44 Earlham St, London, in connection with their Reports to the European Commission concerning Active Solar Space and Water Heating Performance.
[10] See Ref. [5].

References

[11] Information may be obtained from Peterborough Development Corporation, Peterborough, UK who designed the three solar air-heated houses with assistance from the Building Unit at the Polytechnic of Central London. The drawings shown here are reproduced by courtesy of the Development Corporation.

[12] Dodson, C. (1979). 'Passive ventilation alternative'. *Passive Collection of Solar Energy*. London, UK Branch of the International Solar Energy Society, London.

[13] Information may be obtained from Energy Conscious Design, see Ref. [9].

[14] Eagle Sun (1981). 'Advertisement'. *Solar Age*. August.

[15] Schiller, S.R. (1981). 'Freeze protection'. *Solar Age*. August.

[16] Hutchins, M.G. (1982). 'Optical properties of materials for flat plate collectors'. *Helios*, 14, 7, published by the Solar Energy Unit, University College, Cardiff, UK.

[17] Information may be obtained from the Built Environment Research Group at the Polytechnic of Central London.

[18] See Ref. [5], p. 272.

[19] Johannsen, M. (1979). 'Survey of UK collectors'. *Helios*, 7, 1 (see Ref. [16]).

[20] Michelson, E. (1981). 'Optimum sizing of solar water heating systems'. *Helios*, 11, 4 (see Ref. [16]).

[21] Horton, A. & Grove, S. (1979). *Milton Keynes Solar House Solar Heating System, 1975–1979, Performance and Cost Analysis*. Built Environment Research Group, Polytechnic of Central London.

[22] See Ref. [5].

[23] Gillet, W. & Rosenfeld, J. (1981). 'Review of British Standard BS 5918'. *Sun at Work in Britain*, 1213, 12. Published by the UK Branch of the International Solar Energy Society, 21 Albermarle St, London.

[24] Graham, B.J. (1979). 'Evacuated tube collectors'. *Solar Age*, November, p. 15.

[25] Madsen, P. & Goss, K. (1981). 'Non-metallic solar collectors'. *Solar Age*, January, p. 30.

[26] Best, D. (1982). 'New plastics head for higher temperatures'. *Solar Age*, February, p. 51.

[27] Best, D. (1982). 'Innovations with real expectations'. *Solar Age*, March, p. 37.

[28] Best, D. (1981). 'What you should know about phase-change water heaters'. *Solar Age*, December, p. 22.

[29] Sillman, S. (1981). 'Performance and economics of annual-storage solar-heating systems'. *Solar Energy*, 27.6, 513.

[30] Duffie, J.A. & Mitchell, J.W. (1982). 'F-chart predictions and measurements'. US *Annual Meeting of The Association of Mechanical and Electrical Engineers*, April.

[31] Kenna, J.P. (1972–83), in various publications from the Solar Energy Unit, University College, Cardiff, UK.

[32] Augustyn, J. & Schiller, S. (1982). 'A rating system will help consumers and industry alike'. *Solar Age*, May, p. 48.

[33] Information may be obtained from the Solar Energy Unit, University College, Cardiff.

[34] Information may be obtained from the Built Environment Research Group, Polytechnic of Central London.
[35] Duffie, J.A. (1982). Private communication.
[36] Information may be obtained from the Solar Laboratory, University of Wisconsin, Madison, Wisconsin, US.
[37] Information may be obtained from ABACUS, Department of Architecture and Building Science, University of Strathclyde, Glasgow, UK. FLARE is a comprehensive model for the analysis of active solar systems including air, water and photovoltaic collectors.

6

Space heating and ventilation

6.1 Introduction

From the underfloor heating of the Romans to the microbore hydronic installations of today may seem a small way but the path has been circuitous and the neglect of innovation has contributed to squandered resources and lowered living standards. For example, Fig. 6.1 shows an invention which was on the market in the early 1900s but failed to receive the attention it merited.[1] The design is similar to the National Coal Board's present series of room heaters with back boilers. (Current designs include the safeguards of a vent behind the back boiler and a makeup tank.)

Prior to the 1950s, domestic heating in the UK was provided principally by open fires. Then, with rising prosperity, cheaper energy in the form of imported oil and technical advances such as small glandless centrifugal pumps designed to circulate small quantities of water against relatively high heads, central heating began to be adopted. In the past 30 years the percentage of the housing stock with central heating supplied by oil, gas or other fuel, has increased from about three to 50.[2] Electricity for space heating also gained in popularity with the introduction of underfloor heating and storage heaters charged with cheaper night-time electricity.

This wide range of conventional heating systems developed for traditional houses is generally also applicable to low-energy buildings. Since numerous authors have dealt with standard applications,[2, 3, 4, 5, 6, 7] the emphasis of the following discussion will be on a brief but comprehensive presentation of the heating and ventilating systems commonly available and an analysis of their use in buildings and, in particular, homes where the energy demand is significantly reduced by the conservation and supply measures discussed in the previous chapters.

Energy shortages and the consequences they impose on designers should not, of course, be regarded separately from other concomitant changes in society. The question of the choice of a heating system should be intimately related to patterns of work and occupancy. As we have seen (Chapter 1) the BRE estimates that 41–61% of households (that is both occupied flats and houses) are intermittently occupied. Thus, for many houses the heating system required is one that will provide a great deal of heat quickly for several hours in the evening. Even this energy is not required homogeneously. In the earlier part of the evening, heat is required

Fig. 6.1. Hydroradiant fire and single-pipe gravity system designed by A.H. Barker.[1] (Courtesy of CIBS.)

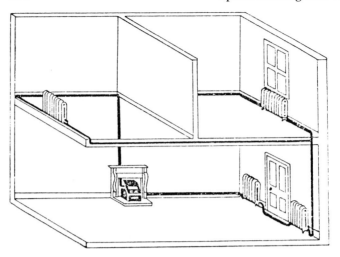

principally in the living and dining rooms and, in the latter part, in the bedrooms.

The overall pattern of demand in a low-energy house is one of moderate demand throughout a somewhat reduced heating season, with some periods of heightened demand during cold spells. If the occupants are in the house during most of the day the energy requirements will be fairly uniform during the daylight hours, with due allowance for, for example, solar gain during spells of bright sunshine; at night, demand will fall considerably. If the occupants are away during the day a morning peak demand from, say, 6.30 a.m. to 9.00 a.m. and an evening peak of, say, 6.00 p.m.–11.00 p.m. are likely.

Each situation must be analysed individually, bearing in mind that during the lifetime of the house, varying groups of people are likely to use it. Considerations of flexibility, in both the immediate and long-term futures; reliability, given the impossibility of assuring any single energy supply; and the difficulty of foreseeing which sources of energy will prove economical argue for the use of various fuels.

Another consideration is that, in new construction or renovating old buildings, the involvement of the users, or potential users, in the design process is important for their satisfaction. It is also essential to energy conservation because even sound design measures such as self-closing doors, shutters for windows, solid-fuel heating systems, thermostats and time switches which people are not willing to accept are worse than useless. For a house then, a meeting with a client to discuss heating-system preferences, desired comfort levels, pattern of occupancy, acceptability of control measures, and so forth, is of great benefit, particularly if the designer remembers that during the lifetime of the house numerous other people will live in it. The result of the meeting will hopefully be a fairly precise determination of the appropriate heating system and an idea of the maximum permissible variation in the heating and other environmental

Introduction

Table 6.1. *Percentage contributions (in terms of delivered energy) to the 1974 UK domestic space heating requirement*[10]

Fuel	Percentage	Fuel	Percentage
Coal	30	Electric — on peak	5
Other solid fuel	8	Electric — off peak	6
Oil	14	Electric — total	11
Gas	37		

requirements, as well as the compromise possible in extreme climatic conditions or emergencies.

Having selected a system, the owners or occupants must be able to control it and thus the energy consumption of the building, whether this be by provision for 'task heating', part-load operation of systems, operation at specific times, partial use of spaces or similar strategies. Additionally, on at least a monthly basis, some way of determining the actual energy used should be provided so that the value of conservation is apparent to those using the space and so that corrective measures can be taken if required. As one of the occupants of the first Milton Keynes Experimental House said, regretting the designers' decision to provide only conventional controls in the house: 'If we could glance at a dial and know how much energy we were paying for . . . I think we could effect even greater savings' (that is, in addition to those afforded by the solar collectors).[8]

The primary consequence of reducing the heat loss in a low-energy house is that smaller heating components may be used. In some cases the result has been to encourage development of appliances specifically designed for the lower load and these are discussed below. A secondary effect is that more flexibility may be possible in siting heat distributors. For example, with double glazing reducing draught and condensation problems at windows, radiators may be placed closer to the heat source and back-to-back in adjacent rooms where possible, thus reducing costs.

For alternative sources of energy such as solar and wind power, the site will be quite important in determining the contribution to the heating demand but for conventional sources the effect is minor. Approximately 80% of the UK is presently served by gas[9] and proximity to coal-mining areas may influence the choice of heating systems, but, apart from these considerations, fuels are generally homogeneously available.

Table 6.1 shows the approximate percentage contributions in terms of delivered energy (this must be multiplied by the efficiency of the heating appliance to determine the useful energy) of various fuels to the UK domestic space heating requirement.

Table 6.2 shows the energy-conversion efficiencies.

Another way of viewing the primary energy (PE) to delivered energy (DE) efficiency, for example electricity, is that 3.69 units of primary energy input are required for one unit of delivered energy. The significance of this is the often-made point, which we strongly endorse, that a rational energy policy must look beyond economics to the primary energy require-

Table 6.2. *Energy conversion efficiencies for primary to delivered and delivered to useful energy for space heating fuels (1974) (after Ref. [10])*

Type of fuel	PE–DE efficiency[a]	DE–UE efficiency[b]	Overall (PE–UE) efficiency
Coal	0.969	0.45	0.44
Other solid fuel	0.831	0.65	0.54
Gas	0.909	0.7	0.64
Oil	0.919	0.7	0.64
Electricity	0.271	1.0[c]	0.27
Domestic sector (all fuels)	0.73	0.66	0.49

[a] The efficiency of conversion of primary energy to delivered energy allows for production, transportation, etc.
[b] The efficiency of delivered energy to useful energy is necessarily approximate because of variation in published estimates of conversion efficiency and because of the wide range of conversion devices.
[c] This figure must be treated with some caution. If, for example, electric heating is used and, because of the 'dead' band of the thermostat, overheating occurs, the DE–UE efficiency will be less than 1.0. Heat wasted by electric immersion heaters in the summer can also be significant.

ments of not only fuels but all the material items (including food) of a society to see how available primary energy resources may best be distributed in both space and time.

In the case of fuels, for numerous historical and technical reasons, there is not always a correspondence between primary energy costs and economic costs. Table 6.3 shows C. Sutherland's valuable table of comparative space heating costs and average thermal efficiencies.

The relative positions of the various fuels tend to change over the years (or months!) according to economical and political realities. Anyone actually choosing a heating system on cost grounds alone will have to obtain current data on both capital and running costs and then estimate how the running costs will change with time. We can only agree that it is more easily said than done.

Returning to Tables 6.2 and 6.3, efficiency is a particularly vexed question since it depends on the appliance and the system. For example, the efficiency of solid-fuel heating varies with the position of the chimney and with the pattern of use of the house since heat stored in the chimney may or may not contribute to usefully warming a space. The BRE[12] provides additional data on system efficiencies. The efficiency also depends on the load and, since a central-heating boiler may spend the majority of its operating time at 20–50% of its rated output, it is important to evaluate its performance under these conditions. Fig. 6.2 shows the relationship between boiler efficiency and load.

British Gas claims that best current practice can provide overall efficiencies in the heating season of 75–80%, depending on whether the installation allows advantage to be taken of casing losses.[13] Modern

Introduction

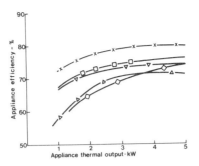

Fig. 6.2. Boiler efficiency as a function of load for five current compact wall-hung gas appliances.[13]

boilers generally have a low thermal capacity and quick response which makes them especially suitable in a lightweight low-energy house designed for intermittent heating.

A common word of caution when selecting any heating system is that too much emphasis should not be placed on quoted full-load thermal efficiencies obtained under laboratory conditions. A unit which is slightly less efficient but which, for example, provides more uniform heating at low level and which responds more quickly to controls, may give lower overall running costs. Nevertheless, it is important to consider some measure of efficiency when selecting any heating appliance since measured efficiencies vary from about 20 to 80%.

Both the EEC and the UK Department of Energy are aware of the enormous scope for energy conservation in this area and are working on directives and legislation to encourage development and selection of high-efficiency convertors. In the US, appliances must now be sold with an efficiency rating and an indication of how the device compares with analogous devices.

Domestic heating in the UK is dominated by gas, in contrast to European countries such as West Germany and Switzerland where oil is the principal fuel. In 1978, of the new central heating systems installed in Britain, over 80% were fuelled by gas.[14] With increasing oil prices the additional demand placed on gas caused British Gas some concern. Former chairman Sir Denis Rooke said: 'It is not that we are running out of gas, but that we have to limit winter peak demand to our total supply capacity in the transmission and distribution lines.'[15] One measure taken was to cancel promotion of central-heating systems and gas fires, which would have aggravated winter demand. Sales of oil-fired central heating systems have been virtually static in the recent past[16] and it is more likely that they will fall than rise as customers shift to gas, solid fuel or wood.

Electricity for heating has profited from increased oil prices, and sales of off-peak storage heaters in particular have risen considerably. Electricity, more than most other fuels, only makes sense economically in houses insulated to a high level. The Electricity Council is encouraging improved thermal standards and generally advocating storage heaters to provide the base load, with any other heating needed being provided by direct-acting space heaters in each room.[17]

Solid fuel is receiving more and more consumer interest, although the

Table 6.3. *Comparative space heating running costs for the UK (south-east area − winter 1982−83)*[11]

Fuel	Sold in units of	Cost per unit sold in pence	Heat content per unit sold on MJ basis	Application
Bituminous coal group 2	50 kg	494	1499	Open fire
Bituminous coal group 2	50 kg	494	1499	Open fire with backboiler
Sunbrite coke	50 kg	624	1400	Room heater
Anthracite Stove Nuts	50 kg	716	1685	Room heater with backboiler
Housewarm singles	50 kg	478	1499	Smoke consuming room heater with backboiler
Anthracite grains	50 kg	633	1627	Gravity feed boiler or backboiler
Electricity general tariff	1 Unit	4.90 4.90	3.6 3.6	Electric fire Electric radiators on day rate
Electricity white meter Economy 7 night rate	1 Unit	1.90 1.90	3.6 3.6	Single storage heater Multiple storage heaters
Gas (general domestic tariff)	1 Therm	33.5 33.5 33.5	105 105 105	Radiant−convector gas fire Wall heater Gas-fired boiler or backboiler
Liquified petroleum gas	15 kg cylinder 1 litre	835 14.40	737 26	Butane heater LPG-fired boiler[d] or backboiler
Heating oil 28 sec. viscosity	1 litre	21.23	37	Oil-fired boiler[e] or backboiler
Heating oil 35 sec. viscosity	1 litre	21.85	38	Oil-fired boiler[e]

[a] Annual heating requirement of a large room. Maintenance and standing charges included when applicable.
[b] Annual heating requirement of an average sized house. Maintenance and standing charges included when applicable.
[c] Difference between white meter and general tariff annual standing charges.
[d] Delivered to 1200 l storage cylinder. Rental shown in standing charge column.
[e] Delivered to 2725 l storage tank.
(Courtesy of C.M.J. Sutherland.)

present domestic consumption of ten million tonnes is well below (and likely to remain so) the 30−40 million tonnes used annually from 1947 to 1964.[18] Appliance sales are both for central-heating installations and supplementary heating where, for example, an open fire may serve as a 'focal point' in a room while providing a comfortable heating source.

Of course, the explanation of these trends is complex, with costs being interwoven with sociological and political factors. For some, central heating will always be preferable for reasons of habit and convenience, and similar considerations will mitigate against the selection of solid fuel in any form. The pattern of occupancy and the thermal response of the building

Introduction

Average thermal efficiency	Cost per useful MJ in pence	Additional annual standing charge (£)	Average annual maintenance cost (£)	Cost of[a] 4.14 GJ for room heating (£)	Cost of[b] 79.9 GJ for central heating (£)
30%	1.10	—	7	165	—
50%	0.66	—	7	—	481
65%	0.69	—	7	106	—
75%	0.57	—	7	—	415
70%	0.46	—	7	—	335
75%	0.52	—	7	—	381
100%	1.36	—	—	196	—
100%	1.36	—	—	—	980
95%	0.56	10.00[c]	—	90	—
95%	0.56	10.00[c]	—	—	410
57%	0.56	39.60	—	120	—
73%	0.44	39.60	18	120	—
75%	0.52	39.60	25	—	369
92%	1.93	—	—	177	—
75%	0.75	36.00	25	—	601
75%	0.77	—	22	—	578
75%	0.77	—	28	—	580

(discussed in Chapter 3) also influence the choice. To avoid discomfort from overheating (or underheating), the heating-system response time should be equal to or less than that of the building. If intermittent heating is required, the response of the heating system should be quick both for comfort and so that the economy in running cost is not lost by the long preheating period of a slow-response system.

Table 6.4 shows the approximate order of response of heating-system emitters; the ranking scale is not linear and in practical terms the difference between any two emitters may not be very great. Other authorities might also choose a somewhat different order and, say, place panel radiators above column radiators. Ultimately, the precise ranking will depend on the exact appliances selected and how they are used in a space.

In general, we believe that intermittent heating and quick-response systems are most suitable for any house designed to take advantage of the highly random nature of passive solar gain in northern Europe.

Table 6.4. *Quickness of response of emitters*[2]

Type of emitter	Order of response	
Fan convectors or ducted warm air	1	(fastest)
Individual natural convectors	2	
Convective skirting heating	3	
Column radiators	4	
Ceiling heating (low thermal capacity system)	5	
Panel radiators	6	
Radiant skirting	7	
Floor heating	8	(slowest)

To conclude this section on a somewhat pessimistic note, the problem of choice under conditions of uncertainty is one we always have with us. Even when there is an element of relative certainty, such as the availability of coal, it is difficult to say whether its use for domestic heating should take the form of solid fuel, substitute natural gas or electricity. Depending on the grades of coal available and consumer preferences, perhaps all three will be appropriate.

It is almost becoming the conventional wisdom to state that the options must be kept open to permit a change over to another fuel or source of energy, but to give specific recommendations is considerably more challenging. The Solid Fuel Advisory Service, not surprisingly, recommends that, no matter what heating system is installed originally, the provision of a chimney is sensible since it provides a long-term fuel option and affords a choice to future occupants. However, on purely economical grounds, there must be an assumption of the probability of the chimney being used during the lifetime of the house since an unused chimney is undoubtedly an investment without a return, as well as an added source of ventilation heat loss.

Systems using hot water or warm air have the advantage of being convertible to coal, gas, oil or electricity. Electric heating, on the other hand, cannot of course be adapted to other fuels. Looking to the future, designing a system for low supply temperatures is likely to facilitate conversion to solar or other sources of ambient energy.

Heating systems may be discussed according to the type of heat source, the extent of the system, that is, central heating or direct space heaters, the medium of heat transfer, the type of emitter or a combination of these. The following material has been divided principally into solid-fuel, water-distribution, forced-air and electric sections with appropriate cross-references made to avoid duplication of material.

6.2 Solid-fuel heating

Table 6.5 shows the characteristics of a number of solid fuels.

Solid fuels may be used for direct heating in fires or stoves, in a central-

Solid-fuel heating

Table 6.5. *Solid-fuel characteristics*

Fuel	Calorific value (MJ/kg)	(GJ/m^3)[a]	Density (kg/m^3)[a]
Anthracite[b]	35	26	750
Bituminous coal[b]	30	22	720
Coke[b]	28	11	380
Wood[c,d]	18	7	420

[a] Allowance for air space made.
[b] Refs. [3, 19].
[c] 12% moisture content; mean of red oak, white elm and chestnut.
[d] Ref. [20].

Fig. 6.3. Wood- and coal-burning stove.[21] (Courtesy of Norcem, UK Ltd.)

heating system or in a combination of both. Open fires are highly inefficient (see Table 6.3) since most of the convected heat emitted goes up the flue. A stove brought out into a room or arranged so that the convection from the back and sides is directed into the room is approximately twice as efficient. Stoves capable of burning traditional solid fuels as well as wood and peat are available in a range of sizes, but in choosing one for a low-energy house care must be taken to ensure that it is not in fact oversized. Wood is generally seen to complement other forms of solid fuel but its future availability is questionable since, on a national scale, the best use of wood may be in construction. On the other hand, in both urban and rural situations, there is, sad to say, wood which is presently wasted and a certain quantity of wood may be grown on-site. Fig. 6.3 shows a small free-standing stove (Jøtul) capable of burning coal or wood.

Many fires or stoves either come with or can be provided with back boilers to supply heat for radiators and domestic hot water. In a low-energy house, especially, it may be difficult to balance the highly variable

Fig. 6.4. Openable stove with a backboiler.[2]

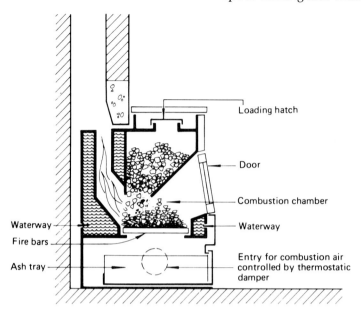

demand for space heating with that for hot water. It may be more reasonable to separate the space heating demand (as supplied by both the appliance and the radiators it feeds) from the hot-water demand by supplying the latter with an alternative source such as an immersion heater (at least during the summer). In this way a situation where a room might be overheated because hot water was required could be avoided. Fig. 6.4 shows a stove with a back boiler.

A number of multifuel cookers are also available, which will provide domestic hot water and hot water for radiators. Similar considerations of sizing and matching supply with demand apply.

Solid-fuel boilers for central-heating systems are now often designed to operate on wood, wood products and peat, as well as on the more traditional solid fuels. New domestic installations usually opt for a gravity-feed hopper which need only be filled about once a day, thus eliminating the inconvenience of the frequent stoking associated with traditional boilers. The clinker formed by combustion must be removed every several days. As with other solid-fuel appliances, consideration must be given to handling the fuel and clinker when positioning the boiler and there is an evident advantage if the fuel store can be located above the hopper. Fig. 6.5 shows a representative solid-fuel boiler.

Typically, these boilers are used in conjunction with hydronic systems (see Section 6.3) but, occasionally, air is used as the heat-transfer medium.

One of the principal reasons for discussing solid fuels in conjunction with ambient energy sources is that we may expect to have coal, or wood, and the sun for some time to come. Others include the economical cost and the low primary energy cost (see Table 6.2). Nevertheless, the marriage is not always ideal for reasons related more to control than convenience. We shall, however, assume, and increasing consumer interest

Solid-fuel heating

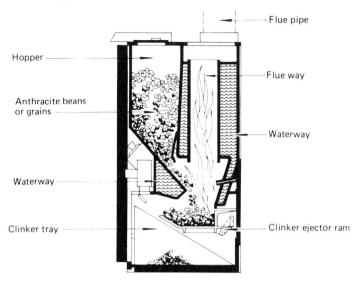

Fig. 6.5. Solid-fuel boiler.[2]

in solid fuel indicates that this is reasonable, that design and handling problems can be minimized. Design considerations include provision of a conventional flue for a solid-fuel boiler and storage facilities. The boiler may draw its combustion air inlet either from inside the space or outside, with the latter having the advantage of minimal interference from the ventilation system in the house. In either case the products of combustion go up the flue which should terminate above roof level.

Storage facilities required vary, with minimum recommendations of 2.7 m^3 for coke and 1.3 m^3 for other smokeless fuels and coal being standard in the past for small houses,[22] but with lower energy demands these might be reduced. On the other hand, the ability to store solid fuel is in some ways a major advantage since it affords a measure of independence from the vagaries of both the weather and national supply systems.

Handling the fuel has become a cleaner and simpler operation in modern appliances but, nevertheless, it remains a dissuasive factor for many individuals who are willing to pay more in exchange for greater convenience. For those who do choose solid fuel, conversion to gas, for example, can be fairly simple as occupants become older and less able to cope with solid fuel.

The control problem is related to the thermal response of the house, the pattern of occupancy and the importance of using passive solar gain in the house. Direct-heating appliances and some central-heating ones are likely to have a slow thermal response because of the heat capacity of the large quantity of metal in the appliance. As the heat demand in the house varies with ambient conditions, the appliance may require frequent manual attention. In the case of central heating, control of the boiler is normally by a thermostat in the water controlling a forced-draught fan. The response is basically similar to that of other hydronic systems using gas or oil but care must be taken to choose a lightweight boiler if a quick

Fig. 6.6. Ground-floor plan of the Vale House (after Ref. [23]).

response is desired. Any system of solid-fuel heating which deliberately or inadvertently stores heat in a chimney will, of course, have a slower response.

Both the problems and potential of solid-fuel heating are apparent in the houses discussed below.

Fig. 6.6 shows the ground-floor plan of the Vale Home, about 20 km to the north of Cambridge. Upstairs there are three bedrooms and a bathroom. The following comments draw heavily on the excellent descriptions of the house by the Vales themselves.[23, 24]

The Vales restored the house and insulated it to a high level (U-value of walls (including glazing) 0.42 W/(m² K); roof = 0.25 W/(m² K); floor = 0.54 W/(m² K); estimated ventilation rate = 0.5 air changes per hour; insulation is inside the brick walls). They spend the greater part of each day at home looking after their smallholding, doing associated research work, taking care of their children and doing the usual housework. As part of what they refer to as their 'alternative approach', they felt that the heating system should be very simple and designed for long life. Consequently, they chose an Aga coke-burning thermal-storage cooker, which also provides domestic hot water, backed up by two solid-fuel stoves. Outside the heating season, cooking is done in a small electric cooker and hot water is provided by an immersion heater.

The cooker is located centrally in the house and connected to a prefabricated insulated chimney (which gives off some heat in an upstairs bedroom). Heat produced is stored in the 500 kg of steel within the cooker and this, combined with a simple thermostatic control which controls the draught, allows the cooker to operate under the constant-load

Solid-fuel heating

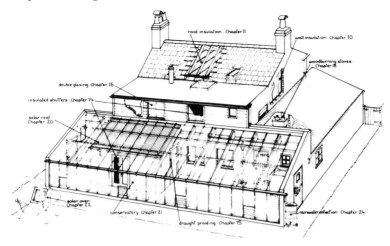

Fig. 6.7. The Vale House.[24] (By permission of Macmillan, London and Basingstoke.)

conditions necessary to high efficiency which, in this case, is about 70%. It is estimated that the Aga operates as a continuous 1 kW direct space heater (no radiators are fed) and over the heating season provides 18 GJ in addition to about 22 GJ for domestic hot water and cooking.

Additional heat is provided by Jøtul and Ulefos stoves with maximum heat outputs of about 8 and 4 kW, respectively, but commonly the Jøtul alone, operating at its minimum output of 3 kW, is adequate — both are used by the Vales to burn wood. Average fuel consumption of the Jøtul is about 20 kg/d (for a 16 h day). Wood is stored under cover and coke for the Aga is stored in a bin in an outbuilding.

No heating has been found to be needed upstairs since warm air rises up the open staircase into the bedrooms whose doors effectively serve as openable vents.

The designers admit that the system does not offer the potential for control or maintenance of exact temperatures of more-conventional systems and that it has the disadvantages of requiring coke to be carried, wood to be sawn occasionally and carried and ashes raked out. On the credit side, however, it makes good use of primary energy, requires little maintenance, is of high reliability and is well suited to the family's continuous-occupancy pattern.

Recently, a large single-glazed conservatory (see Fig. 6.7) was added to the entire south side of the house to provide a sheltered area for plant growth and to permit passive solar gain to play a greater role in space heating.

Another example of an existing house using solid-fuel heating, but this time on a grand scale, is the Bailey House in Surrey.[25] This large, five-bedroom timber-frame house uses a heating system, specified and planned by the Solid Fuel Advisory Service, based on twin-gravity-feed anthracite boilers with a combined maximum output of 53 kW. The boilers feed radiators throughout the house and supply domestic hot water. A twin-thermostat system allows individual control of the burning rate of each fire so that only one boiler need be fired in the summer when just hot

Fig. 6.8. National Coal Board experimental low-energy consuming house design.[26]

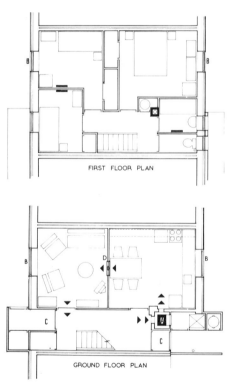

FIRST FLOOR PLAN

GROUND FLOOR PLAN

Solid-fuel heating

water is needed. A control unit, incorporating externally fitted frost stats which will override the thermostats inside the house to compensate for sudden drops in the outside temperature, operates the heating.

A number of experimental designs for low-energy houses using solid-fuel heating exist. Probably the most considered example is the National Coal Board's Terraced House shown in Fig. 6.8. The detailed discussion of the project, given in Ref. [26], forms the basis of the following comments.

In essence the house is a well-insulated volume, designed so that the internal environment remains fairly steady. The calculated design heat-loss rate is only 2.8 kW. (U-values in W/(m^2 K) are: walls, 0.3; roof, 0.25; windows, 2.8; floor, 0.4.) Well-sealed double-glazed windows and air-lock lobbies to the front and rear doors are used to reduce the ventilation heat loss. A structural chimney of high thermal capacity provides chimney heat gain to the house (in this way a house-heating efficiency of about 80% is claimed) and 100 mm of insulation is placed external to a 225 mm masonry wall. The exposed surface of the insulation is rendered to provide protection against the weather and possible damage.

Solar panels are used to supply part of the energy for domestic hot water, with an estimated 1.2 kW contribution from the heating system. For a total maximum heat output of 4 kW at an efficiency of 80% a burning rate of 0.7 kg/h is required and, although this is within the normal range of a conventional appliance burning smokeless fuel, the equipment could not operate satisfactorily for extended periods at the 0.1–0.2 kg/h burning-rate levels required during milder weather conditions. It is also difficult to maintain an attractive visual fire appearance over the operating range of the appliance and, if smoke produced is to be kept below a satisfactory level, the area of water-cooled surface within must be kept to a minimum.

Consequently, heat transfer to the house must be principally by the circulation of warm air rather than by direct radiation from the firebed or from hot-water radiators. A coal-fired boiler (see Fig. 6.9) enclosed within a brick convection chamber heats the air in the chamber which is distributed by natural convection (the stairwell is open) and around the ground floor by a small axial fan located in the partition wall between the kitchen and living room. A proportion of the air is recirculated and provision is made for a carbon filter unit to prevent odours and moisture being circulated around the house. Highly volatile bituminous coal appears to be the most suitable solid fuel because of the low burning rates required. The boiler output is controlled from a room thermostat sited in the kitchen and a second thermostat in the lounge switches the axial fan on and provides filtered warm air from the kitchen when required. The boiler is fitted with an override thermostat to limit the temperature of the warm air leaving the combustion chamber. Additional heating is provided to a bathroom radiator and two small bedroom radiators from the domestic hot-water system.

Fig. 6.10 shows Ambient Energy Design House 1 (see Fig. 3.17 for the floor plans) which was developed independently of the National Coal Board work, by the authors and Ms L. Krall. How the heating system

Fig. 6.9. Experimental low-energy coal-burning appliance.[26]

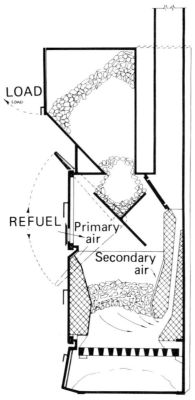

4 kW coal burning warm
air appliance for low energy building

evolved in practice and aspects of the house other than the heating are described in detail in Chapter 12. A 3.7 kW solid-fuel boiler is used for space heating and domestic hot-water heating. The fuel store is filled from the street and has a sloped floor towards the adjacent boiler. Solar collectors on the roof are used to heat air which is ducted to a rock-bin store under the staircase, mixed with recirculated filtered air drawn from the kitchen and then circulated throughout the house. The boiler, which draws its combustion air supply from outside, is used to supply hot water to a coil in the rock store (December to February) or a fan coil unit in the air stream (September to November and March to May). Control is simply by a room thermostat which turns the fan in the fan coil unit on or off and a thermostat in the air stream, which opens or closes the valve to the appropriate heating coil, depending on the month. The combustion air supply which comes from outside is separated from the air supply necessary for ventilation and space heating to permit greater control and efficiency. The rock-bin store is insulated and is not expected to behave like an uninsulated chimney. Wall insulation is internal and the house in these and other ways is designed to respond relatively quickly to passive solar gain.

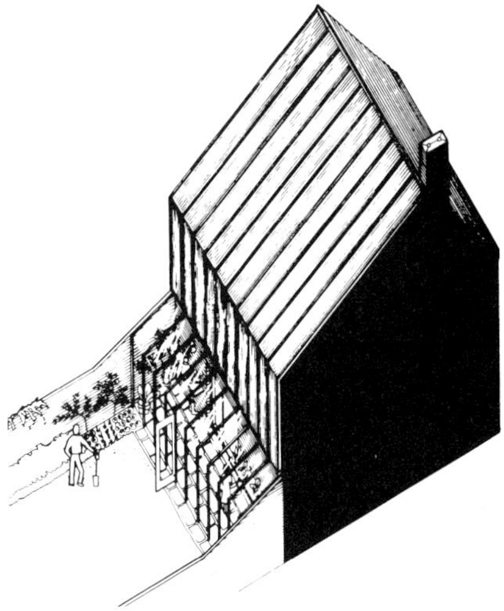

Water-distribution systems 205

Fig. 6.10. Ambient Energy Design House 1.[27]

6.3 Water-distribution systems

Heat sources for water-distribution systems may be solid fuel, gas, oil or electricity, with gas the most popular and electricity the least; Chapter 5 has shown that active solar energy is also easily integrated with a water system. Just as for solid fuels, it is possible with gas to have direct space heaters or a radiant fire with a backboiler but it is more common in central-heating systems to have a floor-standing or wall-hung boiler.

Oil-fired boilers (see Fig. 6.11) use a vapourizing burner or a pressure jet to atomize the fuel for efficient combustion. For smaller domestic installations a vapourizing burner is normally adequate. Control of the boiler is achieved with a valve (actuated by a thermostat in the water) which varies the amount of fuel to the burner. A useful measure to ensure that oil (and to a lesser extent gas) systems are operating efficiently is to analyse the flue gases to ensure that complete combustion is achieved but not at the cost of excessive air carrying the heat away. For smaller installations, periodic use of portable analysers, which give carbon-dioxide content or, less commonly, oxygen content and temperature readings, may be used.

Recent developments in boilers include the incorporation of fluidized beds, provision for multifuels and heat centres in which the domestic hot-water facility is combined with the boiler. In Holland the Stone–Platt Fluidfire domestic boiler has been used in experimental low-energy houses.[28] In the boiler, tubes carrying low-pressure water are passed through a shallow fluidized bed of aluminium-oxide particles, thus giving very rapid heat transfer — this results in greater efficiency and a reduced boiler size. Present models are gas fired but the principle could be applied to other fuels.

Multifuel burners are usually designed with gas or oil as the principal fuel with provision to shift to solid fuels, wood, propane or straw depending on the model. Adaptor kits such as swing-arm pressure jet attachments

Fig. 6.11. Oil-fired boiler.[2]

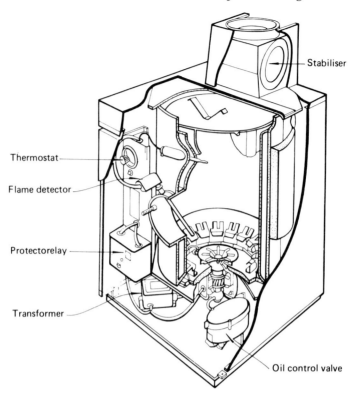

to allow oil to be burned can effect the switch from one fuel to another. Manufacturers claim high efficiencies with all fuels, and sales indicate that many purchasers find the flexibility attractive as an insurance against shortages of a particular fuel, but some criticism of the trend has been voiced. A Swiss report states that dual-fuel boilers (gas or oil plus a solid-fuel option) cannot be designed for maximum possible efficiency since combustion space and heating surface are a compromise between the different fuels.[29] Over the life of the boiler, therefore, a great deal of energy will be wasted and the solid-fuel option hardly ever used, it is claimed. The alternative suggestions are a twin combustion chamber boiler or, better, an independent solid-fuel boiler for those times when another fuel — usually wood — is available.

Heat centres (see Fig. 6.12) are a combination of a boiler, an insulated hot-water cylinder, a circulating pump, an electric programme control, thermostats (boiler, high limit and cylinder) and a flowshare motorized valve.[30] Advantages of the system include reduced piping runs (and associated heat losses), fast recovery on domestic hot water owing to the higher efficiency, elimination of any need to oversize the boiler to ensure adequate temperature since the flowshare valve allows full output from the boiler to go to the central-heating system whenever necessary, and ease of installation.

Standard boilers burning oil usually require a conventional flue,

Fig. 6.12. Interior of a heat centre.[30] (Courtesy of Harvey-Habridge Ltd.)

although some can use a balanced flue (see Fig. 6.13) as most gas boilers do. The great advantage of balanced flues is that they need no chimney.

Current distribution techniques for hydronic systems in standard domestic applications include traditional two-pipe, one pipe with single entry valves, one pipe with conventional connections and microbore. Typical water temperatures are 82 °C flow and 70 °C return.

Two-pipe systems (see Fig. 6.14), which have separate pipes for flow and return, are simple to design, provide water at approximately the same temperature to all emitters and employ either radiators or convectors. Possible disadvantages are an increase in cost and a more obtrusive appearance.

One-pipe systems (see Fig. 6.15) serve radiators progressively with the main flow, not through the emitter but bypassing it. Connections to the emitters are taken off in series and if there is flow through several emitters the flow to the second in a group includes part of the return from the first. Average temperatures in successive emitters are thus reduced and the size must be increased to compensate for this. Also, since the flow through emitters depends on the resistance of the emitter compared to the resistance of the bypass pipe, high-resistance emitters such as convectors may not be suitable. A principal advantage is greater simplicity of piping and, hence, installation. In the past, one-pipe systems were considered

Fig. 6.13. Balanced flue appliance.[31]

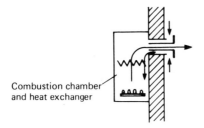

Fig. 6.14. Two-pipe system.

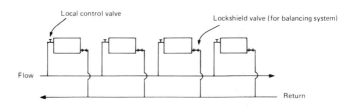

Fig. 6.15. One-pipe system (with conventional connections).

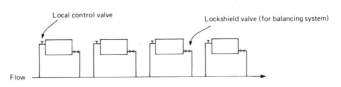

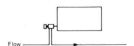

Fig. 6.16. One-pipe system with single-entry thermostatic valve.

most suitable for small installations and, with low-energy houses requiring less heat and fewer emitters, they may be perfectly appropriate.

Local control valves for both two- and one-pipe systems may be either manual or thermostatic (automatic). Special single-entry thermostatic valves (see Fig. 6.16) exist for single-pipe systems which provide a permanent bypass for most of the flow and direct the rest to the radiator.

Microbore systems (see Fig. 6.17) employ smaller pipe sizes (commonly 6, 8 and 10 mm diameter) and feed manifolds (with more-conventional 22 mm diameter pipe) rather than individual radiators. One of their main potential attractions for low-energy homes is the inherent lower water content which gives a faster response to both heat outputs and thermostatic controls.

The most common piping material is, by far, copper in small installations, with steel sometimes used in larger projects. Cross-linked polythene tube, recently introduced by British Steel, may become popular in the future.

Panel radiators, usually of pressed steel, are the most common form of emitter. In spite of their name about 70% of the heat emitted is convective. Principal types available are simple single panel (see Fig. 6.18), double or multiple panel and single panel with a finned back to increase the heat output. In general, radiators respond poorly to control because of their high water content but, of course, in a low-energy house the radiator size will be reduced and thus so will the amount of water. A typical output from a single-panel radiator 1.0 m × 0.7 m and of 40 mm depth, operating at a

Water-distribution systems

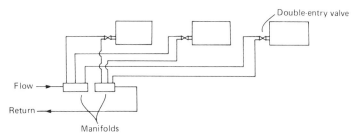

Fig. 6.17. Microbore system (after Ref. [32]).

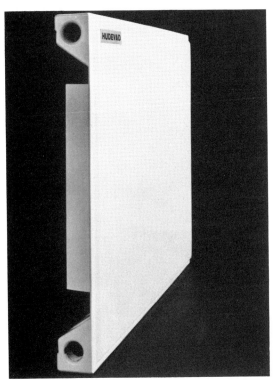

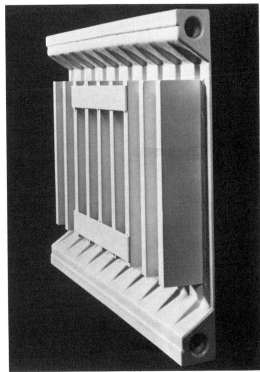

Fig. 6.18. Front and rear view of single-panel pressed-steel radiator.[33] (Courtesy of Hudevad, Gt. Britain.)

56 K differential between mean water and room air temperatures, is 1.9 kW/m^2 ;[33] radiators of increased thickness give higher outputs.

Natural convectors (sometimes known as convector radiators) (see Fig. 6.19) consist of a steel frame inside a casing whose design encourages movement of air over the frame by the stack effect. Since they have both less metal and less water than panel radiators they heat up more rapidly and respond more quickly to control. Heat output is approximately 2.2 kW/m^2 for a 1.0 m × 0.5 m convector of width 89 mm operating at a 60 K differential between room air and mean water temperatures[34] and this can result in significant savings in both cost and space. In spite of these advantages, public acceptance has been slow but, for low-energy houses, particularly those that are intermittently heated, they would seem to be well suited and so interest is likely to increase.

Fig. 6.19. Natural convector.[34]

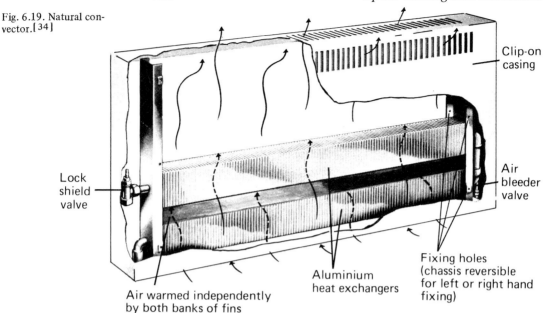

Fig. 6.20. Radiator location in traditional houses.[2]

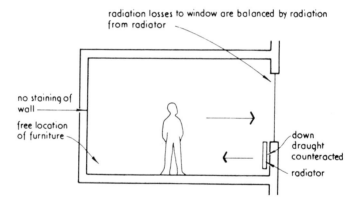

Fan convectors, as their name implies, use a fan to blow air over the heater battery rather than relying on natural convection and so are capable of higher heat-transfer rates. Individual control is very easy since a time-switch and thermostat can be used to govern the fan and hence the output of the device. They have two disadvantages, however. The first is that an electrical supply is required for each unit but this may be compensated for by reduced pipe runs and the smaller number of appliances required due to the flexibility and greater strength of these emitters. The second is that they tend to be noisy.

Radiators and convectors have, in the past, commonly been positioned under windows (see Fig. 6.20), but in low-energy houses with reduced draughts, double glazing and smaller heat loads this is less necessary. Also, some low-energy houses incorporate floor (or close to the floor) to ceiling

Water-distribution systems

glazing on the south wall and here emitters should be brought back into the room to prevent obstruction of incoming solar radiation.

For radiators which are located on external walls, research has shown that reflecting materials behind radiators can significantly reduce the amount of heat loss through the wall, the exact extent depending largely on the U-value of the wall.[35] The material used can vary from inexpensive cooking foil to higher-quality aluminium foil.

Underfloor heating may be used exclusively or in conjunction with other emitters as in the case of underfloor heating on the ground floor and radiators in the first-floor rooms. Either metal or plastic piping may be laid coiled in a screed whose details of construction will vary depending on whether or not it is desired to take advantage of the thermal capacity of the floor. Floor surface temperatures are usually limited to 30 °C to avoid discomfort. Approximately half of the heat emitted may be as radiation and the system has the advantage of maintaining an even-temperature gradient. A number of low-energy houses have adopted underfloor heating because the low temperatures employed (flow temperatures are commonly less than 50 °C) are easily supplied by either conventional equipment or alternative sources such as solar panels and heat pumps. It should be noted, however, that it is also possible to operate radiators and convectors at temperatures as low as 40 °C. Performance will of course be lower and the size must be increased, but not necessarily to an unacceptable point, since the heat load to be supplied in a low-energy house is less.

Other emitters for use with hydronic systems include radiant panels, radiant strips and skirting heating (of three types: convective, radiant−convective and radiant). Details of these and further information on the emitters described above may be found in standard references.[2, 4]

Water-distribution systems commonly use a small, glandless, electrically driven centrifugal pump (see Fig. 6.21) which may or may not be supplied with the boiler. Gravity systems, which require large-bore piping and low-resistance emitters, but have the advantage of being independent of a power supply, are now uncommon.

Allowance for expansion of water is normally by the provision of a cold feed (for replacement of water if necessary) and expansion tank in an open system (see Fig. 6.22). In a sealed system (see Fig. 6.23), which may be required if height in the building is limited, a diaphragm or membrane tank (see Fig. 6.24) together with a safety valve and pressure gauge are used. A sealed system can operate at a pressure above atmospheric and at a temperature higher than 82 °C, in which case, to avoid the possibility of accidents as a result of contact with surfaces, radiators are not recommended − convectors or skirting heating are more appropriate and may be more economical. A detailed comparison of open and closed heating systems is given by Rockhill.[37]

Almost all hydronic systems include provision for a domestic hot-water supply. Fig. 6.25 shows a typical example where the hot-water supply comes from an indirect hot-water cylinder fed from the central-heating boiler. Here an immersion heater in the indirect cylinder heats the water during the summer when the central heating is switched off but, depending on the system, it may be more efficient to simply use the boiler.

Fig. 6.21. Domestic circulating pump (approximate dimensions 130 mm diameter by 130 mm depth).[36] (*a*) Installed. (*b*) Cut-away.

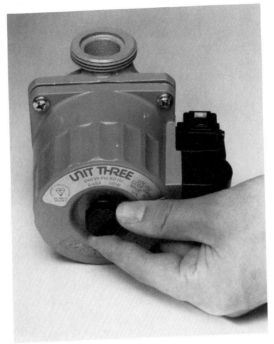

Controls, for any heating system, are an integral, if at times undervalued, part of the effort to save energy. Fisk has pointed out the complex interaction between occupants and controls in domestic heating, which he sees as probably the unique situation in which those who benefit from the controls also pay for them.[38] Consequently, a failure on the part of the controls to maintain comfort conditions may be acceptable if the failure is accompanied by reduced costs. The control system, to be successful, he

Water-distribution systems

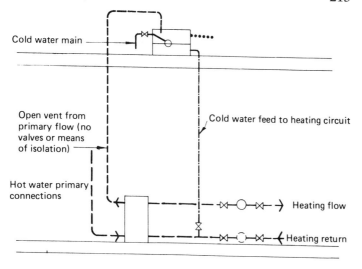

Fig. 6.22. Open distribution system.[2]

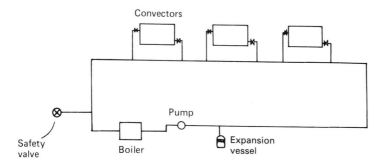

Fig. 6.23. Sealed distribution system (schematic representation after Ref. [3]).

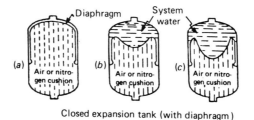

Fig. 6.24. Diaphragm expansion tank.[37] (a) When system is filled, no water enters tank when cushion and water pressure are in equilibrium. (b) As temperature increases, diaphragm moves to accept expanded water. (c) When water rises to maximum, full acceptance of expansion is achieved.

adds, must be matched with the intended system of energy management by the occupants.

In older hydronic central-heating systems it was common for the boiler to remain under the control of its own boilerstat as long as a timeclock or programmer indicated that a need for heat might arise. This involved necessarily higher losses than when internal temperature sensors keep the boiler closed down when there is no demand for heat, and so the next step was to use a single room thermostat to control the system. This, in turn, has the disadvantage of controlling the whole house from one source and so the only room accurately controlled is the one which has the thermo-

Fig. 6.25. Domestic hot-water supply using an indirect hot-water cylinder fed from a central-heating boiler.[2]

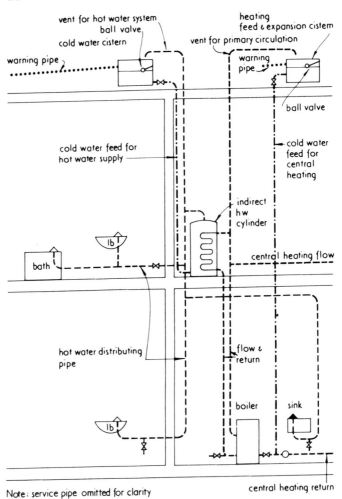

Note: service pipe omitted for clarity

stat, the others either underheat or overheat. However, it has been objected that good design and, especially, correct sizing of the heat emitter, minimized this problem. The argument then went that even the best design could not properly anticipate the effects of other heating influences in the house, whether these be from passive solar gain or casual gains from, say, equipment. To make the best use of these gains, individual room controls are required.

In hydronic systems these commonly take the form of thermostatic radiator valves (see Fig. 6.26) located on the flow pipe just before it enters each radiator (that is, the local control valve of Fig. 6.15). Little information exists on the precise amounts of energy to be saved by the use of these valves (and by alternative control strategies in general)[39] but a number of authorities view them as especially suitable for domestic comfort and economy.[2, 40] (Burberry & Aldersley-Williams[2] provide a valuable detailed guide to a number of control systems.) Because, as mentioned previously, overheating by 1 K results in an approximately

Water-distribution systems

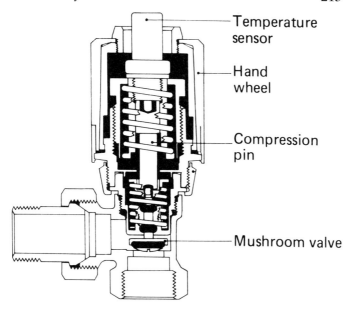

Fig. 6.26. Thermostatic radiator valve.[41]

5–10% increase in energy consumption, accuracy of control is essential and it has been suggested that room thermostats and thermostatic radiator valves with differentials of less than 0.5 K be used.[40] Larger differentials cause the occupants to set the thermostat higher in order to prevent the space temperature falling below the desired level.

Thermostatic radiator valves are best used in conjunction with a time control of the pump and boiler, which may be quite simple, giving, say, one or two periods of operation for the heating and hot-water service per day. A variation can be used to separate heating and hot water if desired.

In the future, microprocessors and other similar small processors are likely to be used to augment the enormous potential of controls.[42, 43] As suggested in Chapter 1, it will be possible to provide considerably more-complicated control programs. A simple application will be the use of the boiler at the beginning of the day for heating only (rather than heating and domestic hot water) to speed response. The memory facilities of 'micros' will permit temperature regulation according to the period of the day to allow, for example, energy savings when the occupants are more active. Programs will also vary from day to day and season to season to allow for weekends and holidays and any other changes.

'Intelligent' programmers will be used for optimum start control. They will learn the heating curve of a house and then, on the basis of external and internal temperatures, will switch the boiler on so that the desired temperature will be achieved at the proper time.

Other capabilities include monitoring the systems for faults and planned maintenance and energy accounting to provide homeowners with records of energy consumption and costs.

The real challenge with microprocessors is with the associated hardware of the control system, that is, the devices that interface with the micro-

Fig. 6.27. System diagram of the BRE Solar Energy House.[44] (BRE, Crown Copyright, HMSO.)

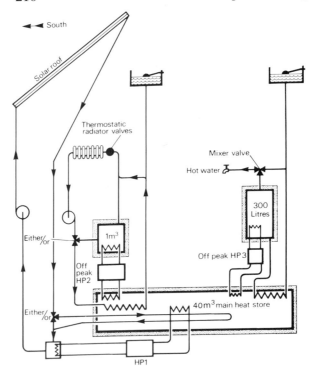

processor. For example, it would be very easy to incorporate in the microprocessor program facilities for turning off a local heating source, such as a radiator with a thermostatic valve, if a window in the space is open. However, the cost of the sensing equipment and the wiring back to the microprocessor might discourage the inclusion of such a facility.

In low-energy houses and particularly those which are separated into zones, intermittently heated and designed for passive solar gain, individual space control is very important. For houses intended to maintain fairly uniform environmental conditions throughout the house and during the day, a single room thermostat (sited properly) may be adequate. Similarly, in such houses, the importance of low thermal capacity heating systems is less. The following examples illustrate the use of water-distribution systems for space heating, in some cases in conjunction with active solar collection.

In the BRE Solar Energy House space heating is provided by hot-water natural convectors equipped with thermostatic radiator valves and fed from microbore piping (see Fig. 6.27).[44] The radiators are of larger than normal size to permit lower water temperatures (from the solar collector and heat-pump system) to be used. When the temperature of the main 35 m³ storage tank is sufficiently high, heat for the radiators is taken from it via a heat exchanger. At other times the supply is drawn from the 1 m³ well-insulated tank which is heated during off-peak hours by a small water to water heat pump that uses the large tank as its heat source.

Underfloor heating is used in one of the Salford Strawberry Hill

Forced-air systems 217

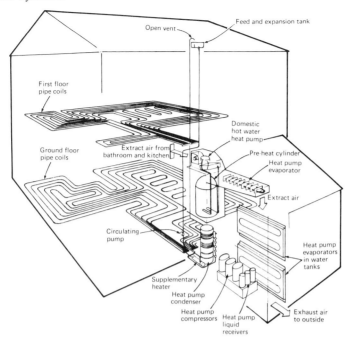

Fig. 6.28. Salford Strawberry Hill House with underfloor heating.[45]

Houses[45] (see also Section 6.6). Fig. 6.28 shows the approximately 180 m of flexible reinforced pvc hosepipe (of 12.5 mm bore) which was set in the screeds over the concrete floors. The house was deliberately designed to be heavyweight and heat is stored both in the floors and walls (brick exterior, concrete block interior and cavity filled with blown polyurethane granules). Although the minimum lifetime of the pipes is not known, it is expected to be long since the screed provides protection from both mechanical damage and ultra-violet radiation and because the maximum anticipated temperature is 40 °C. The pipe was laid using 30 m loops from two headers at each floor level with the geometry adopted permitting standard pipe lengths to be used for the circulation of a large volume of water at low pressure drop. No balancing valves were required.

6.4 Forced-air systems

Forced-air systems comprise the following three principal groups:
(1) ventilation systems;
(2) heating systems which may or may not be combined directly with ventilation systems;
(3) combined heating and ventilating systems which incorporate heat-recovery devices.

In the domestic context Europe has consistently been ahead of the UK in developments in energy conservation in general, and forced-air systems in particular. Over 30 years ago Scandinavia started a trend by equipping blocks of flats with mechanical exhaust systems.[46] With an increasing need to conserve energy and greater awareness of the importance of the

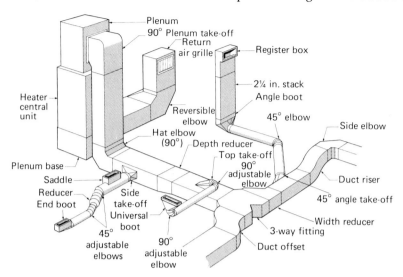

Fig. 6.29. Ducted air heating.[47] (Cavendish Ltd.)

ventilation heat loss in well-insulated buildings these simple ventilation systems are now being supplanted by balanced systems with both mechanical supply and exhaust. In France, in 1977, about two-thirds of all dwellings built had some form of mechanical ventilation.[46] Systems designed solely for ventilation are commonly used with radiators or convectors or electric heating but are applicable to almost all forms of heating systems.

The great advantage of a balanced mechanical ventilation system is that, at least in theory, it permits precise control of the air entering and leaving an enclosed volume and, consequently, the heat loss associated with that air. But, as we saw in Chapter 3, it is necessary to reduce the effective open area for natural ventilation to a small value.

Warm-air heating systems account for just under 10% of the market in the UK for domestic central-heating systems, but this figure may grow in the future as their potential for greater control over air temperature, moisture content and ventilation becomes more valued. A traditional objection to air systems is that they are generally bulkier than hydronic systems and although this remains true in a low-energy house the duct sizes will tend to be smaller than in the past and should pose no serious problems if the duct runs are considered at an early stage in the design. An important advantage is that thermal response is more rapid than in hydronic systems. Fig. 6.29 shows the elements of a ducted warm-air system with the combustion air for the heater control unit coming from within the space heated.

The heat source may run on solid fuel, oil, gas or electricity, with gas being most common in the UK. Warm-air units are divided into two broad categories – direct fired, where heat is generated in the primary source and transferred to the air stream by a heat exchanger that may be as simple as refractory bricks or steel plate; and indirect systems that use water to carry heat from a boiler which can also serve the hot-water system. The

Forced-air systems

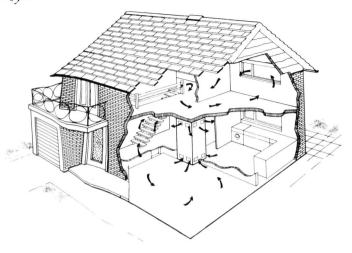

Fig. 6.30. Central heating by natural convection.[49]

efficiency of some direct warm-air heaters recently approved by British Gas is about 75%.[48]

Normally, a centrifugal fan is used to distribute fresh air mixed with recirculated air through the ducts, but some air systems dispense with both ducts and fan and rely simply on natural convection. Fig. 6.30 shows the circulation pattern in a house heated by a gas-fired air heater that warms a chamber which is an integral part of the house and is made of the same materials as is used for the walls, that is, brick, load-bearing block or concrete. Air in the heat chamber is warmed and rises by convection to be discharged by high-level grilles into adjoining rooms. Circulation is set up and cooler air returns to the chamber via low-level grilles, thus continuing the cycle. Upstairs rooms are warmed by conduction through the ceiling and air movement up the stairway. Such a system is extremely simple and may be suitable for an unzoned house of slow thermal response and fairly continuous occupancy.

In systems with fans, attention must be paid to selecting a fan which will distribute sufficient air without creating excessive noise. There is a certain amount of flexibility between fan power and duct size, with a larger fan compensating for a smaller duct in cases where space is limited. Ducts are commonly made from galvanized sheet steel, although glass fibre and certain plastics are suitable, in a circular or rectangular section, and may be fabricated with insulation such as polyurethane or may be insulated afterwards. Duct sizes vary of course but in a small well-insulated house will probably be no greater than 300 mm × 300 mm for the main runs and considerably smaller for most of the others. Air is introduced into spaces by grilles which are preferably located in the floor but may be in the walls or ceiling. Ideally, extract grilles are sited in the diagonally opposed position of the space. Supply grilles are often fitted with a damper which may be controlled manually or automatically to regulate the air flow to the room. This is a consideration of some importance in a low-energy house, in which the maintenance of lower temperatures in unoccupied areas can result in significant savings. Automatic control of the

Fig. 6.31. Warm-air heating with control of the supply air.[50]

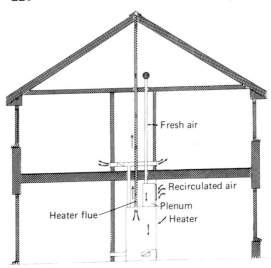

air supply and the temperature in a space may be achieved by motorized dampers connected to a room thermostat, but for a domestic setting this is likely to be both expensive and complicated and hence prone to failure. Typical controls are much simpler (and cruder than in hydronic systems) and consist most often of a room thermostat situated in the living room, which controls the operation of the air heater which then controls the fan; a timeswitch is also often used.

Standard references deal with air velocities in the ducts, the 'throw' from registers and siting of supply, return and transfer grilles.[4, 7] As for radiators in hydronic systems, the reduced heat requirement and more-uniform thermal environment (due to double glazing, reduction of draughts, etc.) of a low-energy house will tend to permit greater flexibility in locating registers.

Fresh-air supply in traditional warmed-air heating systems is mainly from infiltration and enters the warmed space directly. With more-tightly sealed houses and greater control of ventilation and air supply in the house, however, it is preferable to draw the fresh air from outside the heated volume. Fig. 6.31 shows a proprietary system which takes a controlled amount of fresh air from the roof space and mixes it with a varying amount of recirculated air for heating and circulating to the rooms. Combustion air exits by the heater flue and heated air is lost through exfiltration (in an uncontrolled manner).

The scope for energy conservation is considerably increased if a heat-recovery device is incorporated in the system. This is easily done in a balanced system which has inlet and exhaust streams in close proximity at some point. Usually, a plate-type heat exchanger is used for individual houses, with a thermal wheel or heat-pipe unit becoming practical for blocks of flats. Run-around coils are often used in commercial office buildings and heat pumps may be used for heat recovery in a variety of applications. Plate heat-recovery units are passive devices. They act as

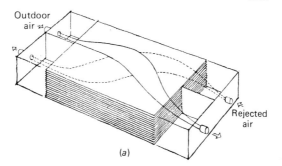
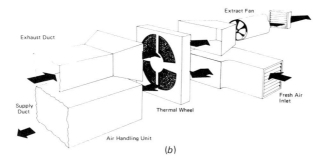

Fig. 6.32. Heat-recovery devices.[53, 54] (a) Plate-type heat exchanger. (b) Thermal wheel. (Courtesy of the Electricity Council).

counterflow heat exchangers by transferring heat from the exhaust stream to, for example, thin aluminium sheets. Incoming air on the other side of the sheets is simultaneously warmed and heat-recovery efficiencies of 60–70%, depending on the differences in temperature, are obtainable.[51] Fig. 6.32 shows a plate-type heat exchanger and a thermal wheel — the latter normally requires an electrical input to turn the wheel but has the advantage, at least in some designs, of facilitating latent-heat recovery in addition to sensible-heat recovery. The efficiency of such devices can be as high as 80%.[52]

In general, warm-air systems and hydronic systems can be equally suitable for low-energy houses, with warm air perhaps preferable in intermittently occupied, fast thermal response dwellings. The scope for controlled ventilation and heat recovery may, on grounds of simplicity of incorporation, favour warm-air systems. Integration with solar heating, whether this be mere preheating of incoming fresh air or an active solar collection system with storage, is also feasible. Recent developments in component design and control have led to a high degree of comfort (that is, in particular, low vertical thermal gradients) and less noise. The following examples illustrate the use of forced-air systems in dwellings.

In the Milton Keynes Solar House described previously (see Fig. 5.16) space heating is provided by a single fan convector fed with water heated by a combination of active solar collection and a conventional 6.5 kW gas-fired boiler.[55] The open staircase of the house permits heated air from the convector, which is located on the ground floor, to circulate through the house. Water as low as 30 °C can supply the convector coil.

Fig. 6.33. Schematic of forced-air heating and ventilation layout in a Cambridge house (scale 1:200). (*a*) Ground floor. (*b*) First floor.

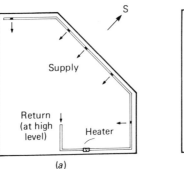
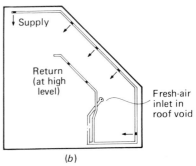

It is of interest to examine in detail the evolution of one component of the above system, the fan convector, because it illustrates the common problems of correctly specifying material and finding suitable equipment commercially. The fan convector produced by the manufacturer contained a belt-driven centrifugal fan for which a 180 W three-phase supply was not available at the house (for houses in the UK, single-phase 240 V supply is standard). Consequently, a 650 W single-phase motor was fitted as the only readily available model; actual consumption was determined to be 500 W. Analysis of the performance of the house indicated that the *primary* energy consumption of the fan was actually greater than the solar space heating contribution.[56] In part because of this, a detailed study of possible replacement of the unit was made and a direct-drive centrifugal fan with a power consumption of 200 W (when suitably modified) was selected; primary energy consumption with the new unit has been calculated to be less than the useful solar contribution.

Fig. 6.33 shows the forced-air proprietary heating system of Fig. 6.31 installed in a home in Newnham, Cambridge (see also Chapter 12).

The design heat load of the house is 7.5 kW and a gas-fired air heater designed for a maximum output of 9.4 kW was selected. Domestic hot water is provided by a circulatory water heater (rated at 4.4 kW) situated inside the air heater, using the same flue and gas connections but operating independently and under separate controls; if necessary, supplementary hot water is from an electric immersion heater in the same cylinder. One-half an air change per hour, or about 190 m^3/h of fresh air, is drawn in by a 300 W centrifugal direct-drive fan in the heater unit, mixed with an amount of recirculating air which varies according to the heat load and distributed to the spaces through ducts insulated with 50 mm of glass fibre; the largest duct size is 275 mm × 175 mm. Downstairs, supply registers were sited, on the perimeter system, in the floor to deliver air reasonably uniformly throughout the space. Upstairs, air is delivered similarly. Return grilles were placed on the opposite wall at ceiling level and return ducts take the recirculated air back to the heater. Space was left in the 'false loft' for a small heat exchanger to be installed at a later date and the roof was designed specifically for future solar preheating of the fresh air. This was achieved by deliberately leaving a path for air flow from the eaves to the ridge between the insulation and tiles.

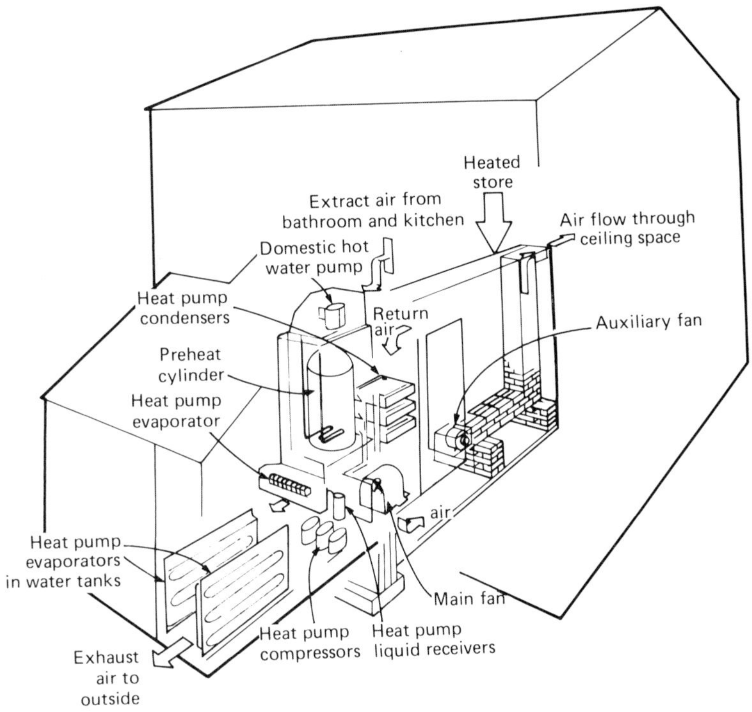

Forced-air systems 223

Fig. 6.34. Salford Strawberry Hill House with forced-air heating.[45]

Forced-air heating was adopted as an alternative choice to underfloor heating in a second Strawberry Hill House[45] (see also Section 6.6). Fig. 6.34 shows the basic features of the system which, instead of relying on the quick-response capability of air systems, opts for a relatively uniform environment and a heat store within the living area. The heat store is a 450 mm cavity in the wall dividing the living room and hall from the dining room and kitchen, with the cavity filled with bricks separated by air passages. The brick stack is heated at night by a closed circulation of air. A second closed circuit is used to circulate air through a ground-floor ceiling void when heat is required in the house, with control of the fan being by a thermostat in the living room.

One of the first low-energy houses to use a forced-air system was the Danish Zero Energy House (see Chapter 3). Because the house is entirely heated by solar energy with low supply temperatures inevitable at the end of the heating season, the radiator sizes required were large and it was judged that this was not compatible with the desired architectural appearance;[53] additionally, it was felt that the relatively large volume of water in such a system would make it slow in response and so waste energy. Floor heating was thought to be ideal because of the low inlet temperature and the general demand for a high degree of comfort, but the floor construction did not permit its use and so a fan-coil system was selected instead. Six identical units (three are located in the half of the house shown in Fig. 6.35) consisting of a return grill, heating coil, fan and outlet grille (see Fig. 6.36) are used for heating and ventilation in the house. Each

Fig. 6.35. Heating-system schematic for one-half of the Zero Energy House.[53]

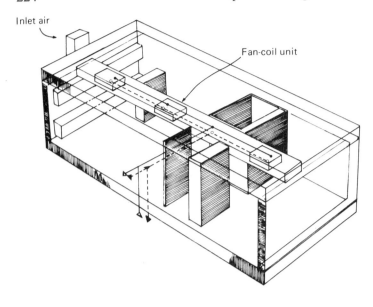

Fig. 6.36. Lengthwise section through a fan-coil unit.[53]

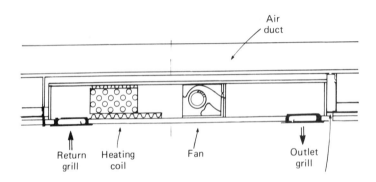

unit supplies approximately 700 W at a water temperature of 45 °C and a room temperature of 23 °C. Air volume per unit is 140 m³/h under design conditions and was deliberately kept low to reduce the noise level. The water supply for the units comes from a water store charged by solar collectors. Individual electric room thermostats, which allow automatic temperature reduction at night, control starting and stopping of the fans.

Ventilation air is introduced to each half of the house through inlet ducts in the utility racks by a separate supply fan and is distributed to the individual spaces. Return air from the rooms flows through the utility racks themselves to the bathrooms where it is expelled. A purpose-made plate heat exchanger (see Fig. 6.32) is used to recover heat from the exhaust air in each half of the house and transfer it to the incoming supply of fresh air. By paying particular attention to pressure losses in the system it was possible to use low-power supply and exhaust fans which together consume only 26 W.

Heating with electricity

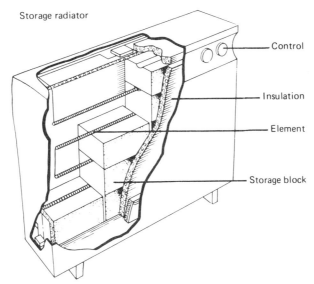

Fig. 6.37. Storage radiator.[57] (Courtesy of the Electricity Council; current models are slimmer and have different controls.)

6.5 Heating with electricity

The advantages of electricity include convenience, low maintenance, cleanliness, relatively inexpensive installation costs and an absence of transportation problems. Direct heating of individual spaces can be by electric fires, convectors, fan heaters and infra-red heaters, and indirect heating can be provided by storage radiators (see Fig. 6.37) which take advantage of the less-expensive rates for night-time electricity produced when demand on the national grid is lowest and when the most efficient stations are in operation.

Standard storage radiators have no moving parts and rely simply on storage blocks of a brick-like material with a high thermal capacity being charged and then discharging heat by radiation and convection from the casing. This lack of control means that storage heaters tend to consume more energy than is necessary, the most notorious case in a changeable European climate being when a warm day follows a cold day and windows are opened to dissipate excess heat. This problem is accentuated in a house designed to take advantage of passive solar gain since, ideally, instantaneous shut-down of the conventional heating system is desired when solar energy can meet the energy demand. The other principal objection to storage heaters in any house is their tendency to be large, heavy, immovable objects.

Storage fan heaters overcome the control problem somewhat by incorporating better insulation and a fan which is controlled by a built-in or remotely sited thermostat. This provides a quicker response but involves greater bulk and higher costs.

A central-heating system called 'Electricaire', based on the storage heater principle, has been developed by the UK Electricity Council (see Fig. 6.38). The heart of the system is a thermal storage unit with a well-

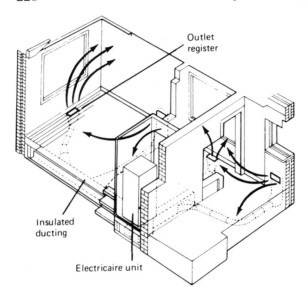

Fig. 6.38. 'Electricaire' warm-air storage heating system.[57] (Courtesy of the Electricity Council.)

insulated core. Air passed through the core by a fan in the unit is heated, proportionately mixed with cool air to maintain a desired temperature and discharged through ducts to each heated area or room. Rising costs of electricity have discouraged adoption of the system since its introduction.

An alternative approach which is now advocated by the Electricity Council is to combine multiple storage heaters for the base load, with any other heating needed being provided by direct-acting space heaters in each room with extensive use of thermostatic control.[17] Such a choice may raise the effective efficiency of the storage system from 70% (for storage radiators producing about one-third of their heat when not required) to close to 100%.[39] It will be appreciated that adoption of any particular system depends on an interplay of factors such as costs of night-time and day-time electricity and meteorological information to correctly estimate the base-load component and that all of these factors are only too variable. What is certain, however, is that electricity is only a sensible choice if the house is very well insulated.

Realizing this, the Electricity Council has instituted awards schemes for homes which make the most economical use of energy for water and space heating. For example, in the Medallion Awards Scheme for private builders the following regulations (among others) must be complied with:[58, 59]

(1) Homes should be all-electric (sole source of energy for heat, light and power).
(2) No fireplaces or flues are to be included in the design.
(3) A high standard of thermal insulation must be met (new dwellings, for example, must have a U-value of 1.4 W/(m^2 K) for the walls including windows, and roofs a U-value of 0.3 W/(m^2 K)).
(4) Design and control of domestic hot water and space heating must be such that at least 50% of the energy required is consumed at the cheaper night rate.

Heating with electricity

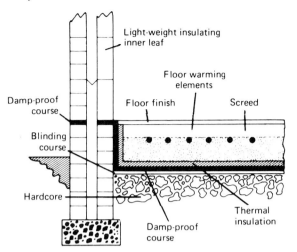

Fig. 6.39. Electric floor warming.[57] (Courtesy of the Electricity Council.)

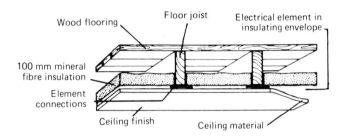

Fig. 6.40. Electric ceiling heating.[57] (Courtesy of the Electricity Council.)

It is a measure of the confusion that reigns in the energy-supply industry that two major campaigns with opposing goals are being waged simultaneously, namely to incorporate chimneys in houses and to exclude them. Since, in general, the approach advocated by the present authors is one of flexibility we find the first two conditions listed above indefensible.

Other ways of using electricity include floor warming and ceiling heating. In the case of floor warming, heating elements are embedded in a concrete floor slab during construction (see Fig. 6.39). The concrete slab, which is insulated, then acts as a storage radiator with control usually either by a room thermostat which varies the degree of charge given to the floor or by a combination of inside and outside sensors which controls the degree of charge more precisely.

Ceiling heating consists of a number of flexible low-temperature radiant elements which operate at about 40 °C. They are usually in sheet form (see Fig. 6.40) and fixed between the ceiling joists and the final ceiling finish which is most often plasterboard. To avoid excessive heat transfer upwards, a high level of insulation is placed above the elements. Control is usually by a wall-mounted air thermostat controlling the elements in the room or space where it is sited.

Houses employing electric heating seem to be associated more with energy conservation than the use of passive solar gain. For example, the

'Guildway Home', selected by the Electricity Council as illustrative of the approach they would like to see furthered, has a floor area of 143 m² and an insulated timber construction which gives U-values of about 0.40 W/(m² K) for the walls (excluding glazing).[60] Background temperatures are maintained by storage heaters and additional heat is provided by thermostatically controlled direct heaters. No specific reference is made to designing for passive solar gain.

Another Medallion Award development of detached two-storey, four-bedroom houses with a floor area of 116 m² uses electric ceiling heating which covers about 70% of the total ceiling area.[61] In ground-floor rooms the sealed elements are stapled to the joists below 80 mm of glass fibre and above the plasterboard; first-floor arrangements are similar but with 150 mm of glass fibre between the joists in the roof space. Each room is under individual thermostatic control. A 2.6 kW storage heater sited in the hallway provides a base heating contribution.

6.6 Heat pumps

Heat pumps extract thermal energy from a low-temperature source, raise the temperature of that energy and then exhaust it to a high-temperature sink, thus increasing its usefulness by the input of work. The most common arrangement for this is an electric motor driving a compressor and in the ordinary household refrigerator heat is removed from the food, raised in temperature in the compressor and exhausted to warm up the kitchen. In principle a heat pump is identical to a refrigerator — only the objectives differ in that the heat pump is used to heat the high-temperature sink and the refrigerator to cool the source. Typically, a refrigerator will deliver about twice as much heat into the kitchen as is put in electrically to drive the motor.[62] Heat pumps can improve upon the figure and therein lies their relevance to energy conservation.

The most common cycle used in heat pumps is the vapour compression cycle shown in Fig. 6.41. The system is sealed and filled with a refrigerant. In the cycle, refrigerant vapour is compressed and pumped to the condenser where heat is given out. The condensed liquid then passes through the throttle (or expansion valve) and cools before going to the evaporator where it takes up heat, evaporates, and goes on to the compressor, so continuing the cycle.

Electricity is most commonly used to drive the compressor via a motor and, although convenient, it has the usual disadvantage of the low ratio of primary to delivered energy supplied by the national grid (see Table 6.2). Alternative power units include internal combustion engines using oil or gas as fuel, diesel engines and external combustion engines.

The source of heat to the evaporator is most often the atmospheric air but it can also be the ground or water; Sumner[62] discusses the merits of these options in detail. The ground and water have the advantage of being less variable in temperature but their use as a heat source usually involves a cost penalty. In the US work has also been undertaken on combining solar energy and heat pumps with preliminary results indicating that, except in

Heat pumps

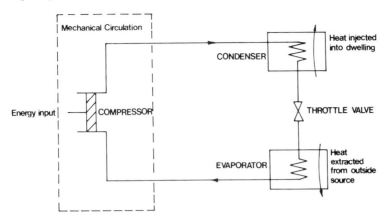

Fig. 6.41. Vapour compression cycle.[63]

special circumstances, the combination is not cost-effective although individually both measures are likely to be.[64]

The ratio of the heat output to the work input is known as the coefficient of performance (sometimes specified as the coefficient of performance for heating) and is given for an ideal (Carnot cycle) machine by

$$\text{COP} = \frac{T_{\text{sink}}}{T_{\text{sink}} - T_{\text{source}}},$$

where T_{sink} and T_{source} are absolute temperatures. (Unfortunately there is disagreement in terminology and this quantity is sometimes called the 'performance energy ratio'.[62]) For practical heat pumps the COP is usually in the region of half the ideal value. The BRE has found that heat pumps in buildings operate with a COP between 2 and 3 averaged over the year although it is quite possible that, in the future, heat-pump values of greater than 3 may be obtained.[65] It should be noted that a disadvantage of heat pumps, evident from the definition of the COP, is that performance falls as the temperature of the source falls — a consequence of this is that supplementary heating is often used at the coldest times. A second parameter of interest is the primary energy ratio (PER), which is defined as the ratio of the heat delivered to the energy content of the fuel consumed to provide the work input required.[66]

Fig. 6.42 shows the vapour absorption cycle which is less commonly used in heat pumps. The principal difference between it and the vapour compression cycle is the substitution of a heat absorption unit for the compressor. In the absorber part of the unit, the refrigerant vapour from the evaporator is dissolved and condensed in a secondary fluid called the absorbent (with the heat evolved being delivered to the building). The refrigerant and absorbent are then separated by the application of heat in the generator, with the evaporated refrigerant passing to the condenser, heat being delivered to the building and the still-liquid absorbent returning to the absorber. The rest of the cycle is the same as the vapour compression cycle. Since no mechanical work is done in the absorption cycle the coefficient of performance tends to be substantially lower but, on the

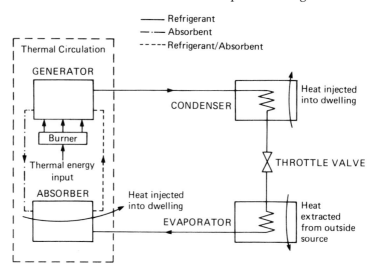

Fig. 6.42. Vapour absorption cycle.[63]

other hand, there is an opportunity for greater simplicity and reliability in the design.

Most heat pumps available in the UK are based on US designs for significantly different conditions. First, the heat requirement in UK houses is often less because of reduced floor areas and more-compact house forms, because internal temperatures tend to be lower (say 18–20 °C rather than 20–23 °C in the US) and because the British tend to wear more clothes and use curtains over windows. Secondly, the UK climate is characterized by moderate temperatures and relatively high humidity. In the US, both heating and cooling are often needed, and partly because it is relatively easy to add a heat-pump capability to an air-conditioning unit, heat pumps have become quite popular there – it is estimated that 1.5 million units have been installed.[63] Because the need in the UK is for a small heating-only unit, though, it is not appropriate to simply import the technology from across the Atlantic. For example, the COP of US heating–air-conditioning units as heat pumps can be improved significantly by doubling the compressor size and designing for heating only.[66]

Research in the UK is geared to best using primary energy, increasing reliability and efficiency and lowering costs. British Gas has studied gas-fuelled heat pumps and concluded that they may have significant advantages over electrically powered versions. Table 6.6 compares the efficiency of a variety of heat pumps and gas boilers.

The main points of note are that heat pumps make better use of primary energy than boilers and that, although the coefficient of performance of an electric heat pump is two-and-a-half times that of a natural-gas-fired heat pump, the latter uses primary energy more efficiently. Electric heat pumps use primary energy significantly better than other electric heating systems but, compared to traditional domestic boilers, their slight performance gain does not at present justify the increased cost.

Technical problems that remain to be overcome in developing a suitable

Table 6.6. *Efficiency of primary energy use of heat pumps and boilers (after Ref. [63])*

Type of system	Primary energy	Losses	Energy supplied to heat pump	Energy from outside source	Energy into building
Electric heat pump (COP = 3)	100	73[a]	27	54	81
Gas boiler (80% efficiency)	100	25[b]	–	–	75
Substitute natural-gas boiler (85% efficiency)	100	44[c]	–	–	56
Gas-fired absorption heat pump (COP = 1.2)	100	6[d]	94	43	113
Substitute natural-gas-fired absorption heat pump (COP = 1.2)	100	30	70	30	84

[a] Comprising generation and transmission losses.
[b] Comprising 6% production loss and 19% flue loss.
[c] Comprising 30% production loss and 14% flue loss.
[d] Production loss.

heat pump for the UK include starting current and defrosting. In the UK there is presently a limit on the use of single-phase motors in domestic premises to prevent starting surges unduly interfering with supply to neighbouring premises, and this limit will affect the size of heat pumps which can be used practically. One estimate is that if the electric motor output could be kept below 750 W there would be little need for electrical reinforcement of the local network.[67] The use of three-phase devices would overcome the problem and could increase efficiency but the extra cost, particularly in the case of modifying the supply to an existing house, would be a disadvantage. The UK Workshop on Heat Pumps concluded that 'further investigations are required to extend the acceptability of single-phase heat pumps'.[66]

Defrosting principally affects air-source heat pumps of both the vapour compression and vapour absorption types. Within a certain range of ambient temperatures, water from the atmosphere, which condenses on the outside coil, freezes on the evaporator surface and reduces the heat-transfer efficiency. In some electric heat pumps the ice is removed by operating in the reverse mode for a short time and it is envisaged that, with certain gas heat pumps, defrosting could be achieved by stopping the flow of central-heating water through the condenser, allowing hot refrigerant to melt the ice.[63] Because of the relatively high humidities in the UK, even at low temperatures, defrosting is a problem. Tests made by the Electricity Council have shown that, on average, 2.4% of the electricity demand of a heat pump used in the UK would be expended on defrosting.[63] Again, the Workshop on Heat Pumps, in suggesting several possible solutions, recommended further research.

Cost was one of the principal reasons for the past lack of interest in heat pumps. The capital cost of a heat pump is greater than that of a boiler of comparable power, thus, the additional cost must be repaid in a limited period by savings in running costs. The Workshop on Heat Pumps concluded that, for devices in the 3–8 kW output range, if the additional capital expenditure did not exceed about £200 (1976 prices) for a whole house-heating system, heat pumps could be competitive and it saw this as the basic question for heat-pump development in the domestic area. At least one manufacturer has recently claimed that heat pumps are now more economical than heating systems using boilers, but manufacturers tend to be an optimistic breed. Long-term performance trials to verify system reliability and seasonal COPs will be needed before firm claims can be made. Any additional back-up heating provided must also be costed carefully.

The Workshop also thought that an alternative approach of small, inexpensive heat pumps for single-room heating merited further research and the BRE has been investigating this possibility. It should be noted that in some countries such as West Germany and France the government subsidizes the initial cost of heat pumps.

The most straightforward domestic application of the heat pump is as a heating device, either in a central-heating system or as a room heater. In low-energy houses, and particularly those using ambient sources of energy, where the complexity of both the space heating and domestic hot-water heating systems is increased and where energy storage is often incorporated, heat pumps can play additional roles as 'energy-management' links between inputs and outputs or between storage and supply.

In general, experts are optimistic about the prospects for the widespread introduction of heat pumps for space heating during the period 1985–2000. It is thought that one of the major markets will be as a replacement for the first generation of central-heating systems, which were less efficient and which become oversized as insulation standards in houses are increased. In the longer term, given the development of suitable devices, it may be expected that every low-energy house will have a heat pump for, say, either space heating or some waste-heat-recovery function. The following examples give an idea of current developments in domestic heat pumps.

The BRE Solar Energy House shown in Fig. 6.27 uses three heat pumps as energy-management links in a system which owes part of its complexity to the need to extract as much experimental data from a given physical situation as possible. The first heat pump upgrades heat from the solar collector into the main heat store when radiation is low. The second is a small water–water heat pump used to upgrade heat from the main store for use by the space heating system and the third upgrades heat from the main store for use in the domestic hot-water system, whose required tem-

Heat pumps

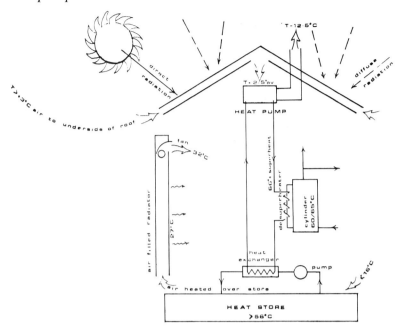

Fig. 6.43. Schematic of a heat-pump system for new and renovated houses.[68]

perature is often higher than that of the store. Other BRE experimental houses use heat pumps to supply all the space heating, to recover heat from the outgoing ventilation air and to recover heat from waste domestic hot water held in a catch tank.[65]

A novel application of heat pumps is found in the Salford Strawberry Hill Houses shown in Figs. 6.28 and 6.34.[45] For these homes it was decided, for reasons of cost and flexibility in the choice of an energy source, to use heat pumps run on night-time electricity only. Exhaust air from the house is used as the energy source. Because the heat pump runs only at night and because it is desirable to store that energy to avoid wasting it when there is no heat demand, cold-side heat storage was employed. Thus, instead of extracting heat from the exhaust air directly, heat is taken from a tank of water, forming ice during the night-time operation of the heat pump. The ice formed is then melted during the day by the warm extract air which passes over the tanks. Such a system which, it will be appreciated from Fig. 6.28, is by no means simple, has nonetheless the important practical advantage of requiring no special provision for defrosting. The expected coefficient of performance of the heat pump has been taken to be 2.5. Heat pumps are also used in the houses for preheating domestic hot water from 10 to 35–40 °C with direct night-time electricity used to supply the additional energy required.

As a final example, J. Keable's use of a standard air-to-air heat pump may be cited since he has demonstrated its applicability and reliability in both a renovated house and a new one.[68] The principles are the same in both, with air being extracted from an air passage formed underneath a roof covering of slate or tiles (see Fig. 6.43) before reaching the heat

Fig. 6.44. Coal-fired hot-water boilers at Basildon New Town.[69] (Courtesy of Ove Arup and Partners.)

pump. Heat from the heat pump is sent to the thermal store directly and, since the machine senses the temperature of the store and the ambient air rather than the indoor temperature, it is easy to avoid operation at unfavourable times of low temperatures. Consequently, the coefficient of performance is improved. Distribution of heat from the thermal store is by hot water to low-temperature radiators in the case of the renovated house and by air through ducts in the new one. The system may be considered to be a combination of a low-cost solar collector and heat pump.

6.7 Group schemes

Combinations of traditional and ambient energy sources similar to those discussed for individual houses in the preceding sections may be used for group schemes.

In the Basildon Housing cited previously, coal-fired group heating was selected. Three boiler houses located in a sunken site each serve 140 houses. Each boiler house has two hot-water boilers rated at 580 kW each and fired by underfed stokers which mechanically extract bituminous washed coal singles from a bunker and feed the fuel direct to the furnace (see Fig. 6.44).[69] In this way, handling is eliminated and a clean atmosphere results in the boiler house. To avoid the high costs of buried mains, the hot-water flow and return pipes are housed in overhead ducts running along covered walkways and hidden beneath the eaves of the pitched roof houses. In the homes, pressed-steel low-pressure radiators fitted with thermostatic control valves are used and the occupants control their (metered) heating supply through the valves and a time programmer.

A central gas-fired boiler is used in tandem with roof-mounted solar collectors in the Lewisham Flats mentioned previously (see Fig. 6.45). The solar collectors provide both preheating of the domestic hot water and underfloor space heating in the lower flat of each block.

The active solar energy system consists of ten 20 m^2 solar panels form-

Group schemes

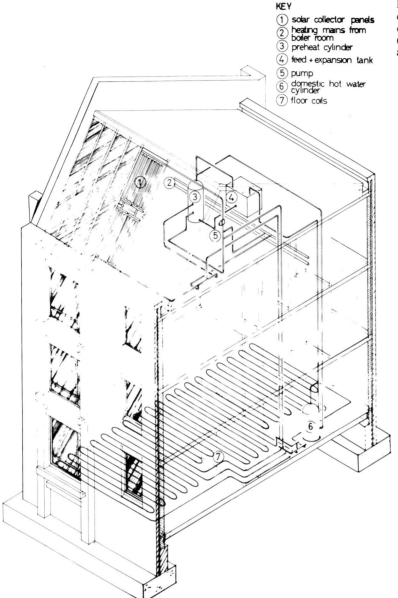

KEY
1. solar collector panels
2. heating mains from boiler room
3. preheat cylinder
4. feed + expansion tank
5. pump
6. domestic hot water cylinder
7. floor coils

Fig. 6.45. Schematic of solar collectors used with gas central heating.[70] (Courtesy of Max Fordham and Partners.)

ing the complete south-facing pitch of each block of the building.[71] Solar heat is designed to be collected at the lowest possible temperature. Thus, since the preheat cylinder is likely to be colder than the building, if the solar collector is warmer than the preheat cylinder there is useful heat to be gathered and the pump is switched on. This pump is set to switch on at a 2 K differential and off at a shade above a 0 K differential.

The preheat cylinder consists of a standard indirect hot-water cylinder of capacity 280 l fitted with a heat-transfer coil able to absorb the output of the solar panel. The preheat tank feeds three hot-water cylinders with preheated water. The pipework from the preheat cylinder to each domestic hot-water cylinder is effectively a common hot-water service cold feed and incorporates the normal design details appropriate to that application.

Each flat contains its own hot-water cylinder and electrical intake. The bathroom and kitchen are not back to back which means that separate plumbing stacks are required for each. However, this saves the need for mechanical ventilation, which reduces both capital and maintenance costs. The simplicity of this part of the installation contributes to the possibility of providing a solar energy scheme within the standard housing costs.

A temperature detector embedded in the floor compares the floor temperature with the preheat cylinder temperature. If the floor temperature is lower than the preheat cylinder temperature then heat can more usefully be employed to heat the floor. When the external thermostat detects an outside temperature below 15 °C, which shows that heat for the building can usefully be used, the water flow is diverted from the preheat cylinder to the floor. The back-up gas-fired heating system is controlled by individual thermostatic radiator valves which will close down when solar heat is provided. The rate of fossil-fuel heat consumption is thus reduced by any solar gain. The control of the radiators in each flat is arranged on a one-pipe loop and a thermostatic valve ensures that there is no flow when the building is up to temperature.

The floor panels consist of 15 mm welded-steel pipework to British Standard 1387 laid in the floor finish. The floor finish consists of 100 mm of screed laid on top of a mineral-wool sound-insulating quilt all laid above precast concrete planks. The pipes are laid at 225 mm centres. The heat output of the floor panels is 100 W/K temperature differential between the water and the room air temperature.

6.8 Conclusion

In selecting a heating system, the resources available to the society, the real needs of the occupants, initial costs, running costs and flexibility must be considered, but, as is evident from the preceding discussion, designers of low-energy houses have arrived at a variety of solutions — every readily available form of heating from solid fuel to electricity has been used — after considering some or all of these factors. In part, the explanation lies in the varying relative importance given to individual factors — if, for example, the predominant concern is efficient use of primary energy resources, solid fuel is preferable to electricity consumed either directly or in storage heaters. (However, if the electricity were to be used in heat pumps or if it were provided by wind or wave power rather than conventional generating plants, heating with electricity could be acceptable.)

Another reason lies in the effect of decreasing the energy demand, which causes both initial and running costs to be lowered. The same percentage cost-difference between, say, electricity and solid fuel, is perceived differently if in one case the annual heating costs are £900 and £600, and in the other they are £225 and £150.

In addition, it is useful to distinguish between houses which are designed primarily with a reduced energy demand in view and those which attempt to combine a low energy requirement with the best use of passive solar gain. In the case of the latter, the heating-system response (both hardware and controls) is extremely important and quick-response systems should be selected. The question is also related to the pattern of occupancy and in houses which are used more or less continuously the choice of heating devices which incorporate some thermal storage capacity, such as Aga cookers or electric storage heaters, may be justified.

All of these factors influence the decision to select central heating or individual space heating, or either an intermediate or combined solution. In a compact, continuously occupied well-insulated house it may be possible to heat the first floor adequately from the ground floor simply by incorporating an open plan and leaving the doors open (or using transfer grilles in the doors). Or a parallel system of, say, gas central heating and a solid-fuel 'focal-point' heater with its attractive amenity value may be used to heat only those parts of the house which are actually occupied. The choice rests with the occupants and the designer.

In any case, it is judicious to design for flexibility and to consider carefully such factors as allowing for low supply temperatures (and it may be noted that both hydronic and air systems are capable of this), provide space for thermal storage (in water, rock beds, eutectic salts, etc.), include a chimney and allow for later conversion to a solar roof. Because individuals have varying criteria for comfort the heating system should also be flexible enough to allow the occupants to easily adjust temperature settings and, particularly, to lower them and thus save energy.

In part due to the lack of experience with low-energy homes, the selection of a heating system remains an art — given the uncertainty surrounding future energy supplies and costs it is likely to remain so for some time.

References

[1] Field, A.A. (1977). 'Anticipations'. *Building Services Engineer*, 45 (1), A22–6.
[2] Burberry, P. & Aldersley-Williams, A. (1977). 'Domestic heating'. *Architects Journal*, 166 (33), 301–25; 166 (37), 453–68; 166 (39), 605–18.
[3] Burberry, P. (1975). *Environment and Services*. London: Batsford.
[4] Kell, J.R. & Martin, P.L. (1971). *Heating and Air Conditioning of Buildings*. London: Architectural Press.
[5] Kut, D. (1968). *Heating and Hot Water Services in Buildings*. Oxford: Pergamon.
[6] Barton, J.J. (1970). *Small-Bore Heating and Hot Water Supply for Small Dwellings*. London: Newnes-Butterworth.
[7] Carrier, W.H. (1959). *Modern Air Conditioning, Heating, and Ventilating*. New York: Pitman.
[8] Tulloch, J. (1978). 'Two-year solar energy experiment produces poor results'. *Building Design*, 22, 7.

[9] Anon. (1980). 'Gas: more changes than problems'. *Domestic Heating +*, March, p. 5.
[10] Romig, F. & Leach, G. (1977). 'Energy conservation in UK dwellings: domestic sector survey and insulation'. *Working Paper*. London: IIED.
[11] Sutherland, C. (1982). Private communication. (Conversion to MJ by R. Thomas: Copyright C. Sutherland.)
[12] Anon. (1956). *Domestic Heating – Estimation of Seasonal Heat Requirements and Fuel Consumption in Houses*. BRS Digest 94. Garston, Watford: BRS.
[13] Green, M.B. & Pickup, G.A. (1978). 'Useful figures for the efficiency of gas central heating'. *Building Services and Environmental Engineer*, 1 (4), 8–9.
[14] Anon. (1979). 'Gas comes top in heating'. *Building Services and Environmental Engineer*, 2 (1), 5.
[15] Anon. (1979). 'Gas: the heat is on'. *Domestic Heating +*, September, p. 6.
[16] Anon. (1979). 'Oil over a barrel'. *Domestic Heating +*, September, p. 16.
[17] Anon. (1980). 'A comeback for electricity?' *Building Services and Environmental Engineer*, 2 (7), 17–19.
[18] National Coal Board, London (1980). Private communication.
[19] Solid Fuel Advisory Service, London (1980). Private communication.
[20] Baumeister, T. (ed.) (1967). *Standard Handbook for Mechanical Engineers*. New York: McGraw Hill.
[21] Anon. (1977). *Jøtul – Magic Warmth from Wood*. Newcastle Emlyn, Dyfed: Simon Thorpe.
[22] Anon. (undated). *The Solid Advantages of Solid Fuel*. London: Solid Fuel Advisory Service.
[23] Vale, R. (1977). 'Low cost thermal upgrading of an existing house'. *Environment and Planning B*, 4, 173–84.
[24] Vale, B. & Vale, R. (1980). *The Self-Sufficient House*. London: Macmillan.
[25] Anon. (1979). 'Saving fuel in comfort'. *Domestic Heating +*, May, p. 41.
[26] Anon. (1980). 'The development of solid-fuel heating'. *Building Services and Environmental Engineer*, 2 (7), 18–20.
[27] Littler, J.G.F. & Thomas, R.B. (1979). *Ambient Energy Design: House Heating for the 1980s*. Cambridge: Ambient Energy Design.
[28] Withers, M. (1977). 'Heating'. *The Architect*, April, pp. 51–4.
[29] Anon. (1979). 'Householder's choice criticised'. *Building Services and Environmental Engineer*, 1 (6), 3.
[30] Anon. (undated). 'Combination gas boiler and cylinder for central heating and domestic hot water'. *Technical Data Sheet*. Telford: Harvey Habridge.
[31] Anon. (1975). *Flues for Domestic Gas Appliances*. Watson House, London: British Gas.
[32] Anon. (1977). *Microbore*. Bilston: The Wednesbury Tube Company.
[33] Anon. (undated). *High Quality Output Radiators*. Epsom: Hudevad Britain.

References

[34] Anon. (undated) Boulter Cam-ray Natural Convector Radiator. Norwich: Boulter Boilers.

[35] Pimbert, S.L. (1977). 'Reflecting materials behind radiators and shelves save energy'. *Building Services Engineer*, 45 (5), A22–3.

[36] Anon. (undated). *The Myson 'L' Series Circulating Pump*. Ongar, Essex: Myson.

[37] Rockhill, H.P. (1967). 'Closed heating systems'. *IHVE Journal*, 35 (99), 1–7.

[38] Fisk, D. (1980). 'Controlling the heat output'. *Building Services and Environmental Engineer*, 2 (7), 21.

[39] Bakke, P. (Chairman) (1975). *Energy Conservation: A Study of Energy Consumption in Buildings and Possible Means of Saving Energy in Housing*. BRE CP 56/75. Garston: BRE.

[40] Vint, M. (1979). 'Controls: the missing link'. *Domestic Heating +*, December, p. 20.

[41] Anon. (undated). *Taco-Constanta Thermostatic Radiator Valve*. Basingstoke: Tacotherm.

[42] Fisk, D.J. (1979). *Microelectronics in Building Services*. BRE CP 12/79. Garston, Watford: BRE.

[43] Ullathorne, D.P. (1980). 'Microprocessors and their possible applications to central heating systems'. *Watson House Bulletin*, 44 (311), 39–45.

[44] Seymour-Walker, K.J. (undated). *BRE Proposals for Experimental Low-Energy Houses*. Garston: BRE.

[45] Randell, J.E. & Boyle, J.M.A. (1979). 'Low energy housing – Salford's answer'. *Building Services and Environmental Engineer*, 1 (10), 18–21.

[46] Field, A. (1978). 'Controlled ventilation for dwellings'. *Building Services and Environmental Engineering*, 1 (2), 6–7.

[47] Zanelli, L. (ed.) (1973). *Central Heating, Plumbing, Electricity*. London: Marshall Cavendish.

[48] Dunstan, G.F. & Green, M.B. (1979). 'Warm air heating in low energy housing'. *2nd International Symposium on Energy Conservation in the Built Environment*, Copenhagen.

[49] Anon. (1976). *Huskiheat Gas Fired Brick Central Heating*. Bishops Stortford: Heatpak.

[50] Anon. (1977). *Controlled Ventilation for Dwellings Combined with Modairflow Gas Warm Air Heating Systems*. Northampton: Johnson and Starley.

[51] de Labbeye, P. & Lebrun, H. (1976). 'Tests effected on a heat exchanger to be utilized in a double flow'. *Electricité de France Research Centre Report HC 312 W 774*. Ecuelles: EDF.

[52] Applegate, G. (1970). 'Heating a factory for nothing'. *The Heating and Ventilating Engineer*, 44 (516), 21–4.

[53] Esbensen, T.V. (undated). *The Zero Energy House*. Technical University of Denmark.

[54] Anon. (1978). *Heat Recovery for Industry*. Technical Information, IND 18. The Electricity Council.

[55] Horton, A., Grove, S. & Lewis, A. (1979). *Milton Keynes Solar House Performance and Cost Analysis of Solar Heating System*. Built Environment Research Group, Polytechnic of Central London.

[56] Horton, A. (1978). *Milton Keynes Solar House Supplement 1: Alterations to the Fan Convector.* Polytechnic of Central London, BERG.
[57] Anon. (undated). *A Guide to Electric Space and Water Heating for Small Commercial Premises.* EC 3487. The Electricity Council.
[58] Anon. (1979). *The Medallion Award Specification.* The Electricity Council.
[59] Anon. (1977). *Medallion Award Homes.* EC 3697. The Electricity Council.
[60] Anon. (1980). 'Attention to energy saving'. *Building Services and Environmental Engineer*, 2 (8), 35.
[61] Anon. (1979). 'Striking a medal for low-cost heating'. *Domestic Heating +*, March, p. 28.
[62] Summer, J.A. (1976). *Domestic Heat Pumps.* Dorchester: Prism.
[63] Jessen, P.F. & Johnston, I.W. (1980). 'The domestic gas-fired heat pump'. *Watson House Bulletin*, 44 (311), 47–56.
[64] Mitchell, J.W., Freeman, T.L. & Beckman, W.A. (1978). 'Heat pumps'. *Solar Age*, 3 (7), 24–8.
[65] Freund, P., Leach, S.J. & Seymour-Walker, K. (1976). *Heat Pumps for Use in Buildings.* BRE CP 19/76. Garston: BRE.
[66] Anon. (1976). *UK Workshop on Heat Pumps.* Didcot: Energy Technology Support Unit.
[67] Anon. (1980). 'Domestic heat pumps'. *Building Services and Environmental Engineer*, 2 (7), 22–3.
[68] Keable, J. (1977). 'The application and economics of heat pumps'. In: *Ambient Energy and Building Design Conference Proceedings*, University of Nottingham, pp. 69–82.
[69] Anon. (1979). 'Coal heating for estate'. *Building Services and Environmental Engineer*, 2 (1), 43.
[70] Anon. (1979). 'Brownhill Road Flats, Lewisham'. In: *Buildings the Key to Energy Conservation*, Kasabov, G. (ed.). London: RIBA Energy Group.
[71] N. Ryding, Max Fordham and Partners (1982). Private communication.

7

Thermal storage

7.1 Introduction

Storage has been discussed in a number of the preceding chapters and, notably, in conjunction with passive solar heating. Here we shall cover the topic more systematically but in some cases in less detail. Storage is often the key to the successful and economical use of solar or wind energy for space and hot-water heating. While it is possible, as we have seen in Chapter 3, to drastically reduce the space heating demand of buildings, the residual demand must either be met from a conventional source of fuel or from energy stored during a period of greater availability. Most buildings using solar or wind energy have been forced to limit storage capacity because of cost and employ a fossil-fuel back-up. Research projects, such as that of the Cambridge Autarkic House Project, are responsible for most of the (rare) designs which rely only on ambient sources of energy. It is nevertheless encouraging to know that even in the UK it is possible to provide in an average year 100% of the space and domestic hot-water heating requirements with solar energy alone.[1]

The size of the store required for any application will depend on the energy demand, the source of energy and the selection of the storage medium. Because the cost per unit of storage falls as the size increases, more and more attention is being given to projects which involve large buildings or groups of dwellings. This in turn permits new equipment to be used more economically and novel means of storage to be examined. An example of the first point is that on a very small scale the additional capital cost of a heat pump in a system may not be economical but for larger schemes this cost becomes relatively small and heat-pump upgrading of stored heat becomes much more attractive. Among the novel large-scale stores being considered are uninsulated rock caverns (similar to those now used for storing oil, for example, in Sweden), confined groundwater aquifers[2, 3, 4] and clay soils.[5] Work is also underway in a number of countries on extracting, usually with a heat pump, the heat that is stored by natural processes in earth, rock and water.

Storage, of course, helps to compensate for daily, monthly and seasonal variations in the supply of energy. In a sense, the problem facing designers is not one of the availability of energy since, even in the UK, for example, the incident radiation is 80 times the present energy demand.[6] The questions are how to economically capture and store ambient sources of

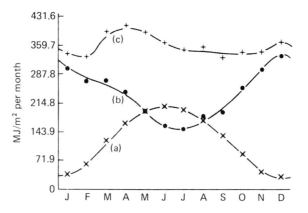

Fig. 7.1. Average monthly totals of output power from a solar collector and wind generator at Aberporth.[7] Curve (a): solar energy collector on south-facing roof at 30° elevation, efficiency = 35%. Curve (b): wind generator at 10 m height with V_{rated} = 20 knots, V_{cut-in} = 10 knots, efficiency = 40%. Curve c: combination of (a) and (b), with equal areas. (Reproduced with permission of the Controller of Her Majesty's Stationery Office.)

energy. Fig. 7.1 shows the pattern of availability of solar and wind energy at one site, Aberporth (52 08 N, 04 34 W, 133 m above sea level), in the UK.

Solar energy is least available during the winter heating season, when it is needed most. On the other hand, wind energy has a much more suitable pattern and, additionally, is quantitatively more important (each site must, of course, be considered individually). Consequently, it is likely that a smaller store could be used with the wind system.

The scope for storage applications is vast and by no means limited to on-site energy supply. Ryle, arguing against nuclear power stations, has said that wind energy should be supplied to the national grid.[8] Because much of the fluctuation in the demand placed on the grid is related to the provision of low-grade space heating for domestic and other buildings and because the central storage systems envisaged would involve multiple conversion of energy and thus low efficiency, he suggests that each home be provided with a small store. Energy would be stored as heat, perhaps by using electric resistance heaters, and storage would be for a period of about 150 h. Of course, a larger store could also be used to improve the seasonal load factor of conventional power stations. Photovoltaic systems also need an inexpensive effective means of storage before they will be used widely. In the US it is predicted that, within this decade, photovoltaics will be competing with oil for power stations in the sunnier areas of the country.[9] Storage will be required for daily and longer periods and the full scope of possibilities from centralized stores to individual ones in homes will need to be studied.

We would suggest that in order to make the best use of ambient energy, space for storage be provided in every building. This would be somewhat similar to the provision of adequate sanitary facilities (often neglected in the past) or the inclusion of chimneys in new houses, as suggested by the UK Solid Fuel Advisory Service. The space could change its use with time – it might start as part of a utility room or garage, become a solid-fuel store and then finish as a thermal store for energy from national wind-turbine power stations. The concept has already taken hold in Sweden

Sensible-heat storage in water

Table 7.1. *Thermal properties of storage media[a]*

Media	Density (kg/m^3)	Specific heat (kJ/(kg K))
Water (liquid)	1000	4.19
Rock (typical)	2500–3500	0.88
Iron	7860	0.50

[a] Ref. [10].

where it is not unusual to incorporate 4–5 m^3 of heavily insulated water storage in the domestic heating system.

The rest of this chapter deals principally with the various thermal storage media available for use in the 30–90 °C range and their applications to buildings.

7.2 Sensible-heat storage in water

For many applications, provision of, say, 20–80% of the space heating and domestic hot-water energy demand will be the object of an ambient energy collector and storage system.

Water has been the traditional choice for low-grade thermal storage because of its high specific heat, minimal cost, safety in operation, convenient use as a heat-transfer agent and storage medium and compatibility with existing heating and water systems. In most cases, thermal energy is stored in water as sensible heat, which simply means that the temperature of the material is raised and the energy is stored internally. The amount of energy that can be stored as sensible heat is given by the formula

$$\text{Energy stored} \ \left(\frac{kJ}{m^3}\right) = \rho \cdot c_p \cdot \Delta T,$$

where
  ρ is the density of the material (kg/m^3);
  c_p is the specific heat (at constant pressure) (kJ/(kg K));
  ΔT is the change in temperature (K).

As mentioned previously, thermal energy can also be stored as latent heat by making use of the heat absorbed by certain substances when undergoing a phase change. Tables 7.1–7.2 provide some data of relevance to the selection of a thermal storage medium.

Although water has significant advantages over other media for sensible-heat storage, it is not without its drawbacks. In particular, the cost of containment can be high (the cost of insulation may also be significant) and there are risks of corrosion due principally to the presence of dissimilar metals in many systems and the possibility of air entering the system. In general, sensible-heat stores tend to be large and heavy and this can pose architectural problems of space allocation and structural loading. When using water for heat storage, attention must also be paid to the possibility of damage due to freezing in cold climates, the risks of overheating in the

Table 7.2. *Energy storage in varying media for a 10 K temperature rise*[a]

Medium	Energy storage (MJ/m^3)
Sensible-heat storage	
Water	41.9
Iron (scrap, 20% voids)	31.4
Concrete (dense)	22.0
Rock (30% voids)	18.5
Brick (fairly dense)	16.7
Phase-change heat storage	
Glaubers salt (Na$_2$SO$_4 \cdot$10 H$_2$O) transition temperature + 10 K rise in bin	392
Ice (heat released)	307

[a] Refs. [6, 10, 11 and 12].

system if either the water reaches boiling point or mechanical failure interrupts flow and the danger, for example, to insulation, if leaks occur.

Determining the size of the store can be as complex a problem as one desires. In the design for the Cambridge Autarkic House a computer simulation of energy supply and demand using hourly weather data from Kew, London, suggested that 40 m^3 of water storage (surrounded by an additional 70 m^3 of polyurethane insulation) would be needed to supply an average annual combined space heating and domestic hot-water load of 7.2 GJ.[1, 13] Provision of such a store is not, however, economical and in most situations space and cost constraints will limit the storage size and simplify the problem of arriving at a suitable volume; conventional fuels then meet the additional energy demand.

For domestic hot-water heating systems in the UK the optimum value of storage has been found to lie in the range 35–120 l per m^2 of solar collector area[14] (see also Chapter 5). Most such systems have 4–6 m^2 of collector and so if we use the two upper figures we have a volume of water of 0.72 m^3. In the US, for domestic hot-water and space heating systems, a similar guideline is 40–80 l per m^2 of solar collector.[15] For space heating, both collector areas and storage volumes will, of course, be much larger with similar ratios of 50–120 l of storage per m^2 of collector being common (see Table 5.5).

Exactly what physical form a store should take has been a somewhat neglected question. More work has gone into collector design than, for example, into suitable stores. Numerous materials have been considered by several authors.[13, 16] Mild-steel tanks are a fairly standard choice because they are readily available at competitive prices and can be designed to withstand pressure should the system be pressurized. A tendency to corrode, particularly at high temperatures, is their principal drawback. At such temperatures galvanizing does not provide suitable protection. For example, in hard waters above 60 °C the zinc becomes cathodic to the steel, thus negating the desired effect.[17] Even with system precautions such as nitrogen rather than air covers to reduce

Sensible-heat storage in water 245

Table 7.3. *Approximate costs of linings and coatings for sealed tanks*[13]

Material	Cost per m² [a] (£)	Maximum temperature (°C)	Lifetime (y)
Aluminium	21	Satisfactory at 95	10–15
Butyl rubber[b]	7–40	70–120	1–20
Soft rubber[b]	27–35	80–100	10–20
Hard rubber[b]	27	105	10–20
Epoxy with glass[b]	7–8	90	10
Bituminous paint[c]	10	95	1
Organic emulsion	1.50	270	20

[a] 1979 approximate costs.
[b] Includes preliminary surface treatment.
[c] Includes preliminary shotblasting of internal surfaces and allows for two coats of paint.

atmospheric corrosion, protection is needed if the store is to have a reasonable life. A common guideline for solar systems is that readily replaceable components should last ten years and those which are not, 20 years. Linings, coatings and water additives (which may be combined with the first two items) are often used for protection but they are not without their problems. For linings and coatings, cost and temperature resistance are the principal difficulties.

Linings fall broadly into the categories of metals, glasses, rubbers and plastics. One possibility is an aluminium lining, itself protected by a vinyl coating, which must be inspected for peeling every three to four years. Its cost and those of several other options are shown in Table 7.3.

Vitreous-glass linings are unsuitable for prolonged periods at temperatures of 90–95 °C. Butyl rubbers can vary in their maximum temperature from 70 to 120 °C with costs mounting at the higher limit. Lifetimes of the cheaper linings are likely to be in the range of only one to five years. General-purpose soft rubbers are suitable up to 80 °C and special formulations will withstand 100 °C. Hard-rubber linings, which have a maximum temperature of about 105 °C, have the disadvantage that cracking may occur if the tank is made of too-thin metal sheet and is jarred or settles unevenly. Probable lifetimes of higher-quality products are ten to 20 years.

Coatings include epoxy coatings, paints and organic emulsions. Manufacturers are reluctant to guarantee epoxy compounds at up to 90 °C with cycling down to 40 °C. It is conceivable that an epoxy with glass system would be suitable but the probable lifetime is no more than ten years. Acceptable polyurethane paints are not presently marketed. There is a bituminous paint which has the required temperature resistance but it requires annual maintenance. A promising recent development is an organic emulsion which, when painted on, combines chemically with the steel surface.[18] Its lifetime is claimed to be the same as that of the tank and the temperature limit is a comfortable 270 °C.

Table 7.4. *Approximate costs of providing 40 m^3 of storage in sealed containers*[13]

Material	Approximate cost[a] (£)
Mild steel	2200
Stainless steel	8000–26 200
Copper	15 000
GRP	15 000
Polypropylene	6800

[a] 1979 approximate prices.

Anti-corrosion additives are common in central-heating systems and can be suitable for use in solar systems. (Wozniak[14] discusses additives in detail.) Manufacturers are reluctant to reveal the composition of their products but it is likely that many are based on mixtures of sodium nitrate and either sodium benzoate or sodium tetraborate. The required strength varies with the system but 1% is a common figure. At such low concentrations, heat-transfer characteristics are unlikely to be affected, nor will the viscosity vary greatly and so the energy consumption for pumping should not be increased. Component lifetimes in protected systems are estimated at longer than 20 years. The cost of protecting 40 m^3 of tank, assuming an additive concentration of 1%, is about £720 (1979 prices). To add a note of caution, however, the Greater London Council has studied corrosion inhibitors for central-heating systems and has found numerous problems ranging from cost to possible toxicity problems to excessive care required in controlling the requisite concentration.[19]

Numerous materials other than mild steel are available for tanks but these tend to be much more costly, as is apparent from Table 7.4.

Stainless steel and copper have the advantage of being highly resistant to corrosion but for large stores costs are excessively high. For smaller applications, it is quite common to use copper cylinders designed originally for domestic central-heating systems as stores. GRP tanks have excellent corrosion resistance but most are not designed to cope simultaneously with pressure and high temperatures. Instead of the normal polyester resins which are not suitable at 90 °C because of degradation of the molecular bond between the resin and glass, a variant such as epoxy must be used. Cylindrical tanks with dished ends are favoured to reduce the stresses induced by pressurization. Periodic inspection is required since, in certain circumstances, water may enter the material itself and eventually cause structural weakening. Cylindrical polypropylene tanks can be pressurized and can withstand 90 °C temperatures; lifetimes are likely to be of the order of 20 years.

Glass reinforced concrete (GRC) tanks are presently excluded from use as major water-retaining structures.[20] Precast concrete can pose problems of porosity and cost, as well as requiring careful consideration of the top and bottom seals for the precast sections. Exotic storage possibilities

Sensible-heat storage in water

abound but tend to be even more expensive. One example is the commercially available evacuated tanks of stainless steel inside and carbon steel outside, used for cryogenic applications. For the Autarkic House, the authors found a more promising possibility in the use of two standard carbon-steel tanks, one inside the other with a vacuum (0.1 Pa) maintained between them. Since no insulation is needed and because the store is compact, the size of any surrounding structure is also reduced. This solution may be economically competitive with more-conventional ones. The great uncertainty, however, is whether the vacuum can be maintained without continuous pumping.

Two other storage possibilities merit mention. The first are the combined solar collector and storage units which, in various forms, are in use in many countries. The Japanese, for example, have developed 'water pillows' made of vinyl (transparent on top and black for the bottom of the pillow), which are used without a covering during the summer and with one in winter.[21] Doubts have been expressed about the suitability of this kind of device in the UK because of their relatively poor performance and the structural problems of installing heavy units on roofs.[22] However, their simplicity is in their favour and in situations where internal space for storage cannot easily be found they may be both practical and aesthetically attractive.

Development along different lines has led to a heat-storage module made of polypropylene.[23] The module was designed to be completely safe in a domestic environment and so is unpressurized and serves only for storage. Heat exchangers link the store to the heating system. The basic module has a storage capacity of 1000 l and the insulation which surrounds the inner vessel limits the heat loss to about 100 W at the maximum operating temperature of 95 °C.

We can now examine how thermal storage in water may be incorporated in buildings. The design for the Autarkic House store[13] is an extreme case but one which illustrates many of the problems involved. The initial conception included a thermal store sited underneath the house for protection from the elements and so that advantage could be taken of the heat loss from the top of the store. In the early stages in the project it was not realized, however, how large the store would have to be. The final volume of 110 m^3 (40 m^3 of water + 70 m^3 insulation) could only be incorporated underneath the house to avoid the aesthetic problem of a massive aboveground object of little architectural interest. But, as we have seen (Chapter 2), in many parts of the UK the water table is quite high. (The site for the Autarkic House was in fact known in the past as Duck's Dole, which gives an indication of the water table.) On such sites, to resist hydrostatic pressures and to avoid moisture infiltration from the surrounding earth, it is necessary to develop quite an elaborate structure. The weight of the water in the tanks cannot be counted on because provision must be made for empty tanks to allow, for example, for inspection. The design solution chosen was a basement made of reinforced concrete. After excavation to a depth of about 4 m a preparatory layer of concrete coated with a damp-proof course of bituthene was to be poured. Then

Fig. 7.2. Insulation of the Danish Zero Energy House thermal store.

would follow a 100 mm concrete screed, a reinforced concrete base 350 mm thick, polyurethane insulation 700 mm thick, a steel tank protected with an organic emulsion to hold the water, 700 mm more of insulation on top of the tank and, finally, polythene again. The polythene was to function as a large impermeable bag to prevent any possibility of water movement from the ground outside to the inside. The insulation was to be foamed in place to provide a better thermal barrier than slabs.

On a similar scale, in the Danish Zero Energy House,[24] designed to be self-sufficient in space heating and hot-water supply, a 30 m^3 cylindrical steel tank surrounded with 60 cm of mineral wool (which was thought to be water repellent) was buried in the ground just outside the house. The water table is well below the tank bottom and so no precautions were taken against ground moisture but, to prevent rainwater from penetrating the insulation, an earth-covered roof was built over it; a mechanically ventilated air space separates the insulation on top of the tank from the earth cover. Fig. 7.2 shows the tank being insulated with the mineral wool.

Not all designers will have the luxury of being able to consider the incorporation of thermal storage at an early stage. Similarly, for the existing housing stock, provision of storage will often be a challenge. In the case of the Milton Keynes Solar House shown schematically in Fig. 5.16, it was decided to install a solar heating system without altering the basic form or plans of one of the standard house types.[25] A combination of 37 m^2 of solar collector (installed on the roof) and 4.5 m^3 of water storage, squeezed into the bedroom cupboards, was provided. The storage consisted of two steel tanks of rectangular cross-section (so that the available space could be best used) arranged vertically. Inhibitors (to BSS 3150) were used to reduce the corrosion hazard and ethylene glycol was added to alleviate the risk of frost damage.

In the more recent (1980) Solar Court Houses at Milton Keynes a combination of 40 m^2 of solar collector and 2.0 m^3 of water storage in standard copper DHW cylinders was used.[26, 27] The cylinders, which are insulated with 50 mm of proprietary board or quilt material, warm air for

Sensible-heat storage in rock

space heating. The indirect store is unpressurized but the collector circuit is pressurized to a nominal 152 kN/m². The collector circuit water is, by volume, 25% proprietary antifreeze with a corrosion inhibitor. An auxiliary gas-fired boiler is used as back-up.

For existing houses, finding space for a large store may be difficult, but a simple small one (of the order of 500 l) for a domestic hot-water system can often be located in a loft space. Wozniak,[14] in discussing the use of commercial hot-water cylinders of copper in solar heating systems, has pointed out that a common fault is inadequate insulation of the storage vessel, which can result in excessive heat losses. As a guideline he suggests that, for a 120 l cylinder, 50–100 mm of glass fibre insulation or its equivalent is economical.

Although metal is by far the most common material for storage containers in the range of 100 l to, say, 50 000 l, other solutions have been tried. To cite but one, in Australia a 20 m³ thermal store of 50 mm thick precast reinforced concrete was incorporated in a solar installation.[28] The system preheats water to warm soft-drink cans. Concrete was selected in preference to mild steel because of lower cost. Space was no problem at the factory and so the store could be located above ground. The unpressurized cylindrical tank was insulated with sprayed polyurethane 125 mm thick and protected from the weather with an external sealer which was sprayed on. Internally, the tank was coated with a synthetic rubberized sealing compound.

Larger applications often rely on stores similar to those above but increased in scale. In Madrid, a new large office building has a 500 m³ water-storage tank in the basement which serves as a heat accumulator.[29] It is designed to bridge a period of 15 days without sunshine – Madrid's longest sunless period on record is 11 days. In Canada, a 120 000 m² office building has a 6000 m³ thermal storage tank which provides water for both heating and cooling.[30] Increasingly in that country, thermal storage is being seen as a way of levelling the electricity demand and avoiding high day-time tariffs. At Studsvik, Sweden, imaginative work is underway on the use of earth pit storage for annual heat storage. Fig. 7.3 shows a demonstration plant with approximately 640 m³ of water storage. If the concept proves practical and economical, it is intended to design larger stations which would serve up to 1000 houses.

7.3 Sensible-heat storage in rock

Sensible-heat storage in rocks has been discussed in some detail previously and so here we shall simply summarize the advantages and disadvantages of this solution and discuss some applications.

Because the specific heat of rock is much lower than that of water and because about one-third of the total volume of such stores consists of voids, their energy storage capacity is lower than that of an equivalent amount of water (Table 7.2). Consequently, a larger store will be needed and where space is limited this will be a disadvantage. Also, in contrast to water systems, heat cannot simultaneously be added to and removed from

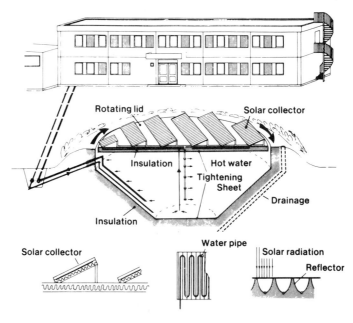

Fig. 7.3. Large-scale storage of solar energy at Studsvik, Sweden.[31]

the storage unit. The following advantages have, however, been noted.[12]

(1) There are no problems of liquid leakage, pipe corrosion or freezing (although there may be problems of air leakage and biological contamination).

(2) Thermal diffusion through the store can be much less than that associated with liquid stores. This is because normal practice in rock beds is to choose stones of an appropriate shape and of uniform size to ensure that minimal (point) contact is achieved and thus minimize conductive heat transfer in the store. A marked division between the hot and cold parts of the store results, with the thermal division moving one way or the other, layer by layer, as heat is added or withdrawn.

(3) A heavy solid store can easily be located in the normal excavated area under a building. Additionally, in areas with a high water table, it will not usually be necessary to design (except at the time of construction) for an empty store as must be done with a water system.

The container for the rocks is commonly concrete but sound wood-frame boxes are also satisfactory and bricks would also seem to be suitable.

In the US one of the best-known rock stores is the Denver Solar House described by Lof et al.[32] There, 10 640 kg of 3 cm diameter rock was used in a store composed of two cylindrical tubes 0.91 m in diameter and 5.5 m high. In Europe few houses have incorporated rock stores. One of the rare examples is a house in Lioux, France, which has a 30 m^3 rock store and a balanced air flow hypocaust (a hollow space under a floor used for heat accumulation and transfer) circulation system (see Fig. 7.4)). A UK example is the rock store in the Peterborough Houses (see Chapter 12) which are being monitored by the EEC and so will provide much-needed data on the performance of such stores in European conditions.

Sensible-heat storage in rock

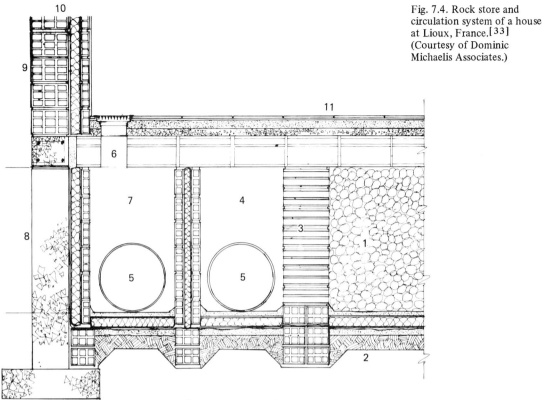

Fig. 7.4. Rock store and circulation system of a house at Lioux, France.[33] (Courtesy of Dominic Michaelis Associates.)

Typical transverse section through ground floor and foundation system, second version showing hypocaust air circulation system.
1 Thermal store: rock. **2** Rough screed/insulation/damp-proof membrane/site concrete/hardcore. **3** Extruded clay bricks laid sideways to allow passage of air. **4** Warm air delivery and collection chamber to and from store. **5** Equalisation duct. **6** Adjustable floor grill. **7** Space heating delivery chamber. **8** Reinforced concrete foundation. **9** External wall of cellular brickwork. **10** Insulation. **11** Insulated floor.

At least two other projects are also underway in the UK to provide additional data. Again, as part of an EEC testing programme, the UK is studying a half-size solar heating system consisting of 19 m² of collector and a 4 m³ rock-bed store.[34] And at the Open University there is a study underway of the potential for interseasonal storage of heat, which involves the development of a prototype combined solar collector and pebble-bed store. Fig. 7.5 shows a section through the intended store design.

One application for the store, which has been studied, is to provide the total annual heating and hot-water requirements for 100 houses.[36] The store would be 280 m long, 10 m wide and 4 m deep and would have a maximum temperature of 110 °C; pebble size is about 4 cm. The insulation is likely to be 0.6 m of glass fibre.

Fig. 7.5. Interseasonal solar energy storage in a pebble-bed store.[35] (With the permission of Dr B. Jones and Mr T. Oreszczyn.)

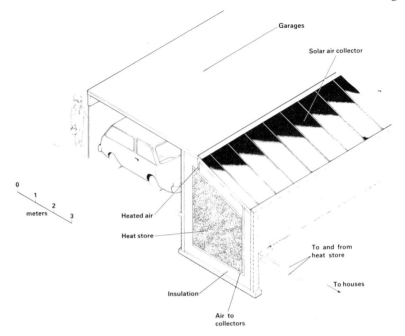

7.4 Phase-change energy storage

As was mentioned in Chapter 4 some materials undergo a phase change, for example from solid to liquid, accompanied by the absorption of a large amount of heat (the heat of fusion) at a temperature which makes them potential storage media. For a substance to be suitable, the phase change must be reversible, that is, the heat must be capable of being released over a large number of cycles without material degradation. To render this abstract description more realistic, many readers will be familiar with phase-change stores in the form of small refrigeration packs for picnics. The pack is taken solid and cold from a refrigerator and put in an insulated container with, if one is fortunate, avocado pears, lobster and champagne. Heat from the food and drink is absorbed by the packs which turn soft. After the picnic, the pack is placed in the refrigerator to which it releases its heat, turning solid in the process. If such a pack could only be used once it would be a rather expensive refrigerant (and heat store) but its attractiveness to purchasers is that it can be used again and again. Most important of all is that such materials can remove (and store) more energy than is possible through sensible-heat storage (see Table 7.2). Another significant advantage is that the process occurs at a specific temperature (the melting point).

Hydrated salts are the phase-change materials that have been investigated in greatest depth with the intention of using them for space heating applications. Telkes[37, 38, 39] in over 30 years of work has described the merits and problems of a number of salts, including $Na_2SO_4 \cdot 10 H_2O$ (Glauber's salt), $Na_2CO_3 \cdot 10 H_2O$ and $Na_2HPO_4 \cdot 12 H_2O$. The character-

Phase-change energy storage

istic reaction for such materials is between the hydrated salt and the anhydrous salt as shown in

$$Na_2SO_4 \cdot 10\, H_2O + \text{Energy} \rightleftarrows Na_2SO_4 + 10\, H_2O.$$

Energy storage occurs by the reaction proceeding from left to right when energy is added. The total energy stored depends on the complete process since sensible heat is involved in heating the hydrated salt to the transition temperature (32 °C in this case), heat of fusion to cause the phase change (this is often referred to as latent-heat storage or, occasionally, as energy storage using latent heat of phase change) and sensible heat, again to raise the anhydrous salt and solution to the final temperature. Energy extraction from storage is the reverse procedure.

Since such salts are mainly water they tend to be inexpensive; additionally, they generally pose no fire hazard. Those proposed for heat storage tend to be non-toxic. Many occupy a smaller volume in the solid state than the solution from which they form, thus there is not necessarily a danger of fracturing a rigid container upon solidification[40] (as there is with ice, for example). They are not, however, without their disadvantages (which is also true for other storage media) and, unfortunately, these disadvantages have severely limited practical applications. The main obstacles have tended to be physical. Glauber's salt, for example, does not actually melt at 32 °C but has an incongruent transformation which is more precisely rendered by

$$Na_2SO_4 \cdot 10\, H_2O\ (\text{solid})\ (\text{at } 32°) + \text{Energy} \rightleftarrows x\ (Na_2SO_4)\ (\text{solid})$$
$$+ [(1-x)\, Na_2SO_4 + 10\, H_2O]\ (\text{solution}),$$

that is, both solid and liquid phases exist at temperatures above 32 °C. Since the solid anhydrous salt is denser than the saturated solution, the two components separate out with the solid settling at the bottom of the container. On cooling it reacts only partially with the solution to reform $Na_2SO_4 \cdot 10\, H_2O$ and, after a small number of cycles, phase separation reduces the heat of fusion to a fraction of its initial value. The second problem, which is also common to other salt hydrates, is that supercooling can occur when heat is extracted from the store. If this happens the latent heat of fusion may not be recovered or it may be recovered at a temperature well below the melting point.[10] These problems have been the subject of a great deal of research. To avoid phase separation, thickening agents can be added, permitting extensive cycling, but the heat of fusion of such mixtures is considerably lower than for pure salt hydrates.[41] Supercooling may be prevented by nucleating additives which have the same crystal structure as the salt hydrate and low solubility in the salt hydrate melt. Thus, the supercooling of melts of $Na_2SO_4 \cdot 10\, H_2O$ and $CaCl_2 \cdot 6\, H_2O$ is reduced to less than 1 °C by the addition of 5% borax ($Na_2B_4O_7 \cdot 10\, H_2O$) and 1% strontium chloride hexahydrate ($SrCl_2 \cdot 6\, H_2O$) respectively;[41] patents have in fact been granted to Telkes for such techniques.[39, 42] Research continued during the past decades and interest grew but technical problems and lack of economic viability meant that few buildings incorporated such storage. Recently, however, complete commercial systems have been marketed in both the US and the UK using Glauber's salt as the salt hydrate. The UK system, developed by Calor, depends on the formation, during preparation, of a finely distributed network of an inert material (authors' note: not identified) which is able to retain in suspension all nucleating crystals and solids associated with incongruent transformation in the salt hydrate.[43] Since the solids are dispersed, they remain available for recombination with the solution and, so, the transformations remain completely reversible. An application of this approach is described below.

The same group has also worked on $Na_2HPO_4 \cdot 12 H_2O$ which has a transformation temperature of 35.5 °C. In Sweden, studies of $CaCl_2 \cdot 6 H_2O$ with a melting point of 29.5 °C are underway[44] with the intention of application to the solar heating of buildings. The Swedes see particular promise in phase-change heat storage combined with heat pumps[45] since the release of heat at a constant temperature permits uniform evaporator temperatures to be maintained (see Chapter 6) — this is, of course, not possible when an external air or ground source is used as a source of heat.

Paraffin waxes are a second type of phase-change material presently being studied, in spite of previous objections to these being a fire hazard in buildings.[39] These materials have relatively high specific heats for organic materials but low thermal conductivities. The consequence of the latter characteristic is that rates of heat transfer are low, thus putting heat in (important if the goal is to charge a store under rapidly changing conditions of solar energy availability) and taking it out (important if, for example, supplying a fast-response heating system is the requirement) can be a problem. Studies at Cardiff[46] concluded that 'it is unlikely that a paraffin wax will supplant water as the storage medium for domestic hot-water applications', but that the possibility of their use in space heating systems was not excluded. Research is continuing on different configurations of heat exchangers and a series of waxes with different melting points (42 °C rather than the 52 °C of the waxes first studied).

Back in 1949, the Dover (Massachusetts) House, claimed to be the first house heated entirely by solar energy, was built and occupied.[47] The solar heating system incorporated a double-glazed air-heating collector and about 13 m^3 of storage bins containing Glauber's salt, and was designed, not surprisingly, by Dr Telkes. Problems in operation of the kind discussed previously led to a loss of efficiency and eventual removal. Nevertheless, the experiment was judged to be of value and theoretical work continued. In 1973, the University of Delaware, where Telkes was working, opened their experimental Solar One House which again uses air collectors and storage in salt hydrates, but this time in a more sophisticated configuration which can be used for heating and cooling.[48, 49] Two outer vertical stacks of pans contain sodium thiosulphate pentahydrate ($Na_2S_2O_3 \cdot 5 H_2O$) and constitute the 'hot' store. A central stack, containing a eutectic, which is mainly sodium sulphate decahydrate, is the 'cold' store.

Developments in storage 255

The total volume is about 6 m³ and storage capacity is about three days' winter heating load or one day's summer cooling.

Closer to home, the Calor solar domestic hot-water and space heating system has been installed in houses in Edinburgh, Dublin and Milton Keynes. The basic system consists of 20–40 m² of solar collectors (available roof area has often proved to be an important constraint) used to heat water whose energy can be stored in one or two 0.4 m³ units, each capable of storing 216 MJ[50]; storage is therefore not for prolonged periods. The storage unit is an unpressurized plastic tank containing, principally, sodium sulphate decahydrate.

Several home-heating systems have also been designed using what may seem to be an illogical and surprising material – ice! The latent heat of fusion of ice is high (333 kJ/kg) (see also Table 7.2) and its cost minimal, although the cost of containment must be considered. Fracture of the container on freezing can be avoided by adding a small quantity of antifreeze (ethylene glycol) or an appropriate salt and by allowing a free space above the liquid.[40]

In the US, where throughout much of the country there is a need for heating in the winter and cooling in the summer, an Annual Cycle Energy System (ACES) operating in the following manner has been proposed for residential and commercial building:[51]

(1) Heat obtained by freezing water to remove the heat of fusion is pumped by a one-direction heat pump to the building.
(2) Ice produced during the heating season is stored and used to provide air conditioning during the summertime.

Fig. 7.6 shows a schematic of the system.

In the UK the use of ice as a storage medium is found in the Salford Houses described previously (Chapter 6). Because of the more temperate climate, for a domestic situation provision need only be made for heating.

7.5 Developments in storage

Ambient energy sources, with their supply variations ranging from the daily to the seasonal, increase an existing need for energy storage due to the variation in energy demand associated with patterns of human activities and weather. There is enormous potential for storage systems, for application at the points of both energy generation or collection and use, which are less expensive, provide more-efficient access to the stored energy and which have a larger thermal storage capacity. Suggestions for storage (which do not necessarily fulfil all of the criteria given) vary from the mundane to the exotic. It is encouraging that most may be viewed as technically feasible and some are so for the very near future.

Starting with sensible heat, at the Ark Project in Canada small-scale storage in a clay mud is being investigated for a greenhouse heating system.[53] On a larger scale, research into solar ponds, which can function as both collectors and stores, continues;[54] initial work in Israel by Tabor[55] has provoked interest in less sunny climates and a proposal has

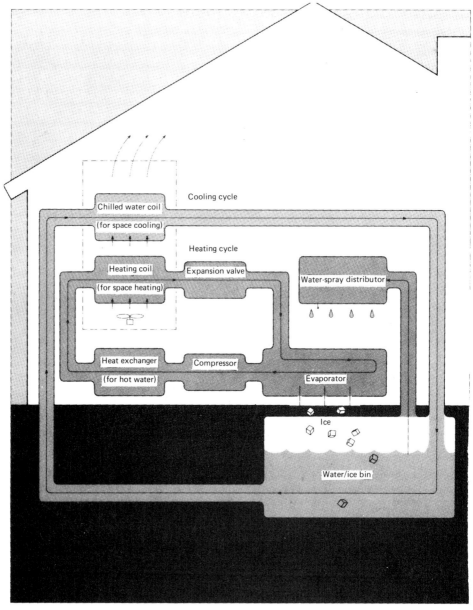

Fig. 7.6. Schematic of the ACES system.[52] (For space heating of hot water: the compressor pumps hot vapour refrigerant through the heat exchanger. The refrigerant continues through the space heating coil where it condenses. The now-liquid refrigerant then moves through the expansion valve and evaporator. As the refrigerant reverts to vapour it extracts the heat from water sprayed on the outside of the evaporator, making ice. The ice is flushed to the water–ice storage bin. For space cooling: ice-water from the bin is circulated through the chilled water coil.)

Developments in storage 257

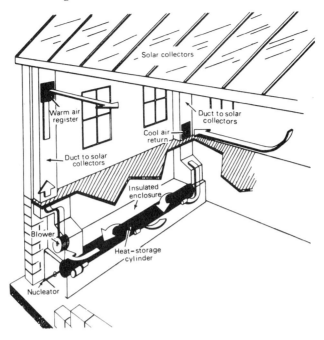

Fig. 7.7. A rolling-cylinder heat store in a solar system.[57]

been made for such a pond in no less a place than London,[56] in spite of the city's image of a capital with an umbrella-carrying populace. In a solar pond, incident radiation penetrates the liquid in the pond and falls on the blackened base which is thus heated. Normally, convection currents would cause the heated liquid at the bottom to rise and transfer the heat to the surface, but in a solar pond this is prevented by dissolving suitable salts (for example, sodium chloride, NaCl) in the lower layers of the water to increase their density while not significantly affecting the transparency. The temperature at the bottom of the pool then increases and the energy may be used, for example, for space heating of a large group of buildings.

Novel developments in phase-change storage include the rolling-cylinder concept shown in Fig. 7.7. The cylinder contains Glauber's salt and by continually turning at three revolutions per minute provides just enough stirring action to prevent component separation.

It is also possible to store latent energy in adsorbent beds of materials such as silica gel and activated alumina.[12, 58] The principle of operation is that suitable materials, in addition to their capacity for sensible-heat storage, remove water from an air stream flowing through them, with a consequent latent-heat gain.

Mechanical storage devices are generally thought to hold little promise for use in solar energy systems because of the losses involved in converting heat to work, but a number of chemical (the term is used loosely since a number of storage techniques, including phase-change stores, involve chemical effects) systems for thermal storage are being actively investigated (chemical stores for electricity are discussed in Chapter 8). One such development is based on the heat of mixing. For example, the well-known reaction when sulphuric acid and water are mixed releases heat but suffers from a relatively low energy density (about 100–200 MJ/m^3 depending on temperature). However, by using it as a heat pump, as in a system being developed jointly by the Rutherford Laboratory and the Open University, the store can be improved greatly.[59] Suggestions for the longer term include metal hydrides, some of which can store more than ten times as much energy as phase-change materials.[60] Work is also underway on photochemical reactions using both organic and inorganic materials. Again, the attraction is that some of the materials involved, for example norbornadiene, have higher storage capacities than phase-change materials.[6, 61]

7.6 Conclusion

As with heating systems, a number of storage techniques are available and are being used in energy-efficient buildings. Sensible-heat storage in water is by far the most common solution but phase-change materials are gaining acceptance as their chemistry becomes understood and as physical problems are overcome. It may be that sensible-heat storage, on grounds of simplicity and cost, will become the preferred store for large-scale applications such as community heating schemes, and phase-change storage will prevail where space is limited as in individual houses.

The factors that will influence such choices are the same as those governing the viability of new means of energy storage — progress in developing and selecting appropriate materials or combinations of materials, material costs, containment costs, safety and the confidence that exists among designers and consumers that the store will perform adequately for a reasonable lifetime. There is a desperate need for more information on storage media and performance over prolonged periods. The US *Workshop on Solar Energy Storage Subsystems*[60] called for research on everything from store geometries in relation to intermittent solar energy availability to basic work on heat-transfer properties of materials.

Long-term storage for single houses, it is generally agreed, is not presently economic by any standard but short-term storage for specific applications such as domestic hot-water heating in the summer is nearly so. In the future, what percentage of the domestic hot-water and space heating load should be met by an ambient source of energy combined with a store, and what the material should be will depend on the factors cited above, with cost being the most notable. We can say with confidence, however, that provision of storage or, as a minimum, space for storage, should be considered for every building being designed now — within the lifetime of most of them it will prove its worth. At the very least the space will be useful to store coal for the future.

References

[1] Littler, J.G.F. & Thomas, R.B. (1977). 'Solar energy use in the Autarkic House'. *Transactions of the Martin Centre for Architectural and Urban Studies*, 2, 93–110.

References

[2] Alexandroff, J.-M. & Alexandroff, G. (1977). 'Stockage Longue-Duree de l'Energie Solaire'. *UNESCO International Conference and Exhibition on Solar Building Technology.* London.

[3] Andersson, O. & Gustafson, G. (1981). 'Thermal energy storage in deep confined groundwater aquifers'. *Swedish Council for Building Research Summary S1: 1981.* Stockholm.

[4] Givoni, B. (1977). 'Underground longterm storage of solar energy – an overview'. *Solar Energy*, 19, 617–23.

[5] Anon. (1981). 'Storage of heat in clay in central Sweden'. *Swedish Building Research News*, 1981: 1. Stockholm.

[6] Long, G. (1975). 'Solar energy: its potential contribution within the United Kingdom'. *Department of Energy Paper No. 16.* London: HMSO.

[7] Caton, P.G.F. & Smith, C.V. (1977). 'Wind and solar radiation – availability in the United Kingdom'. In: *Energy Conservation and Use of Renewable Energies in the Bio-Industries.* Vogt, F. (ed.). Oxford: Pergamon.

[8] Ryle, M. (1977). 'Economics of alternative energy sources'. *Nature*, 267, 111–17.

[9] Tucker, A. (1981). 'Futures'. *The Guardian*, 21 May, p. 10.

[10] Duffie, J.A. & Beckman, W.A. (1974). *Solar Energy Thermal Processes.* New York: Wiley-Interscience.

[11] (1977). *ASHRAE Handbook – Fundamentals.* New York: American Society of Heating, Refrigerating and Air Conditioning Engineers.

[12] International Solar Energy Society – UK Section. (1976). *Solar Energy – a UK Assessment.* London: ISES–UK.

[13] Thomas, R.B. & Littler, J.G.F. (1979). 'Thermal storage in the Autarkic House'. *Transactions of the Martin Centre for Architectural and Urban Studies*, 4, 139–56.

[14] Wozniak, S.J. (1979). *Solar Heating Systems for the UK: Design, Installation and Economic Aspects.* London: HMSO.

[15] Anon. (1977). *Heating and Air Conditioning Systems Installation Standards for One and Two Family Dwellings and Multifamily Housing Including Solar.* Vienna, Virginia: Sheet Metal and Air Conditioning Contractors National Association.

[16] Pickering, E. (1975). 'Residential hot water solar energy storage'. *Proceedings of the Workshop on Solar Energy Storage Subsystems for the Heating and Cooling of Buildings.* American Society of Heating, Refrigerating and Air Conditioning Engineers.

[17] Anon. (1971). *IHVE Guide Book B 1970.* London: Institution of Heating and Ventilating Engineers.

[18] Manufacturers data. (1977). *Neutra-rust.* Middlesex: Double H International Marketing Ltd.

[19] Anon. (1978). 'Control of corrosion in central heating systems – experience gained from a few installations'. *Greater London Council D and M Bulletin 118.*

[20] Anon. (1978). *GRC.* BRE Digest 216. HMSO.

[21] Tanashita, I. (1970). 'Present situation of commercial solar water heaters in Japan'. *International Solar Energy Conference*, Melbourne.

[22] Wozniak, S. (1979). 'Solar heating systems for the UK'. *Building Services and Environmental Engineer*, 2 (2), 16–18.

[23] Field, A. (1981). 'Bridging the load gap from solar and heat pumps'. *Building Services and Environmental Engineer*, 3 (5), 16–17.

[24] Esbensen, T.V. & Korsgaard, V. (1976). 'Dimensioning of the solar heating system in the Zero Energy House in Denmark'. *UK–ISES Conference C8*, pp. 39–52.

[25] Anon. (1975). *Solar Heated House in Milton Keynes*. Milton Keynes: Milton Keynes Development Corporation.

[26] Anon. (1980). 'Solar heated houses, Milton Keynes'. *What's New in Building*, July, p. 5.

[27] Mead, I.K. (1981) Calor Alternative Energy. Private communication.

[28] Morse, R.N. (1978). 'Solar industrial process heating for can warming at Queanbeyan, New South Wales, Australia'. *CSIRO Solar Energy Studies Report No. 10*. East Melbourne: CSIRO.

[29] Richards, R.A.C. (1980). 'Extensive solar energy use in Spanish building'. *Building Services and Environmental Engineer*, 2 (9), 14–15.

[30] Anon. (1981). 'Energy consciousness shows at CIBS/ASHRAE Joint Conference'. *Building Services and Environmental Engineer*, 3 (10), 9.

[31] Roseen, R. (1981). 'A combined solar collector array and heat store for estate applications – assessment of a prototype'. *Summary S9: 1981 of Report R59*: 1981 (in Swedish). Swedish Council for Building Research.

[32] Lof, G.O.G., El-Wakil, M.M. & Chiou, J.P. (1964). 'Design and performance of domestic heating system employing solar heated air – The Colorado House'. *Proceedings of the UN Conference on New Sources of Energy*, 5.

[33] Anon. (1977). 'Solar energy developments'. *RIBA Journal*, 84 (3), 103–10.

[34] Taylor, G. (1981). 'Understanding solar heating'. *Building Services*, 3 (1), 48–9.

[35] Jones, B.W. (1980). 'A solar power system (Prometheus) to provide 100 per cent of low-grade heat needs'. *Applied Energy*, September–October, pp. 329–46.

[36] T. Oreszczyn. (1981). 'The Open University'. Private communication.

[37] Telkes, M. (1974). 'Solar house heating – a problem of heat storage'. *Heating and Ventilating*, May, pp. 68–75.

[38] Telkes, M. (1949). 'Storing solar heat in chemicals'. *Heating and Ventilating*, November, pp. 79–86.

[39] Telkes, M. (1974). 'Solar energy storage'. *ASHRAE Journal*, September, pp. 38–43.

[40] Randell, J.E. (1977). 'Long term heat storage'. In: *Ambient Energy and Building Design Conference Proceedings*, University of Nottingham, pp. 135–51.

[41] Meisingset, K.K. & Gronvold, F. (1981). 'Latent heat storage in salt hydrates'. *Helios*, 11, 9–10.

[42] Telkes, M. (1954). US Patent 2 677 664. May 4.

[43] Anon. (1980). *Calor Thermal Storage*. Slough: Calor Alternative Energy.

References

[44] Carlsson, B., Stymme, H. & Wettermark, G. (1979). 'Low-temperature storage of heat in salt hydrate melts'. *Appendices to Research Report 750016—8*. Stockholm: Swedish Council for Building Research.

[45] Jönsson, A. (1980). 'Heat pump systems using chemical energy storage'. *Report R145*. Stockholm: Swedish Council for Building Research.

[46] Marshall, R.H. (1980). 'A prototype phase-change heat-storage device'. *Building Services and Environmental Engineer*, 2 (11), 12—13.

[47] Nemethy, A. (1949). 'Heated by the sun'. *American Artisan*, Residential Air Conditioning Section, August.

[48] McVeigh, J.C. (1977). *Sun Power*. Oxford: Pergamon.

[49] Böer, K.W. (1973). 'The solar house and its portent'. *Chemical Technology*, 3, July, pp. 394—9.

[50] Anon. (1980). *The Solar Revolution*. Slough: Calor Alternative Energy.

[51] Fischer, H.C. (1975). 'Annual cycle energy system (ACES) for residential and commercial buildings'. *Proceedings of the Workshop on Solar Energy Storage Subsystems for the Heating and Cooling of Buildings*. Charlottesville, Virginia, pp. 129—35.

[52] Arnold, B. (1979). 'House on ice'. *Building Services and Environmental Engineer*, 1 (9), 12.

[53] Caffell, A. & MacKay, K.T. (1981). 'Mud storage: a new concept in greenhouse heat storage'. In: *Energy Conservation and Use of Renewable Energies in the Bio-Industries*, Vogt, F. (ed.). Oxford: Pergamon.

[54] Colbeck, I. (1978). 'A review of solar ponds. Storage in solar energy systems'. *UK—ISES Conference C15*, pp. 31—43.

[55] Tabor, H. (1963). 'Large area solar collectors for power production'. *Solar Energy*, 7, 189.

[56] Bryant, H.C. & Colbeck, I. (1977). 'A solar pond for London'. *Solar Energy*, 19 (3), 321.

[57] Anon. (1978). 'New heat-storage device'. *International Power Generation*. April—May.

[58] Close, D.J. & Dunkle, R.V. (1970). 'Energy storage using desiccant beds'. *International Solar Energy Conference*, Melbourne.

[59] Anon. (1979). 'Chemical heat pump and storage potential on test'. *Building Services and Environmental Engineer*, 2 (1), 8—11.

[60] Davison, R.R. (1975). 'Long-term and seasonal storage group report'. *Proceedings of the Workshop on Solar Energy Storage Subsystems for the Heating and Cooling of Buildings*. American Society of Heating, Refrigerating and Air Conditioning Engineers.

[61] Anon. (1978). 'A chemical way of heating the home'. *New Scientist*, 4 May, p. 295.

8

Wind energy

8.1 Introduction

In the wake of costs for wave power, which have exceeded original estimates, interest in the UK is tending to concentrate on wind power as a means of central electricity generation using renewable sources of energy. In such northern latitudes the use of concentrating solar collectors for power generation is not viable.

For this reason a consortium of companies in conjunction with the North of Scotland Hydroelectric Board is about to construct a 3 MW turbine in the Orkneys, illustrated in Fig. 8.1. There are also tentative plans to place large fields of vertical-axis turbines in shallow parts of the North Sea off the east coast of England. These turbines would be of the Musgrove type shown in Fig. 8.15. Studies[1] show that up to 30% of electricity in use at a given time in the UK could be supplied from variable inputs such as wind turbines, without upsetting the grid network. Southern California Edison expects a 30% contribution by 1991.[2]

However, this chapter seeks to deal with small, local wind turbines. It has to be pointed out that the energy derived from wind-power devices is not unending (the turbines probably should have a lifetime of 20 years or so in the absence of freak weather), is not free (the devices are quite large and thus costly) and that to some, the machines are not handsome, and to others, represent a hazard.

It may help to provide a brief checklist of advantages and problems presented by the home use of wind power, before embarking on a description of the energy and devices available.

Checklist of attributes and problems
— Whilst the free-stream wind speeds can be reasonably high, those in urban areas and near obstructions can be greatly reduced.
— The power in the wind varies as wind speed cubed (V^3).
— To extract, say, 6 kW in a reasonably exposed place requires a turbine of diameter about 5 m, or the width of a terraced house.
— To rise above the layer of more-slowly moving air near the ground implies a tower and the visual intrusion may be unacceptable.
— Legal problems arise associated with damage if there is a mishap, with using the grid as a storage battery and with planning regulations.

Introduction

Fig. 8.1. 60 m diameter, 3 MW turbine to be built on Burger Hill, Orkney, by the Wind Energy Group (a joint venture of Taylor Woodrow Construction, GEC Power Engineering and British Aerospace).

— Aside from the initial capital cost, maintenance must be carried out on an object which is up on a tower.
— Acoustic noise and TV interference can occur.
— Storage batteries produce direct current, whereas many household appliances require AC. The inverter is quite expensive.
— Whilst it might be pleasing to be unaffected by black-outs, unless the costs of electricity storage are drastically reduced, it is not realistic to hope to become independent of the grid except in unusual circumstances.
— In the US, tax credits are available (40% of the first $10 000 via the Federal Government, and up to 55% of the cost via State Government); but there is no such assistance in the UK.
— The credits for power returned to the grid when surplus electricity is being generated are probably going to remain small, if they exist at all.

— In the UK it is not legal to sell electricity to a neighbour across a plot boundary.

8.2 Power extracted by turbines

8.2.1 Power in the wind

In 1972 Betz derived a relationship between the wind speed V (m/s), the area swept out by the turbine A (m^2) and the power P (W) which an ideal device can extract:

$$P = 0.645 \, (A \times V^3)$$

More specifically, Betz found that:

$$P = (0.59) \, (\tfrac{1}{2}\rho \times A \times V^3)$$

where ρ is the density of air. However, although the air density may change by 15% during the year and is about 10% less at 1000 m height, this effect is usually neglected.

This formula gives us four vital pieces of information.

— The power from a device suitable for home use of, say, 4 m diameter, operating at a wind speed of 5 m/s (roughly the mean UK wind speed in non-urban areas) is quite small:

$$P = 0.645 \, (\tfrac{4}{2})^2 \, \pi \times 5^3 \text{ W},$$
$$P = 1 \text{ kW}.$$

— The power varies with the swept area or as the diameter squared.
— The power varies as the cube of the wind speed.
— Only 59% of the power in the wind can be extracted because sufficient momentum must be left to carry the air away from the turbine. Only an ideal device actually approaches this value of 59% and most turbines extract considerably less than this, typical values being around 40%.

The dominant theme then is that the device must be placed in the windiest regime available. If the wind speed is 5 m/s in the unobstructed countryside then, in a neighbouring built-up area, it may well drop to 3 m/s. This drop has a drastic effect, the power falling from 1 kW, in the example above, to 0.22 kW.

8.2.2 Variation of wind speed with location

The wind regimes over the UK have been observed for a long time and we are therefore fortunate in having a large amount of information available. The particular aspect of the wind that concerns us is the duration of wind speeds of different magnitudes over the yearly period. The raw material gathered at meteorological stations, from which such information can be obtained, is the average wind speed for each hour of the year, gathered over a number of years. These data are recorded on an anemograph and can be used in a variety of ways. Perhaps the most useful is to determine the V_{50} for a particular site. V_{50} is the hourly mean wind speed equalled or exceeded for 50% of the time. To find V_{50} for a site, the

Power extracted by turbines 265

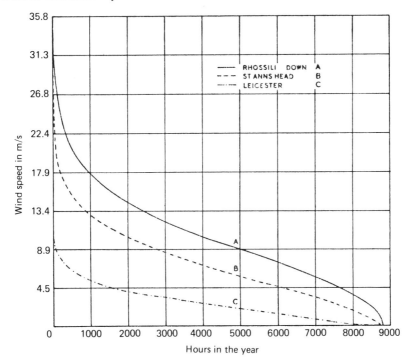

Fig. 8.2. Number of hours in the year when the wind has the indicated speed.[6]

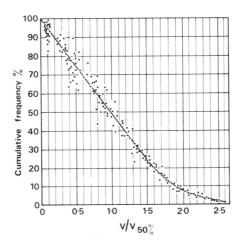

Fig. 8.3. Cumulative frequency distribution of V/V_{50} for 35 observing sites in the UK.[3]

hourly mean wind speeds are plotted as a frequency graph, for example Fig. 8.2. Thus, in Leicester, V_{50} is about 2.5 m/s, that is, the wind blows faster than 2.5 m/s for 50% of the year.

Using the same hourly mean wind-speed data, the ratio of V/V_{50} can be plotted as a cumulative frequency graph as shown in Fig. 8.3, V being the wind speed at any moment. This graph was derived from data for 35 sites around the UK[3] and clearly indicates that once the V_{50} of a site is known, the proportion of the time that wind speeds of different magnitudes will be equalled or exceeded can be determined. For example (see

Fig. 8.4. Hourly mean wind speeds (m/s) exceeded for 50% of the time, 1965–73. Valid for an effective height of 10 m and a gust ratio of 1.60 and for altitudes between 0 and 70 m above mean sea level.[4] (The gust ratio is defined in the text.)
(Reprinted with permission, HMSO.)

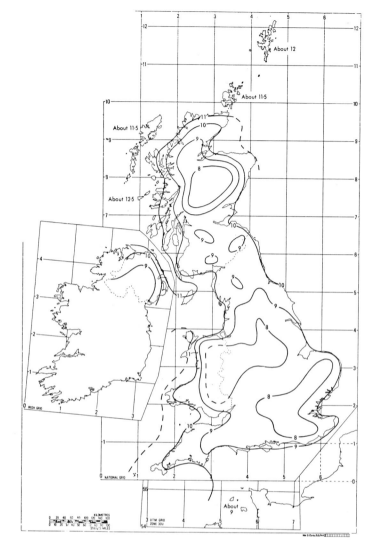

Fig. 8.3), for a site with a V/V_{50} value of 5 m/s, windspeeds of 5 m/s will be equalled or exceeded for 50% of the time since $V/V_{50} = 1.0$. Wind speeds of 10 m/s, corresponding to a V/V_{50} value of 2.0, will be equalled or exceeded only 7% of the time.

Ideally, the value of V_{50} should be determined by long-term observations at the site in question. However, a reasonably reliable short-cut to finding the distribution of different wind speeds is available by using a map of V_{50} (Fig. 8.4). This map[4] applies to open inland sites and refers to the V_{50} at an effective height of 10 m. For sites and heights other than the ones to which the map refers, modification factors must be applied.

For sites other than the open inland site to which the map refers, the map value is modified by considering the gust ratio at the site in question. Since January 1973, anemograph stations have measured for each clock

Table 8.1. *Variation of gust ratio with terrain*[8]

Type of site	Gust ratio (G)	Modification factor for map value 1.6/G
City	2.10	0.76
Suburban	2.00	0.80
Open inland	1.60	1.00
Coastal	1.45	1.10
Hill	0.80	2.00

Table 8.2. *Variation of wind speed with height (from Caton[4] and Rayment[3])*

Type of site	Power-law exponent	Modification of map value
City[a]	0.23	$0.4476\, H^{0.23}$
Suburban	0.22	$0.4820\, H^{0.22}$
Open inland	0.17	$0.6761\, H^{0.17}$
Coastal	0.13	$0.8154\, H^{0.13}$
Hill	—	—

[a] Heights measured upwards from prevailing roofline.

hour the maximum gust speed of about 3 s duration and, simultaneously, the hourly wind speed. The maximum gust speed for each hour has been divided by the mean speed for each hour and the average value from several years' data has been taken. This is the gust ratio (G). It has been found that sites with similar terrains have similar gust ratios. For example, the gust ratio for city sites is about 2.1:1 whereas the ratio for open inland ones, to which the map applies, is about 1.6:1. To find the relevant V_{50}, one therefore has to multiply the map value by $1.6/G$, where G is the gust ratio characteristic of the site (see Table 8.1).

8.2.3 Variation of wind speed with height

For heights other than the standard effective height of 10 m, the value can be modified using the power law:

$$V = c \times H^p,$$

where V is the wind speed, c is a constant, H is the effective height and p is an exponent characteristic of the underlying terrain. For an open site, H can be taken to be the height above ground. For a city site, where buildings are densely packed, it has been suggested that the effective height should be the height above the general level of rooftops. For a suburban site, the effective plane will be dependent on the extent of screening by buildings and trees, but for a low-density, low-rise region, it seems reasonable to assume the zero plane to be the ground level as long as the prevailing wind is not blocked by nearby trees or buildings. Exponents and constants for modifying the map values of Fig. 8.4 are shown in Table 8.2.

We are now in a position to judge V_{50} for any site in the UK. Rayment[5] has shown that the total annual energy in the wind (E_{wt}) in the UK is such that

$$E_{wt} = 43.43\,(V_{50})^3 \text{ MJ/(m}^2\text{ y)},$$

when V_{50} is measured in m/s.

Using the wind map and the modification factors for V_{50}, it is therefore possible to determine the annual energy available in the wind for any site and height. For example, suppose we wish to consider a site in the Midlands in a non-urban open inland area with no major obstructions, a value of V_{50} from Fig. 8.4 of 4.5, and a tower height of 25 m. The terrain modification from Table 8.1 is 1.0. The height modification from Table 8.2 is:

$$(0.6761)\,(25\text{ m})^{0.17} = 1.17.$$

Thus, V_{50} for the site is

$$(4.5\text{ m/s})\,(1.0)\,(1.17) = 5.26\text{ m/s}$$

and

$$E_{wt} = 43.43\,(5.26)^3 \text{ MJ/(m}^2\text{ y)}$$
$$= 6.32 \text{ GJ/(m}^2\text{ y)}.$$

Thus, a turbine with a diameter of 5 m and a swept area of 20 m² encounters an annual wind energy of 127 GJ/y.

As we have seen, even the most efficient device could only extract 59% of this, or 75 GJ/y and, unfortunately, the energy actually available for use will be very much less than this.

Before moving on to a discussion of further losses, a city example is considered. If the site is in London, in a congested position with a mast at 25 m but roof tops at 8 m, the following data might apply:

$$V_{50} \text{ (Fig. 8.4)} = 4.0 \text{ m/s},$$

Terrain modifier (Table 8.1) = 0.76,

Height modifier (Table 8.2) = 0.86,

$$E_{wt} = 43.43\,(4.0 \times 0.76 \times 0.86)^3 \text{ MJ/(m}^2\text{ y)}$$
$$= 774 \text{ MJ/(m}^2\text{ y)}.$$

Thus, even our ideal device of 20 m² could only extract 9.14 GJ/y compared with 75 in the country.

Fig. 8.5 illustrates the energy in the wind across the UK.

8.3 Power extracted by real turbines

The actual power extracted by a turbine depends on three factors, namely the cut-in speed, the rated speed and the power coefficient.

The lowest wind speed which will overcome friction in the turbine bear-

Power extracted by real turbines 269

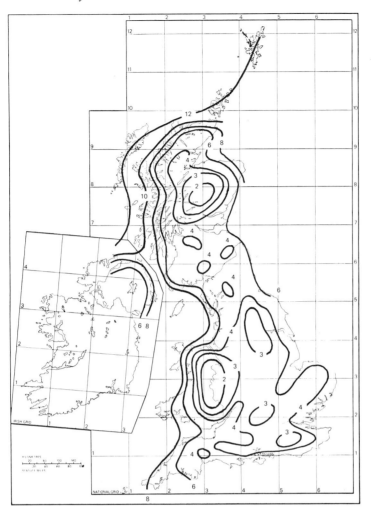

Fig. 8.5. Energy in the wind at an effective height of 10 m in open countryside.[5]

ings and actually cause the device to produce useful power is called the cut-in speed and is typically 3–4.5 m/s (a survey of commercially available devices with rated outputs of less than 50 kW gives a mean cut-in speed of 4.2 m/s).

Above the rated wind speed, the turbine, gearbox, generator and mast would be overstressed unless wind is spilled, thus output is constant above the rated speed until at even higher speeds the turbine is shut down to avoid damage.

The power coefficient is the ratio of power extracted by the blade system to the power available in the wind and thus has a maximum value of 59%.

Returning now to the rated speed, Fig. 8.6 illustrates data derived from Fig. 8.3 and shows the value of $(V/V_{50})^3$ (which is proportional to the power in the wind) against the number of hours in the year.

Suppose V_{50} is 4.5 m/s, then a high wind speed of 12 m/s would have a

Fig. 8.6. Power duration curve typical of the UK.

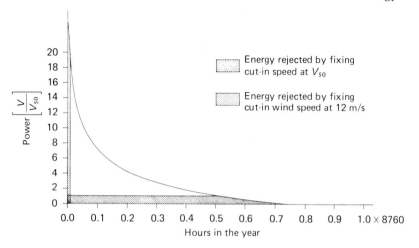

value of $(V/V_{50})^3$ of 19. But this speed is only available for 0.01 of the time during the year. The area under the curve in Fig. 8.6 is proportional to the energy per year in the wind. It is clear that the fraction of power for wind speeds above 12 m/s is small, yet to be able to absorb these wind speeds the structure of all the components must be adequate. A cheaper and less-robust system could be built if the rated speed were only 10 m/s, for example. Optimization of the performance suggests that rated speeds should be about $2.25(V_{50})$.

The cut-in speeds can then be decided. For example, if a cut-in speed of V_{50} is chosen, then $(V/V_{50})^3 = 1.0$. The value of $(V/V_{50})^3$ at the rated speed is 2.25^3 or 11.4, thus at V_{50} the device is only operating at 1.0/11.4 of its maximum capacity. The part load efficiency of components such as gearboxes, bearings and generators can be very poor. Moreover, the power available annually, which is rejected by not using winds below V_{50}, is small as can be seen in Fig. 8.6. Typically, then, cut-in speeds should be between V_{50} and $1.3(V_{50})$.

It should be noted that 80% of the wind energy normally lies between speeds of $1.125 \times V_{50}$ and $2.5 \times V_{50}$, and if our 'perfect' device of 20 m² operating in the Midlands has a cut-in speed of 1.125 (4.5) or 5 m/s and a rated speed of 2.5 (4.5) or 9 m/s it would capture 58 GJ/y.

8.4 Types of turbine

Earlier in this chapter the Betz relationship was quoted, which suggests that a device can only extract 59% of the wind power. There is some disagreement about the theoretical maximum and Sabinin has suggested a figure of 69%.[6]

There is, however, little doubt that practically speaking the best figure currently achieved is an extraction rate of about 45% of the energy in the wind, by aerodynamically designed blades. (There is a little confusion about power coefficients since some authors have quoted values as a percentage of the Betz maximum, for example 60% of the theoretical maxi-

Types of turbine 271

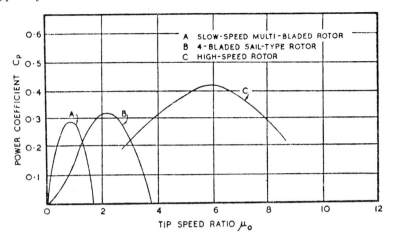

Fig. 8.7. Power coefficients for different types of rotor.[6]

mum value of 59%. In this text the C_p is quoted as the fraction of power in the wind, that is, 36% in the above case of 59 × 60%.) Golding[6] quotes the values in Fig. 8.7.

The power coefficients are plotted against the ratio of the speed of the blade tip (that is, $2\pi rN$) to the upwind air speed, where r is the radius and N the rate of revolution; for example, a large device such as Growian II in Fig. 8.8 has a radius of 73 m and rotates at 17 rpm, giving a tip–speed ratio of 6 at a wind speed of 22 m/s.

It is clear from Fig. 8.7 that the power coefficient, although fairly broad for well-designed airofoil blades, falls off quite quickly on either side of the peak. A good design would be one chosen so that, at a particular site, the wind speeds would be such that power output over the year was maximized and this implies keeping the tip–speed ratios within fairly narrow bounds to preserve a good power coefficient.

In the Midlands example quoted previously then, the output will fall from 58 to about 36 GJ/y because of the actual value of the power coefficient of even a well-designed rotor.

Types of rotor

A large number of forms have appeared over the course of time including the vertical-axis Chinese windmill of Fig. 8.9 and the horizontal one described by Hero of Alexandria in the third century BC, shown in Fig. 8.10.

To extract power from the wind, the turbine must first intercept it as it passes through the area swept out by the blades or vanes. To do this at low wind speeds when the turbine is rotating slowly requires a large number of vanes. Thus, the multibladed pumping mill is able to perform efficiently when the tip–speed ratio is low.

However, as the wind speed increases, the large area of vanes exerts a considerable rotational force on the air column and efficiency falls. This can be seen in Fig. 8.11. If the turbine consists of two or three thin aerofoil-section blades, at low wind speeds the blades do not rotate fast

Fig. 8.8. Growian turbine[7] with a 125 m tower. (Reprinted with permission © 1981, *Popular Science*, Times Mirror Magazines Inc. Photograph by C. De Groote.)

enough to intercept all the wind passing through the swept area and the efficiency is low. At higher wind speeds, the blade tips may be rotating at about six times the wind speed and can extract power efficiently. This is because the thin blades do not present a large area to the wind and so drag is reduced and the air column is not caused to rotate. The power coefficient of such turbines is in the region of 0.45 when the tip–speed ratio is about 6 (Fig. 8.11). Since turbines with aerofoil blades work more efficiently than other types, and do so at high rotational speeds, they are ideal for combining with an electrical generator because the amount of gearing-up will not need to be as great as with other types.

Figs. 8.12 and 8.13 illustrate the Savonius and multibladed devices also shown in Fig. 8.11, both of which are more suitable for pumping irrigation water than generating electricity.

The two categories of aerofoil turbines that can be used for electricity

Types of turbine   273

Fig. 8.9. Primitive vertical-axis Chinese windmill.[6]

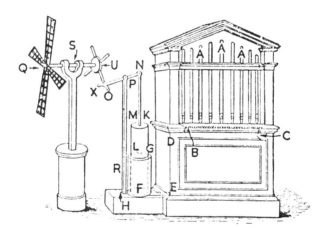

Fig. 8.10. Small horizontal-type windmill blowing an organ (from *The Pneumatics of Hero of Alexandria*[91]).

generation are the horizontal-axis propellor types and the vertical-axis types. The horizontal-axis machines have a turbine with a rotating shaft that is parallel to the ground and supported on a tower, for example Fig. 8.8. It is immediately apparent that the turbine must cope with changes in wind direction. If the turbine is mounted downwind of the tower, as is usually the case with large aerogenerators of this type, the turbine may be self-directing. If it is designed to be upwind of the tower, as is often the case with smaller aerogenerators, a tail vane is required. There are advantages and disadvantages associated with each strategy.

If the turbine is downwind, the blades can be coned. This means that the blades are inclined at a small angle so that their tips are downwind of

Fig. 8.11. Power coefficients for various turbines.

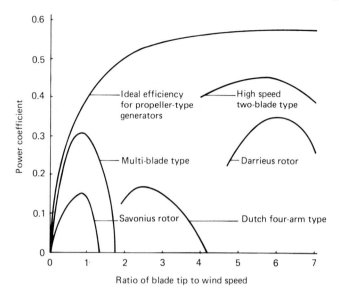

Fig. 8.12. Savonius rotor and air flow through the blades.[8]

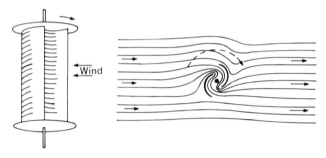

the hub about which they rotate. This is done to reduce stresses. Since the turbine reduces the axial momentum of the wind, it must experience an axial force which the blades and tower have to resist. This axial force is independent of the number of blades in the turbine and is solely determined by the power that the turbine extracts. Since aerogenerators have 'low-solidity' turbines (the ratio of the total frontal area of blade presented to the wind to the area swept by the blades has a low value, typically 0.05), the stress that the axial force produces per unit area of blade will be high. If the blades are coned, the centrifugal force at high rotational speeds will make the blades attempt to assume an upright position. This will tend to counter the effects of the downwind axial force.

The disadvantage of a downwind turbine is that once every revolution each blade must pass through a patch of relatively calm air in the lee of the tower. This will momentarily change the stresses in it and can have highly undesirable effects. For example, the NASA 100 kW machine in Fig. 8.14 was found to have blade stresses twice as great as was predicted due to the 'shadow' cast by a staircase on the tower.

To avoid such problems, the supporting tower for any downwind-

Types of turbine 275

Fig. 8.13. Multiblade turbine.

Fig. 8.14. ERDA/NASA 100 kW wind turbine at Plum Brook, near Cleveland, Ohio.

mounted device should be as 'transparent' as possible while still being designed to withstand the axial force transmitted to it from the turbine.

If the windmill is mounted upwind of the tower, a tail vane will be required, which will add to the expense of the machine. This vane can also be used for feathering the blades. Although upwind mounting considerably reduces the problem of transient stress changes in the blades it does not eliminate it. There will be a patch of calmer air that builds up in front of the tower. Moreover, it is now the tower that has to withstand fluctuating stresses as each limb shields it from the wind once per revolution. If the blades need to be coned, they will have to be arranged so that their hub is some way upwind of the tower to prevent the blade tips hitting it. This will lead to a considerable eccentric load on the tower.

Apart from the detailed aerodynamic characteristics of the blades, which will not be considered here, the question of how many blades to have in the turbine, and what the solidity should be, arises. For maximum efficiency, drag and air-column rotation should be minimized so a low solidity is desirable. However, as previously described, the axial force is independent of the number of blades. It is therefore impractical to aim for a very low solidity unless very strong blade materials can be used to withstand the stresses. The number of blades between which the solidity is divided is usually two or three. Two blades mean that static and dynamic balancing is simpler so tower vibrations and blade stresses are likely to be reduced. Since the turbine will probably have to be balanced where it is manufactured, a two-bladed turbine presents less of a transportation problem. However, the main difficulty with two-bladed turbines is their behaviour when the wind changes direction. The moment of inertia of the blades about the yaw axis changes every $90°$, and so at every revolution the blades and tower experience two periods when there is considerable resistance to tracking the wind. This will cause unwanted stresses in the system. With three blades this is no problem although the difficulties of balancing and transportation apply.

With horizontal-axis aerogenerators, the turbine shaft must be connected to the gearbox and alternator at the top of the tower. Although it would be possible to use a right-angle drive shaft to connect the turbine shaft to a gearbox and alternator at ground level, the torsion in such a system would appear to be a problem. The tower then has to withstand the considerable loads imposed on it by the turbine, the gearbox, and the alternator, as well as that imposed by the axial force on the turbine and the wind loads on the tower itself. With the alternator at the top of the tower, some means must be found to lead the electricity generated to ground level, while still allowing the turbine to track the wind. Clearly, a simple cable would become twisted. This is normally achieved by the use of slip rings which are liable to cause maintenance problems.

The second type of aerogenerator is the vertical-axis machine. In the Savonius rotor (Fig. 8.12), two half-cylindrical blades are offset about a vertical axis in such a way that the wind can pass between them. In addition to the pressure on the concave surface facing the wind, the wind is directed through the machine so that it produces an additional pressure,

Types of turbine

assisting rotation, on the back side of the other half of the rotation. Although this type of machine is very simple to construct, often from an oil barrel cut in half, its performance characteristics (see Fig. 8.11) are not suitable for the generation of electricity due to its inability to operate at high speeds.

The major advance in vertical-axis machines was made by Darrieus in 1926, although it was not until the late 1960s that South and Rangi rediscovered his patent when they tried to file their own for a similar machine. The first type of Darrieus machines employ curved aerofoil-section blades rotating about a vertical shaft. In order to reduce bending stresses, the blades are usually shaped to a 'troposkein' curve, the configuration that a skipping rope assumes. This means that when rotating at high speed, the only major stresses in the blades are tensile ones.

The aerodynamic principle that applies to such machines is the same as that for horizontal-axis aerofoil machines. With a horizontal-axis machine a torque is produced continually as long as the turbine is facing the wind. With the vertical-axis machine, however, power is extracted from the wind when the blades are crossing the path of the wind. When travelling directly up- or downwind, no torque is produced. However, the cyclical nature of the power extraction is minimized by the momentum of the turbine. With a three-bladed turbine, there will be no part of each revolution during which the blades do not extract power. As with horizontal-axis machines, the blades should be of the lowest possible solidity for maximum efficiency. If this is done, the power coefficient of the turbine can be comparable to a horizontal-axis machine when the tip–speed ratio is about 4 (see Fig. 8.11). The Darrieus-type machines have a large proportion of the blades a considerable distance from the tower so the 'shadow' effect of the tower is largely avoided. Moreover, because the shaft of the machine is vertical, the alternator and gearbox can be at ground level which avoids the need for a sturdy tower. Another advantage of the Darrieus-type is that the machine can extract power from the wind without needing a directioning system. This means that no tail vane or slip rings to carry current from the alternator are required. However, troposkein machines do have several drawbacks. The curved aerofoil blades are expensive to manufacture and it is only the outermost portion of the blades that sweep an appreciable area of wind thereby extracting power.

The Musgrove variable-geometry vertical-axis windmill has upright aerofoil blades attached to a cross-member that rotates about a vertical axis. For a given diameter and height, the cross-section of wind swept out by the blades is therefore much greater than that swept by a Darrieus machine. This leads to a greater power output. Although the blades have to cope with bending stresses, being straight they are cheaper to make. To reduce bending stresses in high winds and to reduce the power output, the blades are allowed to incline towards the horizontal as centrifugal forces overcome the restraining force of a wire attached between the vertical shaft and the blades. This reduces the swept area and hence the power output, as well as allowing the blades to take stresses increasingly in tension rather than in bending. Fig. 8.15 illustrates a Musgrove turbine built at the

Fig. 8.15. Musgrove turbine, diameter 6 m, blade length 4 m.

Rutherford High Energy Physics Laboratory and designed to provide electricity for the Cambridge Autarkic House.

8.5 Electrical generation

8.5.1 Electrical generators

Various kinds of generators can be attached to a wind turbine. The choice of system will generally have been made by the manufacturer of the wind-power system and a consumer will usually elect a system based on the following questions:
- How large an electrical output is required?
- Is the output to be used only for heating?
- Is the output to be stored in batteries?
- Is the system to be connected to the electricity grid?

There are three classes of generator in common use for turbines, two producing alternating current and one direct current. A survey of about 50 commercially available wind systems in the range 0–50 kW shows that small systems (0–5 kW) tend to use DC generators. Intermediate systems supply AC from alternators and large AC systems use induction generators. A fourth type, the synchronous generator is as yet uncommon.

In the past, large DC systems have been proposed, as Golding[6] points out. Table 8.3 summarizes some of the properties of generators.

The first problem with designing the system lies in matching the power output from a turbine which varies as $C_p \times V^3$, with the power converted by a generator which varies at best as the (rotational speed)^2. The second problem concerns matching the load with the generator output.

Electrical generation

Table 8.3. *Selection of generators*

	Relative cost/kW	Efficiency at full load (%)
DC generator		
wound field	5	90
permanent magnet	4	93
AC alternator		
wound field	3	90
permanent magnet	3	93
vehicle alternators	–	20–70
AC induction generator	1½	90

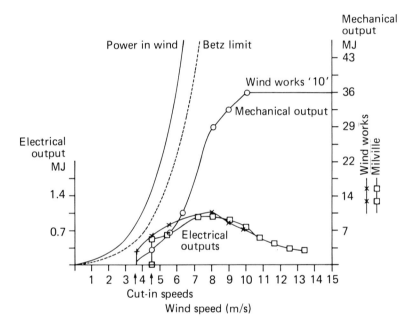

Fig. 8.16. Overall power coefficients of two 10 kW commercial systems (diagram plotted from data in Marier[10]).

The wind turbine, if terminated by a fixed resistive load, such as an immersion heater, will face a very large starting torque at low wind speeds (thus delaying the onset of generation). It will tend to stall in fluctuating winds, even when, for much of the time, generation would be possible. The overall performance will thus be markedly improved with the use of suitable control.

At low speeds a system of switching resistive loads in and out allows the load to be very small when the turbine starts up; allows higher rotational speeds, thus moving to a more efficient part of the C_p curve; and is able to bring in more and more load as the wind rises to simulate a load varying as V^3. The field current may also be varied (in field-wound generators) which over a limited speed range allows the electrical output roughly to vary as V^3.

Power coefficients for the turbines were shown in Fig. 8.11 but the concept can be extended as in Fig. 8.16 to cover the turbine plus generator. In this case then, C_{po} represents the fraction of power in the wind which is converted to electricity. Whereas power coefficients for the turbines reached values of 0.45, overall power curves only reach values of about 0.3. (The curves shown are for a Windworks 10 kW system, with 10 m diameter, three-bladed horizontal-axis rotor and a 240 V DC alternator, and a Millville '10-3-IND' with 7.6 m diameter, three-bladed horizontal-axis rotor and a 220 V AC induction generator.)

It is clear then that the generators and gearboxes cause additional power loss. This loss appears as heat, and if the equipment is within the heated volume of the house, such heat can be valuable.

8.5.2 Gearboxes

The electrical generators with small numbers of poles all require high shaft speeds of 1000 rpm or higher. The turbine shaft is rotating typically at 50–100 rpm for a 10 kW device and the gear ratio is about 15:1. Losses in gear boxes or timing belts or chains to overcome friction are significant particularly at low loads and Fig. 8.17 indicates typical values. Similarly, except for the very highly developed rare earth permanent magnet generator[16] mentioned in the previous section, major losses also occur in the generator. Thus the overall power coefficients of Fig. 8.16 now look reasonable. At the rated power of 36 MJ, the percentage recovery factors are as follows:

Power in the wind	100(%)
(1) Turbine power coefficient	40–45
(2) Gears	92
(3) Generator	80–85
Overall power coefficient (product of 1, 2 & 3)	30–35

Most generators require a field current. At low wind speeds, residual magnetism will be insufficient to allow electrical generation. Thus the generator curve of Fig. 8.17 is depressed still further in many cases by the provision of field power.

8.5.3 Electricity storage

During calm periods, users must rely for electricity on the grid, a back-up supply such as a diesel generator, or batteries.

Estimates of the duration of calm periods vary rather widely, but when the authors examined five years of hourly wind speeds recorded in East Anglia, the longest calm period (defined as one incapable of generating electricity from a 5 kW system with its attendant inefficiencies) was 42 days.

The average electricity consumption in the UK for needs other than space and water heating is about 6.5 GJ/y plus about 6.5 GJ for cooking. To store the energy for one day's use of lights and appliances (about 18 MJ) requires about 110 kg of lead acid batteries, at a cost of about

Electrical generation

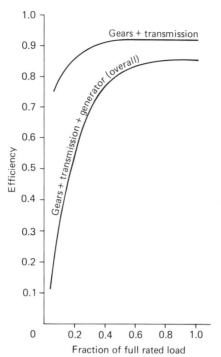

Fig. 8.17. The part-load efficiencies of gears, transmission and alternator, excluding any field current supplied to the generator.

£350. To store 18 MJ/d for 42 days would cost almost £15 000 (at 1982 prices) for the batteries alone.

In most cases then, the turbine system will be used either to supply only heat (in which case thermal storage can be used) or to replace grid electricity supplies when the wind is blowing. In isolated cases connection to the grid may be very expensive and batteries may be used. Car batteries are unsuitable, being designed for short-term but heavy demands, but industrial truck cells, whose discharge is normally less than 50% should last more than ten years. Lead antimony units will be less expensive than lead calcium ones but will require more frequent topping up. A large number of cells requires suitable access space, ventilation and temperatures kept between about 10 and 25 °C.

Exotic batteries

Very large amounts are being spent on battery research for load handling in grid systems and vehicles. Useful reviews of new systems are to be found in the *Electrical Review* and *US Popular Science*.

Table 8.4 illustrates some of the cells under evaluation. At present there is no real alternative to lead acid cells for the storage of electricity on a small scale.

The cells require a charging unit which tapers the energy supplied, to prolong battery lifetime. An inverter is also needed in most cases, to transform the DC supply from the battery to AC of the correct frequency. Many motorized appliances will not work with a DC supply, televisions

Table 8.4. *New cell systems (adapted from Ref. [11])*

General data	Pb/PbO$_2$ H$_2$SO$_4$	Ni/Fe KOH	Ni/Cd	Zn/air	Fe/air	H$_2$O$_2$
State of development	Proven	Traction batteries	Small batteries	Lab. cells	Prototypes	Demonstration cells
Operating temp (°C)	30	30	30	≈ 40–50	≈ 40–50	≈ 100–200
Overall efficiency (%)	71	55	53	43	47	42
Cycle life	2000	4000	2000	1500	1500	1500
Cost (£/kWh)	25	65	250	75	23	–

and other electronic systems demand a supply close to 50 or 60 Hz. Since motors call for a starting current of about five times their running current, quite large inverters are needed and a 3 kW device will cost about £1500.

When the turbine system is used as a grid substitute in windy conditions, an induction generator will provide a constant voltage and frequency output at the values of the grid and a current which varies with the power abstracted from the wind. Thus the output is suitable for direct connection to the house system and when more power is produced than can be absorbed, the energy could, with CEGB agreement, be fed back into the grid. When the mill is not producing power, unless automatic disconnection is provided, the grid would supply the generator turning it into a motor.

If a DC generator has been installed it may still be connected to the grid (thus avoiding the use of batteries) via a Gemini synchronous inverter, which normally converts the DC to AC for use by appliances in the home but in time of surplus sends the excess down the line. The Gemini does not produce a perfect sine wave and may cause some problems with sensitive devices such as microcomputers.

8.6 Use of wind-generated electricity

8.6.1 AC or DC?

Many household appliances demand AC. Power devices generally use induction motors, since they are robust and cheap, and these will not work on DC. Electronic equipment demands AC with a closely regulated frequency and a reasonably sinusoidal wave form. Transformers fail at markedly low frequencies and synchronous clocks and phonographs will turn at incorrect speeds if the wrong frequency is supplied.

There are always alternative pieces of equipment, often of lower power demand, designed, for example, for boats and caravans; but in general it will be highly inconvenient to have only DC power supplied to a house. Furthermore, if low voltages are used, extra costs will be incurred due to

Use of wind-generated electricity 283

the heavier cables needed. DC demands extra robust switches and thermostats. Normal fluorescent light fixtures should not be used on DC. Any appliance which relies only on heating will work equally well on AC or DC.

8.6.2 Inverters

Our recommendation then is that 240 V AC (UK) or 120 V AC (US) be used. There is unlikely to be enough power to charge a thermal store and, unless grid connection is made, cooking would have to be by other means. To convert the DC power from batteries an inverter is required (or a motor-generator set). At peak load an inverter has an efficiency of about 90%. Any losses appear as heat which may be useful. At part loads, since power is required to run the inverter, the efficiency drops to, say, 80%. When not in use the power drawn is typically 50 W, which over 24 hours would represent a large power drain on the battery. Thus the device should switch itself off when no power is demanded. A more extensive review of problems associated with inverters can be found in *Wind Power for the Homeowner*,[10] and Cox[12] reviews the problems caused by present-day inexpensive inverters which may feed reactive power into the grid if such connection is made.

8.6.3 Thermal storage

In many cases, those using wind power will connect to the grid via a Gemini[13] synchronous inverter or use the power for space heating or domestic hot-water heating. Such uses avoid most of the control problems associated with batteries.

Surplus energy may be stored as heat. In the UK, storage heaters operating on off-peak electricity are common and can be used to absorb power produced by a turbine and generator. According to the type of generator, the current, and possibly the voltage, will vary with the strength of the wind; a simple control system which connects more and more heating elements in the circuit will be necessary to match the power uptake to power output. One great advantage of the heating mode of use of wind power is that even short bursts of energy can be used and the system will not be upset by a varying source of power.

Instead of heating the bricks (which, being made essentially of iron ore, are of high thermal capacity) in a storage heater, water may be heated. As discussed in Chapter 7, water cannot practically be used much above 80 °C. This means that the heat capacities of the two kinds of store are rather different, as shown in Table 7.2.

To reduce the surface temperature of a storage unit at 500 °C to a safe value of 30 °C would require about 300 mm of high temperature insulation. This degree of insulation would ensure that less than 15% of the heat would escape via the insulation over the first five-day period if the store were left unused. Hot water could be provided via an air loop to a heat exchanger for domestic use. Watson discussed such storage further.[8]

8.7 Economies of scale

Generally, the high costs of electrical or thermal energy storage militate against sole reliance on wind power at the one-house scale for either electrical or heating demands. For example, an electrical store to cover the long calm periods found even in the UK would cost between £10 000 and £20 000. A thermal store using high thermal mass bricks would have a volume of about 10 m^3 including insulation (6 m^3) and would cost about £2000 including a control system for charging and an air loop for heat extraction, on the assumption that all the space heating and domestic water needs for 40 calm days would be stored.

If smaller storage units are used the savings on network energy to the consumer are only the marginal ones of electricity, gas, etc. at the prevailing rate, and not the metering or installation charges.

Watson[8] has looked exhaustively at the economies of scale and found that if electricity prices rise at 12%/y for 20 years, then for sites with wind speeds such that V_{50} is $\geqslant 4$ m/s, a turbine heating system with grid back-up is cheaper than an all-grid heating system. The fraction of heat supplied by wind power is 70% for the optimally designed system. Watson indicates how to size the system and in this particular case finds that a ten-day store and a turbine of 60 m^2 swept area ($V_{50} = 4$ m/s) is the cheapest solution. (Areas vary according to V_{50}, for example at $V_{50} = 5.0$ m/s the area is 31 m^2.) However, such a system competes with gas only at sites for which $V_{50} \geqslant 6$ m/s.

By contrast, Watson finds that even at the lower fuel-inflation rate of 8%/y for a *group* of 20 houses, a turbine system and central store with grid back-up is less expensive than an electrical system solely supplied from the grid, for all sites with $V_{50} \geqslant 4$ m/s; *and* is less expensive than a gas system, for all sites with $V_{50} \geqslant 5.5$ m/s. Table 8.5[8] illustrates this point.

8.8 Installed systems

8.8.1 Conservation House at NCAT

Fig. 8.18 illustrates the Conservation House at the National Centre for Alternative Technology (NCAT) (Machynlleth, Powys, Wales).

The calculated specific-heat loss for the house is 66 W/K. The peak demand is 1.2 kW and the space heating requirement with incidental gains taken into account is only an average of 29 MJ/d in December and January and a total of 3.4 GJ over the heating season. This compares with 47 GJ over the heating season for an equivalent sized conventional house.

Heating system

An air-to-air heat pump was chosen for space heating and the low-power requirement enabled a small and cheap system to be employed. The compressor is the same type as that used in a domestic deep freeze. The evaporator heat exchanger in the roof space is fed with outside air mixed with ventilation air from the kitchen extractor fan and the condenser is mounted with the compressor in a cupboard on the upper floor. Warm air

Installed systems

Table 8.5.[6] *Total[a] present values (PV) of wind-powered system for a group of 20 houses. Discount rate 4%, fuel inflation rate 8%/y*[8]

Present value of items in each house	(£)	
Back-up energy bought	1728	
Thermal store	243	
Heat distribution	284	
Radiators, etc.	1068	
Present value of aerogenerator per house V_{50}(m/s)	(£)	Total present value (£)
3.5	7481	10804
4.0	5426	8749
4.5	4089	7412
5.0	3180	6503
5.5	2524	5847
6.0	2050	5373

Present value of electricity and heating system per house (no turbine system) £9885
Present value of gas and heating system per house (no turbine system) £6064

[a] Prices at May 1979.

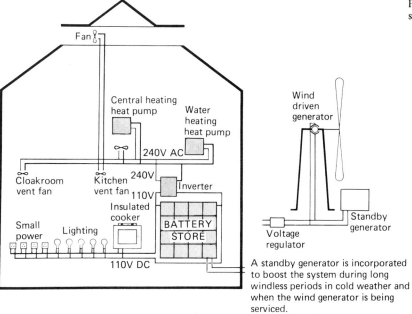

Fig. 8.18. The NCAT Conservation House.[14]

A standby generator is incorporated to boost the system during long windless periods in cold weather and when the wind generator is being serviced.

is fed to all rooms via the underfloor spaces and returned via ceiling-level vents. The refrigerant circuit can be reversed by a changeover valve for automatic defrosting of the evaporator and for summer cooling should this prove necessary. Experience of the hot summer of 1976 suggests that, in this location, cooling will not be needed, as the long thermal time constant of the house provides a very stable temperature with little diurnal variation.

In 1978 a propane-fired heating system was installed experimentally to facilitate measurements of the building's performance. Preliminary results show a heat loss close to the calculated value and suggest that heating costs on low-priced fuels should average around £1 per week (at 1978 prices).

Hot water

A similar, small heat pump is used to heat the 200 l hot-water tank. A large volume of relatively low-temperature (50 °C) water is used to maximize the COP while maintaining an adequate hot-water supply (for some purposes cold water will not need to be mixed with it). The evaporator of this heat pump is immersed in a 275 l waste-water tank below the ground floor. By discharging water at a similar temperature to the cold-water-feed temperature, or colder if necessary, sufficient energy can be extracted to maintain the temperature of the hot-water cylinder. It has been estimated that heat lost from the water to the house is approximately balanced by the electrical energy supplied to the heat pump. A COP of 2 has been measured for this system. Electrical consumption is estimated at 16.2 MJ/d or 5.90 GJ/y.

Energy supply

The house described, so far, as having a total energy demand of about one-fifth of a similar sized conventional house, is suitable for normal mains electricity powering, but as the emphasis at the Centre is on ambient energy use and the site is not connected to the national grid, a 2 kW Dunlite aerogenerator was installed next to the house, together with 72 MJ of lead acid battery storage. On a good site this would supply most of the energy required, except for long windless periods combined with cold weather. With the relatively low wind speed near the house, there is insufficient wind energy for heating but adequate electricity for lighting and cooking is supplied at 110 V DC. It cannot be claimed that the wind–electric system is generally a cost-effective option, mainly because of battery replacement cost. However, the initial cost of the system (about £4000) is similar to the cost of connecting isolated premises to the national grid, and in such situations a wind-power scheme is much more attractive financially. However, a house designed specifically for wind-power heating should incorporate thermal storage (rather than batteries) equivalent to at least a week of winter load.

Problems experienced
- The 2 kW wind generator was poorly sited being next to the house and in the shadow of surrounding ridges.
- As a result the energy output has been less than expected and

Installed systems 287

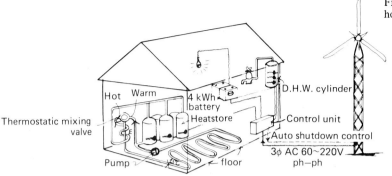

Fig. 8.19. Wind-powered house at NCAT.[15]

because the battery store (72 MJ) is too small, not all the output is available. It has been estimated that by resiting the device on a neighbouring hill top, the annual electricity output would rise to 9.7 GJ/y. Using the method outlined earlier in this chapter the following data would be expected:

Rotor diameter for Dunlite 2000	= 4.11 m
V_{50} from Fig. 8.4	= 4.5 m/s
Gust ratio, Table 8.1	= 1.6
Gust ratio, modification to V_{50}	= 1.0
Correction for height from Table 8.2	= $0.6761 H^{0.17}$
(Above 10 m quoted in Fig. 8.4)	= 1.05
V_{50} for the site	= 4.7 m/s
E_{wt} (wind energy/y encountered by blades)	= 61 GJ/y
Power coefficient of the Dunlite[10]	= 0.284
Combined efficiency of gears and generator	= 70%
Power between cut-in and rated speed	= 80%
Electrical power output/y	= 61 (0.284 × 0.7 × 0.8)
	= 9.7 GJ/y

— The system of a 110 V DC output driving a 240 V AC heat-pump compressor via batteries and an inverter is very inefficient.
— The heat pump iced up very readily in cold weather thus impairing its COP.

8.8.2 Wind-powered cottage at NCAT

Attempts were made at the NCAT to provide most of the power for a small cottage, using a 5 kW Swiss Elektro machine.[15]

Fig. 8.19 indicates the scheme, in which most of the energy is used via immersion heaters to provide domestic hot water and space heating via a seven-day thermal storage unit and underfloor heating. The heaters are switched automatically in series and parallel to provide a suitable load for the output voltage which varies between 60 and 220 V.

The major problem encountered[15] has been the poor output which has been only 50% of the rated output at the rated wind speed.

References

[1] Whittle, G.E., Bossanyi, E.A., Maclean, C., Dunn, P.D., Lipman, N.H. & Musgrove, P.T. (1980). 'A simulation model of an electricity generating system incorporating wind turbine plant'. *Third International Symposium on Wind Energy Systems, Copenhagen*, BHRA Fluid Engineering Cranfield, Bedford, UK.
and
Bossanyi, E.A. (1982). 'Wind and tidal integration into an electricity network'. *Cranfield Wind Energy Workshop*, Cranfield, Bedford, UK.

[2] Article in *Electrical Review*, 13 March 1981, Vol. 208, no. 10, p. 19. No title or author known.

[3] Rayment, R. (1976–77). 'Wind energy in the UK'. *Building Services Engineer*, 44, 63.

[4] Caton, P.G.F. (1976). 'Maps of hourly mean wind speed over the United Kingdom'. *Meteorological Office Climatological Memorandum, no. 79*.

[5] Rayment, R., see Ref. [3].

[6] Golding, E.W. (1976). *The Generation of Electricity by Wind Power*. Wiley, London.

[7] Anon. (1981). *US Popular Science*, January 1983.

[8] Watson, M. (1979). *Wind Turbines for the Provision of Domestic Heating*. Dissertation of the Department of Architecture, University of Cambridge; and (1983), 'Use of wind energy for space and water heating'. *Rural Power Sources, International Solar Energy Society*, May 1983, London.

[9] Bennett Woodcroft (1851). *Pneumatics of Hero of Alexandria*. Charles Whittingham, London.

[10] Marier, D. (1981). *Wind Power for the Homeowner*. Rodale Press, Emmaus, Pennsylvania, US.

[11] Talbot, J. (1981). 'Rechargeable cells'. *Electrical Review*, 208, 18.

[12] Cox, C.H. (1981). 'Making the utility connection'. *Solar Age*, December, p. 39.

[13] Information on Gemini synchronous inverters is available from Windworks, Box 329, Route 3, Mukwonago, Wisconsin, US and from Upton Ltd, East Bergholt, UK.

[14] Willoughby, J. & Todd, R. (1981). *Ambient Energy Utilization at the National Centre for Alternative Technology*. National Centre for Alternative Technology, Powys, UK.

[15] Todd, R. (1980). 'Energy and buildings at the Centre for Alternative Technology'. *Technical Information Report 1*, National Centre for Alternative Technology, Powys, UK.

[16] A very efficient mult-pole rare earth generator has been designed by Bolton and Ferraris at Imperial College, London, and built at the Rutherford Laboratory, Chilton, UK. It is driven at low rpm.

9

Water-supply systems

9.1 Introduction

Water-supply and waste-disposal systems are related in two ways. Firstly, in the UK, WC (toilet) flushing consumes about 36.5 l/p d or roughly one-third of the average household water use.[1] Secondly, the point of final waste disposal may also be a source of fresh water. About one-third of the public water supply is from rivers, yet 60% of the local-authority sewage discharges to these rivers do not comply with the recommended 20/30 standard.[2] (The Royal Commission on Sewage Disposal (1898–1915) made recommendations which were later adopted on the quality of sewage effluents to be discharged to water courses. The maximum concentration of suspended solids was limited at 30 mg/l and Biochemical Oxygen Demand (BOD) was limited to 20 mg/l for a dilution of 1.8 with river water; the Biochemical Oxygen Demand is a measure of the oxygen consumed during the oxidation (stabilization) of organic matter by a mixed microbial population and under aerobic conditions.)

In this chapter we shall briefly deal with water-supply systems and in the following one with waste disposal. Chapter 11 gives some water (and energy) conservation measures.

Virtually all homes in the UK are connected to mains water supplies and are likely to be so in the foreseeable future since planning permission is extremely difficult to obtain for sites that are not serviced. This tends to concentrate development and one of the interests of a group like the Autarkic Housing Project was to study potential changes in land-use patterns if housing was not constrained by availability of mains servicing systems.

In other countries, for example the US, private water supplies are often much more important. Even in the UK, occasions arise where alternative supplies must be considered – one of the present authors has acted as a consultant for a group setting up a holiday centre for inner-city youngsters on an off-shore island without any form of servicing.

The use of wells, infiltration galleries, springs and streams is covered in standard works on water supply and will not be discussed here. Instead we will outline the water-supply system we proposed for the Cambridge Autarkic House, which was based on rainwater collection, recycling and heat recovery on waste water.

Fig. 9.1. The Cambridge Autarkic House water-supply system.

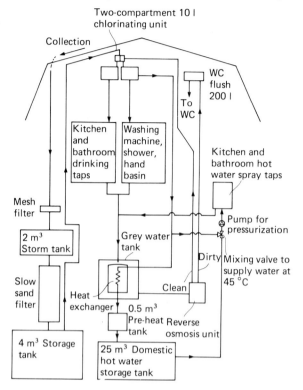

9.2 The Cambridge Autarkic House water-supply system

Fig. 9.1 illustrates the scheme. Water is collected from the roof and sloping walls of the house in a storm tank before transfer to a storage tank via a slow sand filter.

A previous calculation[3] had shown that in Cambridge (which is in a comparatively dry area of the UK) a storage tank of 25 000 l and a collection area of 145 m² would be required to guarantee a daily water supply of 100 l. This is considerably less than typical present water-consumption rates for dwellings and so extreme conservation measures would be required. Moreover, neither the required volume nor collection area was available and so recycling of water was necessary.

The slow sand filter used is a standard item which consists of a vertical cylinder containing fine sand at the top, coarser sand in the middle and gravel below. As the water trickles through the filter the turbidity should be reduced from 50 ppm (maximum permissible) to 5 ppm, the bacteria content reduced by 85–99% depending on the initial count and the colouration should become clearer.[3] Sufficient bacteria may remain to cause problems and so it is safer to further purify the water. With time, turbidity in the filter builds up and backwashing is required to clean it.

In this system, further purification is accomplished by then pumping the water to a 10 l chlorinating unit and two 100 l storage tanks in the roof. The chlorinating unit uses calcium hypochlorite tablets to disinfect

The Cambridge Autarkic House water-supply system

Table 9.1. *Roof areas and storage volumes for water production (101.7 l/d) with a solar still*[3]

Fraction of roof facing south and acting as a still	Size of storage reservoir m^3			
	5	10	15	25
	Roof areas m^2			
0	174	163	155	145
½	160	97	92	88
1	150	69	66	63

the water. Disinfection refers to the adequate destruction of waterborne pathogenic microorganisms including bacteria, viruses and protozoa.[4] Disinfection varies from sterilization in that the latter also destroys certain bacterial spores. One storage tank supplies the kitchen and bathroom drinking taps. The other serves the washing machine, shower and hand basin, as well as the hot-water-system feed.

The hot-water-system feed passes through a heat exchanger where it is warmed by discharged domestic hot water, then through the small preheat and large thermal storage tanks which are both solar heated. It is then pressurized and supplied to spray taps in the kitchen and bathrooms. After giving up its heat in the heat exchanger the waste water goes to a reverse osmosis unit. This unit eliminates suspended solids and removes almost all organic matter, bacteria and spores, as well as dissolved salts from, for example, soap. It is placed after the heat exchanger so that the entering temperature is not excessive. Reverse osmosis units use electrical energy to reverse the normal osmosis pattern and increase the concentration gradient to produce clean water from the grey water fed into them.

Clean water from the unit is pumped to the non-drinking-water storage tank in the attic and dirty water is pumped to a 200 l WC flush tank. Other 'autonomous' water systems are well described by Rump,[1] Smith[5] and the Vales.[6]

Use of a solar still was also considered for the Autarkic House.[3] An inexpensive but durable form of still might consist of a series of self-supporting polyurethane foam trays in a roof, each glazed with a single pane of glass at as low an angle as 6.5–10°. The joints must be well sealed, the inside of the still painted black on the bottom with, for example, a black-pigmented silicone paint, and the internal sides painted white. A shallow layer of water (9–10 mm) is maintained in the tray and when hot it evaporates, condensing on the cover glass and running down into a collecting gutter.

Table 9.1 shows the results of a computer program developed to estimate the roof area and storage volume needed to satisfy the previously cited daily water demand of 101.7 l. The table is based on a calculated daily output, from March to September for a poor year, of 1.6 l/m^2 still area. In all three cases the entire roof area is used to collect rain. Thus, the table states, for example, that if a storage reservoir of 10 m^3 is chosen, the roof area required, if one-half of the roof is a solar still (and all of it collects rain) is 97 m². In the event, the required collector area was not available and so recycling with the reverse osmosis unit was selected.

References

[1] Rump, M.E. (1978). *Potential Water Economy Measures in Dwellings: Their Feasibility and Economics.* BRE CP 65/78. Garston, Watford: BRE.

[2] Anon. (1971). 'Out of sight out of mind'. *Report of the Working Party on Disposal of Sludge in Liverpool Bay.* London: HMSO.

[3] Littler, J.G.F. (1975). 'Solar still and water supply calculation'. *Autarkic Housing Project Working Paper 20*, University of Cambridge, Department of Architecture.

[4] Leckie, J., Masters, G., Whitehouse, H. & Young, L. (1975). *Other Homes and Garbage.* San Francisco: Sierra Club.

[5] Smith, G.E. (1973). 'Economics of water collection and waste recycling'. *Autarkic Housing Project Working Paper 6.* University of Cambridge, Department of Architecture.

[6] Vale, R. & Vale, B. (1975). *The Autonomous House.* London: Thames and Hudson.

10

Waste disposal and utilization

10.1 Introduction

Surprisingly enough, several years ago one of the highlights in California Governor Jerry Brown's politics was the introduction of a law forbidding the sale of any lavatory which flushes more than 14 l of water. Waste disposal does not often receive such publicity although the problems it entails in both industrialized and non-industrialized countries, albeit for different reasons in the two groups, are impressive.

Here we shall deal with wastes in greater detail than we did with water systems but not because the central network is less extensive — in the UK 94% of all households are connected to mains sewers[1] (in the US, on the other hand, the comparable figure is about 67%[2]). Rather, it is because wastes can be a source of on-site energy if methane digesters are used. Although this tends to be less practical at the level of a single home, groupings of houses and other building types such as schools should not ignore the energy potential of the wastes they produce.

Mains servicing in the UK, as elsewhere, consists of a cistern-flush toilet connected to a network of underground sewers which transport sewage and domestic waste water to a treatment or disposal facility. While to those of us who use such systems almost nothing seems more natural, it is of interest to note that the concept of using storm sewers for human wastes is only about 140 years old and the first integrated system only came into full use about 1870 in London. The first sewers merely conveyed the wastes to bodies of water where they were discharged with often disastrous consequences — sadly this practice continues today in many areas. Present waste-treatment plants vary in size and operation but in the UK almost all incorporate a primary treatment which removes most of the suspended solid matter, such as paper, by means of sedimentation, settling or septic tanks; the sludge which accumulates is removed periodically and often treated — 40% is applied to agricultural land, 20% is tipped and 40% is dumped at sea.[3]

Secondary treatment involves bacterial digestion of the organic wastes and leaves a residue containing nitrogen and phosphorus. When this residue is discharged into lakes or streams it can result in eutrophication, with simpler organisms like algae growing too quickly to be eaten by the more slowly growing organisms higher up the food chain. Algal blooms remove the oxygen from the water, causing it to decay and become stagnant.

Additional, but expensive, tertiary treatment can reduce such dangers. In the UK, one-quarter of the rivers and one-half of the canals have been described as of 'poor', 'doubtful' or 'grossly polluted' quality[4, 5] (sewage is not, of course, the only source of pollution). There have also been indications that the common practice of dumping sludge at sea may have adverse effects on phytoplankton communities and even fisheries.[6] Water supplies might also be considered to be at risk since, as was mentioned in the previous chapter, one-third of the public water supply is from rivers.

The disposal techniques presently available (throughout the world) are conveniently divided into aerobic and anaerobic systems.

10.2 Aerobic systems

The common aerobic systems vary from the primitive, such as the pit latrine, to the fairly sophisticated, such as the Clivus Multrum. Kalbermatten *et al.*[7] provide a useful introduction to and bibliography for many of the techniques. Here we will consider only the Clivus Multrum since it would seem to be the aerobic method most likely to be socially acceptable in European areas of high population density.

The Clivus Multrum[7] (see Fig. 10.1) is an aerobic decomposition chamber designed, primarily, for single-family houses, to dispose of human excreta and organic kitchen wastes. The prototype was constructed in 1939. The device is divided by vertical partitions into three chambers which are connected at the bottom. A specially designed toilet bowl is situated above the excreta chamber and the garbage chute leads to the refuse chamber. The addition of kitchen wastes or other organic material permits the correct carbon–nitrogen ratio for composting (the device is in fact a continuous composting toilet) to be achieved. The decomposing material travels slowly down the container to the storage chamber. Aerobic decomposition (by the microorganisms in the wastes) is achieved by sucking air through the wastes from channels running along the container. A draught is created by the presence of a chimney which is sometimes equipped with a fan. Since the inside pressure is lower than the outside no odour can escape and in a properly operating Multrum it is claimed by the manufacturers that there are no odour control problems although this does not seem to be a claim universally accepted by users.[8]

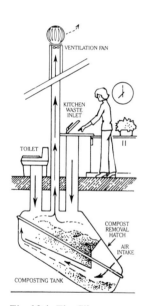

Fig. 10.1. The Clivus Multrum.[9] (Courtesy of Clivus Multrum USA, Inc.)

In particularly cold climates electric heaters may be needed to control the temperature so that decomposition, which is temperature dependent, proceeds. Care should also be given in the design of the installation to ensure that cold air is not drawn into the house.

The great advantages of the Clivus are that no water is needed and the volume of wastes is only 5–10% of the original volume, the rest being given off principally as non-noxious carbon dioxide and water vapour. The end product is a sludge amounting to approximately 10–40 l of humus-like soil per person per year; on average the sludge is removed every 15–24 months. A bacteriological analysis of the sludge revealed that no fecal coliform bacteria (*E. coli*) were present in the samples studied nor

Anaerobic systems

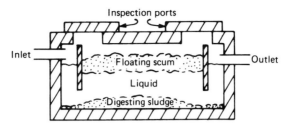

Fig. 10.2. Typical septic-tank design.[2]

were any species of pathogenic Clostridium found.[10] Only about 0.5% of the bacteria present in the sludge were pathogenic and these are known to be common in soils. The sludge contains nitrogen, phosphorus and potassium plus a number of minor plant nutrients and so will be of some value as a fertilizer and soil conditioner.

A major difficulty with the Clivus is that it does not provide a means of disposing of waste water from the sink, shower and clotheswasher, the so-called 'greywater'. Conventionally, this is done by discharging the water to a septic tank followed by a drainage field or seepage pit, but Clivus has proposed a number of simpler solutions such as leaching chambers with no pretreatment, based on the less harmful nature (lower BOD, virtual absence of pathogens if toilet wastes are voided to the Multrum) of greywater compared to typical sewerage.[11]

10.3 Anaerobic systems

Septic tanks are the most common anaerobic systems. Conventionally, the tank is a small closed chamber (or group of chambers) sited just below ground level, which receives both excreta and waste water (see Fig. 10.2). Solids settle to the bottom where they are digested anaerobically. Digestion is reasonably good, as evidenced by a reduction of approximately 50% in the BOD but enough sludge still accumulates so that the tanks must be desludged every one to five years.[7] The continuous effluent from the tank is low in suspended solids but has a high BOD; it is essentially untreated and is not suitable for direct discharge to a water course. Normally, the effluent is treated aerobically on a percolating filter, by irrigation on land (which may result in odour problems) or by subsurface irrigation (which may pollute water supplies).

Septic tanks do not normally make any special provision for the production or collection of the methane gas which results from anaerobic digestion. Methane digesters, on the other hand, are designed specifically with this in mind and so offer a potential source of energy for cooking, heating or lighting in the form of methane. They also deal with all of the wastes, giving a single final product, sludge, which can be used as a soil conditioner and fertilizer. The smell of digested sludge is a bit like that of tar and not at all similar to that of untreated sewage or septic-tank effluent.

Historically, digesters evolved from septic tanks and digestion has proven itself to be of value in waste disposal worldwide — which is not to say that problems have not been encountered. In the UK, anaerobic

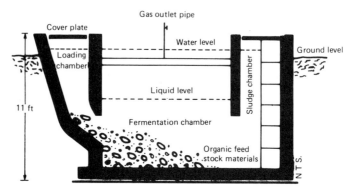

Fig. 10.3. Chinese methane digester.[17]

digestion is by far the most widely used single process for treating sewage sludge in municipal plants and its controlled use for this purpose dates back to the last century.[12] In these plants it normally suffices to raise the temperature of the sludge by 20 °C at all times of the year (very roughly, from 7—27 °C in winter and 15—35 °C in summer) for adequate performance.[13] The methane produced is often used to drive the plant's machinery.

In Europe, agricultural applications of digesters are exciting interest and it has been reported that at least a dozen systems are already operating successfully in Britain and France, with payback periods (at 1978 energy prices) of around six years.[14]

In Asia, numerous countries have developed small-scale methane digesters, or bio-gas plants as they are often called.[7, 15, 16] In the Chinese province of Szechwan, it was estimated in 1975 that 400 000 digesters were built or under construction and that 17 000 000 people used the gas produced for cooking and lighting.[17] Fig. 10.3 shows a typical 10 m^3 digester. As the gas is generated, it pushes down on the fermentation liquid which, to relieve the pressure, rises and flows through a small hole near the top of the sludge chamber into the area on top of the device. As the pressure drops, or as gas is drawn off through the outlet pipe, the liquid level recedes.

10.4 Methane digestion

Methane digestion is complex and only a brief introduction can be given here; numerous works,[8, 18, 19, 20, 21] from which much of the following discussion is drawn, treat the subject in greater detail.

Fig. 10.4 shows the basic process, Fig. 10.5 summarizes the biochemical reactions and Fig. 10.6 shows the sequential mechanisms schematically.

In the first stage of anaerobic waste treatment the 'acid-producing' bacteria, a heterogeneous group which includes facultative anaerobes, predominate and aerobes degrade the organic fraction of the raw waste to substrates which can be used by the methane-producing bacteria. The generation, or doubling times, of members of the acid-producing bacteria

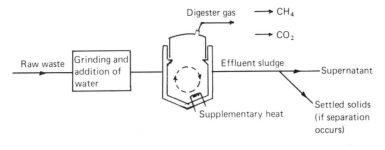

Fig. 10.4. The methane digestion process.

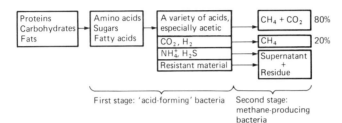

Fig. 10.5. Biochemistry of anaerobic waste treatment. (Courtesy of C. Freeman.)

are believed to be very much shorter than for the methanogenic group — minutes or hours as opposed to days. Thus, if a 'shock' loading occurs, the acid producers are able to respond with an increase in population size much more rapidly than the methane formers, with a consequent common result being an increase in the volatile acid concentration and possible inhibition.

The methane-producing bacteria are strictly anaerobic — even small amounts of oxygen are harmful to them. There are several different types, with each being characterized by its ability to convert a relatively small number of organic compounds into methane. Thus, for complete digestion, several different types are needed.

Biologically, successful digestion depends upon achieving and (for digesters which are loaded continuously) maintaining a balance between acid-forming and methane-producing bacteria. This can be accomplished by supplying raw wastes of a suitable composition and by proper maintenance of the pH and temperature.

The composition of the feed should have a carbon to nitrogen ratio (C/N) of about 30:1 for digestion to proceed at an optimum rate;[20] Freeman & Pyle[21] give a broader range of between 10 and 30 for 'good operation'. The 30:1 ratio is based on the digester bacteria using up carbon (for energy) 30 times faster than nitrogen (which in part is used for building cell structures). If there is too much carbon in the raw wastes, nitrogen will be used up first, carbon will be left over and the digester will slow down. If there is too much nitrogen, on the other hand, the carbon soon becomes exhausted and fermentation stops. The remaining nitrogen is lost as ammonia gas and the reduction in nitrogen content decreases the fertility of the effluent sludge. Table 10.1 shows the approximate carbon and nitrogen values of some wastes. (The figures must be used cautiously since chemical analyses do not necessarily represent the carbon and

Table 10.1. *Typical carbon–nitrogen ratios of some feeds*[21]

Material	N (% dry weight)	C–N ratio
Night soil	6	6–10
Cow manure	1.7	18
Chicken manure	6.3	7.3
Horse manure	2.3	25
Hay, grass	4	12
Hay, alfalfa	2.8	17
Seaweed	1.9	79
Oat straw	1.1	48
Wheat straw	0.5	150
Bagasse	0.3	150
Sawdust	0.1	200–500

Table 10.2. *Typical reported yields from anaerobic digesters*[21]

Material	Gas yield (m^3/kg volatile matter fed)	Gas composition (% methane)
Cow dung	0.1–0.3	65
Chicken manure	0.3	60
Pig manure	0.4–0.5	65–70
Farm wastes	0.3–0.4	60–70
Elephant grass	0.4–0.6	60
Chicken manure–paper pulp	0.4–0.5	60
Chicken manure–grass clippings	0.4	68
Sewage sludge	0.6	68

nitrogen available to the bacteria and also because the carbon and nitrogen contents can vary greatly with age and growing conditions of the plant, diet and age of the animal, etc.)

Gas production is often given as a function of the volatile solids in the wastes (see Table 10.2). (Volatile solids is that portion of a solid sample which volatilizes when the sample is heated to 600 °C; it is a measure of the organic portion of the sample since the ash remaining is the inorganic or mineral portion. Total solids is the weight proportion of solids in a sample; 10% total solids means that 100 g of a sample would yield 10 g dry solids on evaporation at around 100 °C.) Table 10.3 gives a detailed breakdown of the gas composition.

Another aspect of the feed is particle size. Feeds such as straw, newspaper and refuse may require grinding to, say, a diameter of 10–15 mm if they are to be pumped or piped.

Feed concentration (total solids percentage) is important because if too much digestible material is fed in, acid accumulation and inhibition can result. Experience with operating digesters has shown that a feed slurry containing 7–9% solids is optimum for digestion.[8]

The concentration will also affect the ease of mixing and pumping the

Table 10.3. *Representative composition of digester gas produced from farm wastes*[19]

Component		Percentage volume
Methane	(CH$_4$)	54–70
Carbon dioxide	(CO$_2$)	27–45
Nitrogen	(N$_2$)	0.5–3
Hydrogen	(H$_2$)	1–10
Carbon monoxide	(CO)	0.1
Oxygen	(O$_2$)	0.1
Hydrogen sulphide	(H$_2$S)	trace

digester material. Mixing is important because it promotes the interaction of the microorganisms with their substrate (the digester contents) and because it disperses any local pockets where acids might have concentrated. Large units often aim at continuous mixing but the machinery and energy consumption implied makes this less feasible for a smaller plant. Infrequent mixing (say once or twice a day) has proved adequate for many of the Indian village plants (where a typical problem seems to be controlling the scum which can form at the surface).

The term pH refers to the amount of acid or base present in solution (it is defined as the logarithm to the base ten of the reciprocal of the hydronium ion concentration expressed in moles per litre). McCarty[18] has stated that anaerobic treatment can proceed quite well with a pH varying from about 6.6–7.6 with an optimum range of about 7.0–7.2; Freeman & Pyle[21] give a wider operating range of 6.0–8.0. When starting off a digester, a neutral pH of 7.0 should be aimed for. The initial acid phase of digestion may cause the pH to drop but care must be taken that it does not fall below approximately 6.2 because at this point efficiency drops off readily and the acidic conditions can become quite toxic to the methane bacteria; this control can be achieved either by decreasing the waste feed to the digester or by adding neutralizing materials such as lime, or by both. As digestion proceeds the pH rises (see Fig. 10.6) until it reaches a point when it is well buffered and the digester contents can stabilize themselves even when large amounts of acid or alkali are added. At this point, raw feed may be added periodically and a constant production of gas and sludge achieved.

Temperature affects both the rate and course of digestion; the reactions themselves are slightly exothermic overall. Two optimum temperature levels for anaerobic treatment have been reported. The first is in the mesophilic range from 29 to 38 °C and the second in the thermophilic range from 49 to 57 °C. Although treatment proceeds much more rapidly in the thermophilic region (with a consequently shorter detention time, that is, the average period of retention of sludge in the digester), the additional energy requirement needed to maintain the temperature tends to limit such applications to municipal digesters. At below 15 °C digestion proceeds very slowly although the process can function at a temperature as

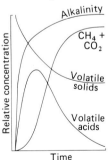

Fig. 10.6. Sequential mechanism of digestion. (Courtesy of C. Freeman.)

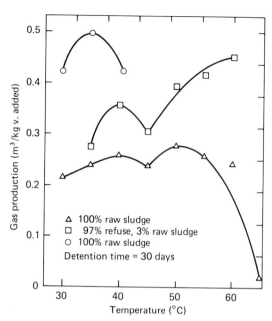

Fig. 10.7. Effect of temperature on gas production.[22]

low as 5 °C.[21] Maintenance of a constant temperature is important for optimum gas production. Fig. 10.7 shows the effect of temperature on gas production using selected feeds.

The problem that arises at this point is that almost all of what is known scientifically about methane digestion is due to studies of large-scale processes. Domestic-scale digesters with mixed feeds supplied at irregular intervals have received very little attention and indeed the number of such devices operating in western countries is probably very small indeed. Consequently, we can only supply estimates based on our work and the work of C. Freeman for the Autarkic Housing Project.

Fig. 10.8 shows a design (developed in conjunction with Farm Gas Ltd) for a domestic-scale digester. The annual feed to such a digester in the context of a four-person home and garden might consist of human wastes, kitchen wastes, paper and garden wastes.

A digester which has only human wastes available in limited amounts is not practical. The average weights of faeces and urine produced by a mixed population are 0.086 and 1.055 kg/(d p)[24] (wet solids basis) and it was estimated that this would give 85 kg/y dry weight and might produce 40 m^3 of digester gas per year.[24, 25] If, as a conservative estimate, we assume that only 50% of this is methane and using a calorific value of 37.4 MJ/m^3 for pure methane, we have about 0.8 GJ. Since four-person families using conventional cookers (cooking being the most likely household application for methane) annually use between 4.3 and 6.8 GJ (see also Chapter 11), it is evident that other sources of waste are required if methane is to contribute significantly to the energy demand of a house.

It was estimated that kitchen wastes and paper might contribute 310 kg dry weight and supply 2.2 GJ/y.[25, 26, 27] Vegetable garden wastes from a

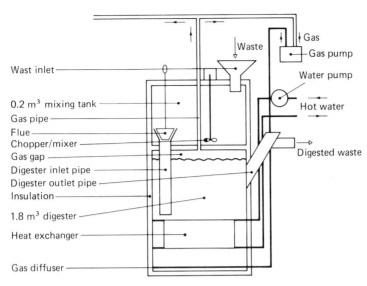

Fig. 10.8. Schematic of an anaerobic digester for the Autarkic House.[23]

400 m² area were calculated to be about 90 kg dry weight and might provide 0.7 GJ/y.[25, 27, 28, 29] Thus the total energy supply would be approximately 3.7 GJ/y and would be capable of contributing significantly to the energy requirement for cooking. (From this, one would have to subtract any energy needed to maintain the digester at a working temperature but this will depend on such factors as design, location of the digester in the house and climate.)

It must be emphasized, however, that work on domestic-scale digesters is only in its infancy — both theory and practice need development. To close on a note of caution: at a concentration of 5–14%, methane is flammable and so poses a serious safety problem in the home. (To keep this in perspective, though, North Sea gas currently supplying many UK homes is 95% methane.) Consequently, burying a gas store in the garden may be a reasonable safety precaution. Similarly, the danger of explosive mixtures of gas and air also means that digestion tanks and the gas-collection systems should always be kept under a positive methane pressure.

References

[1] Anon. (1970). 'Taken for granted'. *Report of the Working Party on Sewage Disposal, Ministry of Housing and Local Government.* London: HMSO.
[2] Barley, J. & Wallman, H. (1971). 'A survey of household waste treatment systems'. *Journal Water Pollution Control Federation*, December, pp. 2349–60.
[3] Coker, E. & Davis, R. (1978). 'Sewage sludge — waste or agricultural asset?' *New Scientist*, 78 (1101), 298–300.
[4] Anon. (1970). *Report of River Pollution Survey of England and Wales.* Department of the Environment, London: HMSO.

[5] Anon. (1972). *Report of River Pollution Survey of England and Wales*. Department of the Environment, London: HMSO.
[6] Anon. (1971). 'Out of sight out of mind'. *Report of the Working Party on Disposal of Sludge in Liverpool Bay*. London: HMSO.
[7] Kalbermatten, J.M., De Anne, S.J. & Gunnerson, C.G. (1980). *Appropriate Technology for Water Supply and Sanitation*. Washington, DC: The World Bank, p. 18.
[8] Leckie, J., Masters, G., Whitehouse, H. & Young, L. (1975). *Other Homes and Garbage*. San Francisco: Sierra Club, p. 215.
[9] Anon. (Undated). *Clivus (Multrum)*. Tyreso, Sweden: A. B. Clivus.
[10] Anon. (1977). *Bacteriological Studies of Clivus Multrum End-Products*. Cambridge, Mass: Clivus Multrum USA, Inc.
[11] Fagel, M. & Lindstrom, C. (1976). *The Treatment of Household Water in Homes Equipped with the Clivus Multrum Organic Waste Treatment System*. Cambridge, Mass: Clivus Multrum USA, Inc.
[12] Anon. (1974). 'Anaerobic treatment processes and methane production'. *Notes on Water Pollution No. 64*. Department of the Environment. London: HMSO.
[13] Escritt, L.B. (1971). *Sewers and Sewage Works*. London: George Allen and Unwin.
[14] Martin, D. (ed.) (1980). *Specification 80*. London: Architectural Press.
[15] Prasad, C.R., Krishna Prasad, K. & Reddy, A.K.N. (undated). *Bio-Gas Plants – Prospects, Problems and Tasks*. Bangalore: Indian Institute of Science; New Delhi: Management Development Institute.
[16] Subramanian, S.K. (1977). *Bio-Gas Systems in Asia*. New Delhi: Management Development Institute.
[17] Smil, V. (1977). 'Energy solution in China'. *Environment*, 19 (7), 27–31.
[18] McCarty, P.L. (1964). 'Anaerobic waste treatment fundamentals'. *Public Works*, September, pp. 107–12; October, pp. 123–6; November, pp. 91–4; December, pp. 95–9.
[19] Anon. (1973). *Methane Digesters for Fuel Gas and Fertilizer*. Woods Hole, Massachusetts: New Alchemy Institute.
[20] Stafford, D.A. (1974). 'Methane production from waste'. *Effluent and Water Treatment Journal*, February, pp. 73–9.
[21] Freeman, C. & Pyle, L. (1977). *Methane Generation by Anaerobic Fermentation – an Annotated Bibliography*. London: Intermediate Technology.
[22] Diaz, L.F., Kurz, F. & Trezek, G.J. (undated). *Methane Gas Production as Part of a Refuse Recycling System*. University of California, Berkeley. Department of Mechanical Engineering.
[23] Cheshire, M.J., Farm Gas Ltd. (1977). Private communication.
[24] Snell, R.J. (1943). 'Anaerobic digestion III. Anaerobic digestion of undiluted human excreta'. *Sewage Works Journal*, 15 (4), 679–701.
[25] Imhoff, K.M., Muller, W.J. & Thistlethwaite, D.K.B. (1972). *Disposal of Sewage and Other Water Borne Wastes*. London: Butterworths.
[26] Skitt, J. (1972). *Disposal of Refuse and Other Wastes*. London: Charles Knight.

References

[27] Klein, S.A. (1972). 'Anaerobic digestion of solid wastes'. *Compost Science*, January–February, pp. 6–11.

[28] Thomas, R.B. (1980). 'Biological aspects of an Autarkic House'. Ph.D. Thesis, University of Cambridge.

[29] Rosenberg, G. (1952). 'Methane production from farm wastes as a source of tractor fuel'. *Agriculture*, 58, 487–94.

11

Domestic-energy saving

11.1 Introduction

Results from the recent Better Insulated Housing Programme[1] are summarized in Table 11.1 which shows the fractions of energy consumed on-site for various purposes.

The experimental houses at Bebington (Fig. 4.48), which are electrically heated, indicate that only about half the electricity supplied annually is used directly for space heating.

Much of the energy provided to the lights, fridge and so on, contributes usefully during the heating season to warming the space. Capper[2] suggests the fractions shown in Table 11.2.

One might wish to argue a little with the low figures (Leach[3] assumes 0.8 in all cases and Siviour's conclusions are shown in Table 3.1) but the fact remains that a substantial fraction of the heating in houses arises from electricity (or gas) used in appliances. Since this energy *is* wholly or partly electrical (generated with an efficiency of roughly 30%), the cost to the country in terms of primary fuel, or to the consumer in terms of cost, constitutes a large part of the annual fuel bill for each house. Thus, one would expect to find regulations concerning the efficiency of appliances. There are almost none in the UK, although in the US some goods must be marked with an efficiency indicator.

In the Bo'ness study,[1] nearly all households have a clothes washer, one-third a tumble drier, all have a fridge and 40% have a freezer. Most have a colour television. Two-thirds of the households have electric fires, one-third have calor gas and one-fifth have paraffin heaters as subsidiary heating devices.

11.2 Domestic hot water

Typically, of the water used in houses, 25% is fed directly from the mains (that is, cold water for drinking, cooking, etc.), 40% is cold water from the storage tank and 35% is hot water. A reduction in demand is highly desirable nationally, for the water industry is extremely capital intensive (costing, even in 1967, £50 000 to create one job compared with a national average of £4000). Table 11.3 indicates how warm water is consumed.

The energy required to produce 80 m^3 of water at 45 °C from a mains

Table 11.1. *On-site energy consumption*

	(%)
Space heating	50
Appliances	20
Immersion heater (DHW)	20
Cooking	7
Lighting	3

Table 11.2. *Fraction of energy which contributes to space heating*

Lighting	1.0
Water heating	0.5
Cooking	0.4
Appliances	0.4

Table 11.3. *Use of warm water in households*

Use	a (l)	b	c (m^3/y)
Washing hands	5	14	4
Washing hair	20	4	4
Washing face and hands	10	14	7
Bath	110	8	44
Shower	40	—	—
Dishes	15	14	11
Clothes washer	50–300	2	10
Total			80

[a] Ref. [4].
[b] Estimated number of uses for a family of four per week.
[c] Projected annual consumption for a family of four.

temperature of 10 °C is about 12 GJ, which corresponds well with the mean use of the immersion heater in the Bo'ness houses (12 GJ/y).[1]

Clearly, the principal activities for which economies may be made are bathing and washing clothes.

Front-loading washing machines and cold-water washing powders are mentioned below. Showers use less than half as much water as baths. Part of the high load ascribed to washing hands arises from the dead-legs between the basin and the tank (which typically hold 1–2l). Insulating such pipes only marginally affects the problem, since the water is likely to cool off between each draw off. Smaller pipes help but may lead to blockage with deposits. Careful design reduces the pipe run, of course. The care lavished on water economy in the Minimum Energy Dwelling in Cali-

Table 11.4. *Major domestic electrical consumers other than cookers and immersion heaters*

Appliance	Typical rating in UK (W)	UK annual consumption (GJ/y)	US annual consumption (GJ/y)[b]
Lighting	–	1.8[a] 0.9[d]	7.2
Fridge	100	2.7[a,d]	2.7
Freezer	300	1.8[a] 3.6[d]	4.3
Tumble drier	2000	1.4[c]	3.6
Clothes washer	2500	0.7[d]	0.2
Dish washer	3000	3.1[d] 1.3[a]	1.2
Kettle	3000	0.9[c]	
Colour TV	300	0.9[a] 1.8[d]	1.8
Central-heating pump	80	0.5[c]	–
Overblanket	100	0.4[c]	–
Toaster	1200	0.2[c]	–

[a] Data calculated from NEDDO Report on Energy HMSO1974 and the Better Insulated Housing Programme.[1]
[b] *New York Public Service Commission Report on Appliance Efficiency*, 1973, published by the New York State Energy Commission.
[c] Estimated.
[d] Leach.[3]

fornia,[5] must represent the ultimate in energy reduction. This house features the 'Ultraflow' system in which push-button control at each point of use activates a solenoid valve and mixer at the hot-water tank, allowing delivery through *one* flexible pipe of small diameter. There are savings on installation and energy costs.

11.3 Electrical power

It commonly seems to be thought that a large array of electrical gadgets produces a great consumption of electricity but, since the power of such devices and the time for which they are switched on is low, the energy used may be small.

The items which consume by far the largest amounts of energy are shown in Table 11.4.

Manufacturers are responding only very slowly to the implications of these figures. The most obvious economies which are available now concern colour televisions (whose power rating has come down rapidly in the last few years with solid-state techniques) and lights (where fluorescent

Cooking

tubes are now commonplace). There are also proximity switches (at present for offices) which detect people and switch off lights in empty rooms. It seems quite possible that the consumption for lights and TV could be halved. The same is not so true for the other appliances.

Energy use in clothes washers is considerably lower for front-loaders (which use less water), and great economies and added convenience can be achieved using ultra high-speed spin-drying cycles (of at least 1200 rpm). Clearly, the water spun out of clothes does not need to be removed by heating in a tumble drier, or by evaporation into the house from clothes strewn about radiators, thus reducing the possibilities of condensation. The weight of clothes to be carted out to a clothes line is also ameliorated!

Economies can be made by insisting that the dish washer or clothes washer accepts hot water from the hot storage tank without mixing in cold water unless necessary (that is, avoiding unnecessary use of the water heater in the appliance). Clearly, too, cold-water washing cycles and powders, which are common in the US, should reach the UK ultimately. The only other obvious areas to attack concern food preservation. Fortunately, 'well-aired pantries' (adding considerably to the heat load of a house in winter) are less common now. But fridges and freezers all too frequently have skin condensing coils, which seems insane unless the coils are mightily well insulated from the cooled space. The insulation, in practice, is poor, door seals quickly wear out and magnetic strips cease to close, all adding unnecessarily to the electrical load.

In the absence of a rating system, consumers should perhaps consult the *Which Guides* which have dealt with appliance efficiencies. Obviously, front-opening freezers should be avoided, and chest freezers kept full. Market penetration by freezers seems likely to be about 40% by the year 2000[3] and thus, both in domestic and national terms, the energy consumed is considerable. Even chest freezers are poorly insulated and, provided they do not have skin coils and if one can seal on added insulation in such a way that interstitial condensation does not occur, energy may be saved.

11.4 Cooking

Table 11.5 indicates the consumption by an electric cooker and hob. Gas consumption would be higher by at least a factor of 1.3.

Obviously, microwave ovens can save energy but the sacrifice in quality may be too great. Grilling relies on a fierce heat and free movement of air to carry away water vapour. It is difficult to enclose a grill. At present, ovens must merely be insulated to ensure that the user does not burn himself by touching the casing but, clearly, more could be done to reduce heat loss. Gas ovens, in which the burning is inside the oven, require ventilation whose cooling effect must be counteracted by using more gas. Induction heating certainly directs heat very positively into the pans on a hob but although induction hobs have been made, they remain very expensive.

Leach suggests that cooking efficiency will rise considerably (specific

Table 11.5. *Energy consumption for cooking per household*

Function	Rated power (W)	UK annual consumption (GJ/y)	US annual consumption (GJ/y)
Oven	3000		
Grill	2500	4.3^a–5.4^b	5.9^c
Hob	1000–2000/ring		

[a] Bo'ness Better Insulated Housing Study, Ref. [1].
[b] NEDDO, *Report on Energy*, HMSO, 1974.
[c] *New York Environment Research Centre Energy Index*, 1974.

fuel index: 1975 = 1, 2010 = 0.5). He also suggests that the amount of cooking will fall (1975 index = 1, 2010 index = 0.86).

In 1974, about 55% of dwellings used gas for cooking, with the production of a good deal of water vapour (about 1 kg per 0.04 GJ). The vapour is in addition to that arising from the cooking itself (about 0.05 kg/h).[6] Since much of this water vapour, plus that from showers, etc. (also about 0.05 kg/h) and people, must be removed by ventilation, it is important to reduce the amount of water vapour created in order to economize on the ventilation heat loss.

11.5 Space heating

Comfort levels are discussed in Chapter 3. Very often, people saddled with electric heating in poorly insulated houses use portable devices such as paraffin or liquified gas heaters. Inevitably, there are then complaints about condensation, since each kilogramme of fuel burned produces slightly more than 1 kg of water vapour. To remove the vapour, windows must be opened! (Dew-point calculations are addressed in Appendix 3.)

In highly insulated houses where air-change rates are likely to be low, it is highly undesirable to use such portable heaters. It may be worth pointing out that an electric overblanket rated at roughly 100 W consumes very little energy, and retains comfort levels in bed, whilst allowing room temperatures to sink to, say, 16 °C. Naturally, the room temperature ought to be kept above the dew point at room surfaces, with the possible exception of the window.

11.6 Transportation

The growth of energy use in transport stems mainly from the increasing use of cars in the UK. (Currently the ratios are roughly as shown in Table 11.6.) The average car consumes about 45 GJ of fuel per year. There are about 15 million cars compared with 20 million households. Thus the total energy consumption for all cars is about 7×10^8 GJ, whilst that for space heating in all households is about 10×10^8 GJ (total energy use in houses being about 15×10^8 GJ.

Table 11.6. *Ratios of passenger km (after Leach[3])*

Bicycle	1
Motorcycle	2
Rail passenger	100
Bus and coach passenger	150
Car passenger	1000

Clearly, the incentive to save fuel for a household is as strong in transport as in space heating and the methods are well known and excellently summarized by Leach.[3] They obviously include selecting a car with a good km/l rating for the type of travel the household commonly enjoys, the use of steel-core radial tyres at the correct pressure, proper maintenance and gentle driving. Simple precautions such as these, following the choice of a car, may reduce consumption by anything up to 50%!

References

[1] Capper, G. (1981). *Interim Report Number Three, Better Insulated Housing Programme.* Building Department, Heriot Watt University, Edinburgh, UK.
[2] Capper, G., ibid.
[3] Leach, G. (1979). *A Low Energy Strategy for the UK.* International Institute for Environment and Development, Percy Street, London.
[4] Littler, J.G.F. (1975). 'Solar still and water supply computation'. *Working Paper 20*, Martin Centre for Architectural and Urban Studies, Cambridge, UK.
[5] Burt Hill Associates (1977). *Minimum Energy Dwelling.* Burt Hill Associates, Butler, Pennsylvania, US.
[6] Burberry, P. (1975). *Environment and Services.* Mitchell's Building Construction. London, Batsford.

12

Housing case studies

12.1 Introduction

This chapter takes the reader through some of the design processes which occurred during the planning of several low-energy houses. The Peterborough Houses are complex, having large air-heating solar collectors, a sophisticated ductwork system and microprocessor control. The Newnham Houses are more conventionally heated, but highly insulated. These new dwellings are discussed in Section 12.2, and in 12.3 reference is made to the rehabilitation of existing houses, including one with a passive roof-space solar collector.

12.2 New houses – three solar air-heated houses in Peterborough

The authors won a competition at the 1979 National Energy Show, for a terraced low-energy house, designed with Lucy Krall, shown in Fig. 12.1. Peterborough Development Corporation became interested after seeing the model on television and suggested that the same sort of principles could be applied to one of their houses.

In the end, a terrace of three was built, as shown in Fig. 12.2. Because a lot of domestic hot water was likely to be produced by large collectors, we felt that six-person houses should be built. Larger units also yielded a greater roof area than would smaller terraced houses. In fact, in order to provide a reasonably large area (32 m² gross), collectors were also placed on the first-storey walls.

In retrospect, this idea of dual slope collectors was probably influenced by the Autarkic House (Fig. 12.3) in which the architectural design forced the provision of collectors on three azimuths and two tilts. The whole of the southerly facade in the Peterborough Houses is glazed, since the ground floor consists of a conservatory opening into the living room. This idea of wholly glazing a facade was turned into an extremely elegant system by the Energy Design Group, who won an award for their rehabilitation scheme at the First European Passive Solar Design Competition. The solar skin is illustrated in Fig. 12.4.

The severe problems faced by the architects at Peterborough were to incorporate the proposed heating system into a workable house. This is rather the converse of the normal procedure! It led to great compromises on all sides.

Three solar air-heated houses in Peterborough 311

Fig. 12.1. The initial design in 1979 for the Peterborough houses.

Fig. 12.2. The homes at Peterborough in 1983.

Fig. 12.5 illustrates the site plan, and indicates a major problem of siting. Overshadowing was avoided by placing the three houses on the perimeter of the development but at the cost of an orientation 40° east of south.

It remains to be seen how drastic this poor orientation will be. Of course, with the shallow roof pitch (30°) and the prevalence of diffuse radiation in the UK, the harm may, comparatively, be less than in a region with more beam radiation.

Inside the house, the architects solved the problem of providing lobbies to exterior doors, as shown in Fig. 12.6. The 4 m³ heat store was placed on a reinforced slab, forming part of a core in the centre of the house, containing all the dampers, etc. needed to direct the air stream past, in or out,

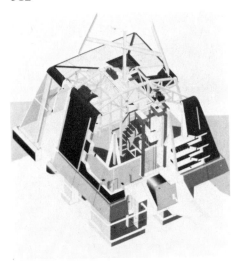

Fig. 12.3. The Cambridge Autarkic House (architect: Alex Pike; technical design team: D. Forrest, J. Littler and R. Thomas, 1977).

of the store. The slab is insulated throughout and thickened under the store as shown in Fig. 12.7. The volume was chosen to provide only one day's storage — larger stores would scarcely ever be completely filled during the heating season.

Building the store itself constituted a problem, since the house is timber framed and it was undesirable to bring in a wet trade just to build a box for a pile of loose rocks. Thus, lateral pressure was avoided by using stacked bricks. Calculation quickly shows that pressure drops are likely to be small at sensible flow rates across stores of this size (1 m × 2 m × 2 m) using bricks or even 50 mm stones. Nor is the rate of heat transfer a problem (see Chapter 4). The store will inevitably be stratified. The difficulty arises in ensuring an *even* air flow through the pile.

On the skin of the house the architects had to incorporate windows between collector panels to the fully glazed first floor, and to cope with extensive thermal movement of the collector panels. The section in Fig. 12.8 shows that panels stretch from the ridge to eaves (about 5 m) and under conditions of control failure, with no air flowing through the panels, they will reach at least 150 °C. Even under low flow conditions, temperatures above 130 °C have been observed (June 1983). Fig. 12.8 indicates the solution adopted to cope with extensions up to 20 mm.

The architects encountered major problems in incorporating the ductwork (which is very large, for example bottom feeder 300 mm × 250 mm × 6000 mm), the collector panels and glazing, and the heat store. Compromise stretched only so far on the structure — in fact, to double-glazed windows, slab insulation and the normal standard of wall insulation. This normal standard at Peterborough (nominally 75 mm of glass wool) is fairly high, corresponding to a wall U-value of about 0.4 W/(m² K) but, as an advisor, one would have sought a value of, say, 0.3 W/(m² K). However, efforts were made to incorporate an internal vapour barrier to reduce infiltration; but it was clear during construction that many procedures had to be rethought and design changes made in order that the barrier be easily

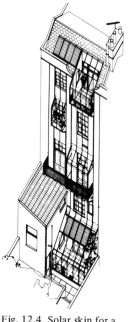

Fig. 12.4. Solar skin for a Georgian terrace in Bath, by Peter Clegg and Richard Feilden (1980).

Three solar air-heated houses in Peterborough

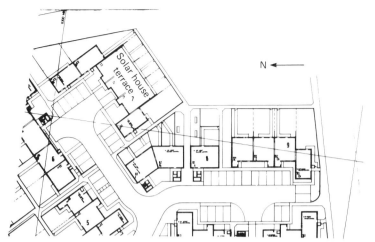

Fig. 12.5. Site plan of the Peterborough Houses. (Figures 12.5–9 by Peterborough Development Corporation, Chief Architect K. Maplestone. Design and drawings by M. Rodgers, R. French and A. Newman.)

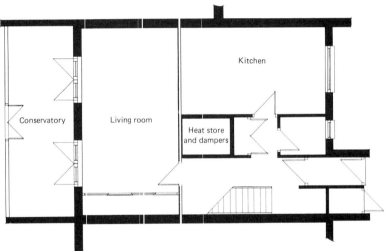

Fig. 12.6. Ground floor of the Peterborough Houses.

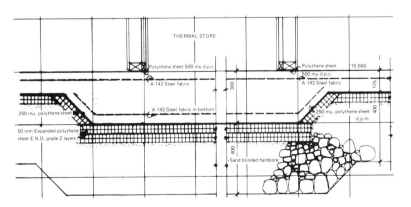

Fig. 12.7. Slab thickened and insulated under the heat store.

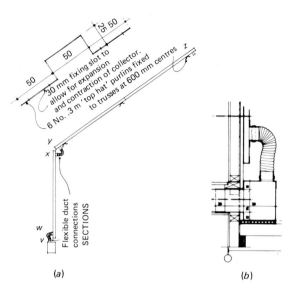

Fig. 12.8. (a) Uniform size of collector panels (2.4 m) and their linkage by flexible ducts. (b) Flexible link to bottom feeder duct at the level of the conservatory roof.

installed and remain without breaks. The book[1] concerning such design, to be produced for the International Energy Agency, should help in this regard.

The major problem for the Peterborough engineers was coping with a totally alien method of heating or preheating the air. The collectors may overheat and require fail-safe venting. The hot water is heated by a transfer coil in the air stream (see Fig. 5.2) but this sort of transfer from air to water, with a low pressure drop in the air stream and a strange set of constraints, often takes one off the end of a manufacturer's performance curve. The three-way dampers (Fig. 5.2) had to be specially made, although complete units with a fan and both dampers are available off-the-shelf in the US. The desirable rate of air flow via the collectors (for high collection efficiency) is about three times that needed for heating the house. Thus, a two-fan system was adopted.

Finally, the design called for up to ten modes of operation (see Table 5.1). Such control complexity requires a microprocessor and it remains to be seen how successfully the control scheme works. One of the main advantages of a solar air-heating system is that the fluid heated in the collectors subsequently heats the house, without the intervention of heat exchangers. Since the solar heated air may require further warming by a gas coil, the system is necessarily of the 'parallel' type. This is by distinction from a system in which solar heated water is further heated by gas and then passed into radiators, in a 'series' heating arrangement. From the point of view of collector efficiency, the latter is highly undesirable.

A second advantage which the designers saw for the system was the potential use of very low-grade heat. Cold makeup air to the house can be drawn in via the collectors (Mode 6B, Table 5.1), and even on fairly dull days can be warmed before passing over the auxiliary heater. In this way the house is pressurized against infiltration by cold air. It remains to be

Energy-efficient houses in Newnham, Cambridge 315

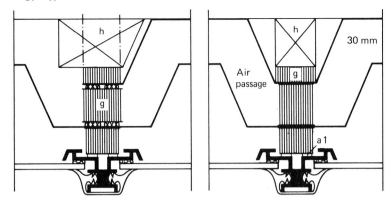

Fig. 12.9. Section through single-glazed collector. Standard profiled steel sheets.

seen whether this mode is useful. In practice, solar air systems tend to leak badly. An example of a Peterborough panel was pressure tested[2] and found to be almost gas tight at 100 Pa. The original design called for an optimum width of 30 mm between the two layers of metal (the front being painted black and exposed to the sun, and the back being sealed to insulation) and a fluted pattern (Fig. 12.9). The fluted shape achieved a higher surface area for heat transfer but proved too expensive for implementation.

Computer simulation using 'f-chart' (see Chapter 5) suggests that the active system should provide 30–40% of the space and hot-water heating for the year. Monitoring is underway to confirm or refute this calculation. However, a further boost to the solar fraction will arise from the single-glazed conservatory linked to the living room by two pairs of double-glazed doors. The doors will be opened during the heating season when the conservatory temperature exceeds that in the living room. Fans may subsequently be fitted if necessary. The contribution from the conservatory is likely to be about 25% of the space heating load.

12.3 New houses – energy-efficient houses in Newnham, Cambridge

For two houses in Newnham, Cambridge, one problem the architects faced was how to take advantage of passive solar gain on a restricted site where the surrounding homes are orientated along a street running approximately north-east–south-west. The solution adopted was to abandon the traditional rectangular shape by cutting off the south-facing corner of the rectangle, thus giving a south-facing wall (see Fig. 6.33). The approach introduced some complexity in space planning and construction but this was more than outweighed by the spacious and visually attractive interior achieved. Glazing was concentrated on the south, south-east and south-west facades (see Fig. 12.10) and kept to a minimum elsewhere. In addition to the compact form and the south-facing facade for passive solar gain, energy conservation was achieved in a number of other ways including high insulation levels and careful selection of the heating system (see

Fig. 12.10. View from the south-east. (Architects: Lyster, Grillet and Harding.)

Fig. 12.11. Fixing of expanded polystyrene insulation to the inner wall.

Chapter 6). Provision was also made for incorporating future developments in energy conservation and the use of ambient energy.

Perhaps the most interesting aspect of the houses was that this was all achieved by constant interaction between clients and architects and by one of the present authors acting as energy consultant. For example, one suggestion which was adopted was to alter the roof space to reduce the heated volume and allow for the installation of an air-to-air heat-recovery device at a later date. On the other hand, closing off the staircase to separate the house into zones of varying temperature was not accepted for reasons of aesthetics and convenience. But what was important was that the occupiers of the building had participated in its thermal design and so were much more likely to use it properly and economically.

The wall construction from inside out was 12 mm plasterboard, 100 mm lightweight concrete block, 100 mm cavity with 50 mm of expanded polystyrene on the inner half, and 100 mm brick, giving a U-value of 0.39 W/(m² K). Fig. 12.11 shows the expanded polystyrene boards being fixed to the concrete block with wall ties.

Renovation of a farmhouse to include a sun space 317

Roof construction was basically tiles on battens above 150 mm of glass fibre insulation with 12 mm plasterboard, giving a U-value of 0.22 W/(m² K). The floor consisted of 22 mm floorboards over 80 mm of glass fibre insulation suspended from the floor joists, giving a U-value of 0.27 W/(m² K). For reasons of reduced heat loss and increased thermal comfort it was decided to use multiple glazing. Triple glazing (with two vacuum-sealed gaps) was selected (see Fig. 12.10) rather than double glazing because of the better thermal performance and also because the higher quality of the windows (wood, finish and ironmongery) was judged to outweigh the disadvantage of the somewhat higher cost.

It was anticipated that high insulation levels, multiple glazing, provision of a lobby and attention to quality of workmanship on-site would produce a well-sealed house with considerably reduced ventilation heat losses. To guard against the possibility of condensation on internal surfaces, extract fans were fitted in the utility room, kitchen and bathrooms. The estimated on-site energy consumption for space heating is about 300 and 400 MJ/(m² y) for both space heating and domestic hot water, which compares favourably with an extremely well-insulated house of the near future (see Chapter 1).

In addition to allowing for future solar preheating of the inlet air and incorporation of a small air-to-air heat exchanger, as mentioned in Chapter 6, the following measures were planned for:
(1) increased wall insulation by filling the cavity;
(2) use of approximately 2 m³ in the utility room as a thermal store;
(3) incorporation of a unit to recover heat from waste water by grouping services in one area of the house.

12.4 Rehabilitated houses — renovation of a farmhouse to include a sun space (contributed by Peter Clegg)

In 1978, work began on the subdivision into four dwellings and renovation of a farmhouse and outbuildings in the lower Cotswolds. One wing of the main farmhouse faced due south and offered opportunities for the use of passive solar design (see Fig. 12.12). About one-third of the house was completely rebuilt using reclaimed natural materials and following the pattern of vernacular details. This was done primarily as an exercise using traditional construction but built to a high insulation standard. New roofs contain 150 mm glass fibre, and the walls consist of 200 mm of rubble stonework built up against 50 mm of 'Styrofoam' with an inner skin of dense concrete blocks which also contain foam insulation. Internal thermal mass of the building was therefore kept relatively high.

One of the other precepts of vernacular design was that window openings were kept relatively small, and roofs had low eaves, with dormers to illuminate the upper floors. Taking into account these restrictions, and the owner's interest in plants, the obvious way to make a better solar contribution to the building was to construct a conservatory.

A bronze anodized aluminium glazing system was used, in order to match the colouring of the traditional material, to reduce maintenance

Fig. 12.12. The converted farmhouse (Feilden Clegg Design, Bath).

costs and for the provision of a light glazing bar. The system, manufactured under the name 'Florada' is of US origin and uses a curved bar with polycarbonate glazing under the eaves to simplify construction and reduce the structure. Using typical sections and lengths of bar from different free-standing greenhouse models, a 13 m long structure was built; some of it forms an entrance porch to the house, another section is used as a greenhouse proper and a third section is intended as additional living area.

The glazing bars are tilted over at an angle, primarily for aesthetic reasons, but also to admit more direct sunlight in winter. The foot of the glazing bar rests on the lip of a glazed clay drainage channel to collect rainwater, on top of a 750 mm high insulated retaining wall (the building is partially underground).

It was intended that excess heat should be vented into the house via windows, doors and clay pipes which form openings through the wall. A return-air system was envisaged, and a duct was constructed to collect air from the ridge of the house and discharge it under a suspended timber floor over part of the ground floor of the building. The mechanical part of this system has never been fully completed and is unlikely to be, in view of the fact that dissipation of heat into the house seems to work satisfactorily by simply opening doors and windows. A considerable amount of heat is also stored in the masonry walls on the outside of the house, which helps to keep night-time temperatures in the greenhouse above freezing for the plants inside.

Typical of many houses constructed by architect–clients the project is part of the continuing experiment and never seems to get finished! A num-

House conversion incorporating roof-space collector 319

ber of problems have been found with the greenhouse system in that, whilst it is basically relatively cheap and simple it does not lend itself particularly well to being tilted away from the vertical on the south face and, despite the avoidance of laps in the glazing details, the structure leaks air and even driving rain, particularly around the roof vents. The thermal performance of the conservatory is difficult to assess, though it did make a noticeable contribution to heating the home in the first winter, but a substantial proportion of this may have been due to the reduction of cold-air infiltration on a particularly windy site. In the exceptionally cold winter of 1981–82, the conservatory temperature was maintained between 5 and 10 °C above the outside temperature, which was higher than expected.

The conservatory performed exceptionally well as a plant house, with the first cucumbers being harvested early in May 1982. It appeared to require little or no shading, and summertime temperatures in the greenhouse and the house can be reduced to acceptable levels using through natural ventilation.

12.5 Rehabilitated houses – house conversion incorporating a roof-space collector (contributed by Peter Clegg)

Any passive solar measure incorporated into a building – particularly in renovating an existing building – must either be very simple and economical to install, or must serve a dual function (for example, a conservatory), in which case the entire cost of the passive solar measure need not be justified in purely energy-saving terms. In our opinion, if neither of these options is applicable, the measures may not be economical when considered against simple high-insulation alternatives.

In the case of renovating existing buildings, it is often possible to justify additional expense at a time when other work to the building is being done, since the overcost of the passive solar measures will be much less. This is the case in a recent conversion which incorporates a simple roof-space collector into a former agricultural building.

The existing building is 25 m long and 6.0 m wide, single storey, and faces south 30° east. The ratio of external envelope area to floor area is very high, and the building form is thus bound to generate a higher heat requirement than a more compact plan. The solar exposure was also very good, and it therefore seemed appropriate to investigate various forms of passive solar heating.

The amount of direct solar gain through windows was limited by a desire to use the existing window openings and avoid excessive areas of glazing directly into rooms. Conservatories were not regarded as appropriate because of the low eaves height, and the desire to maintain the simple form of the existing building. Trombe walls were rejected because of the difficulty of adapting an existing masonry structure while retaining the present windows.

The most appropriate option was to use the roof space itself as a collector, and the house, with heavyweight masonry partitions and floors as thermal storage, see Fig. 12.13(*a*, *b*). The roof of the building needed strip-

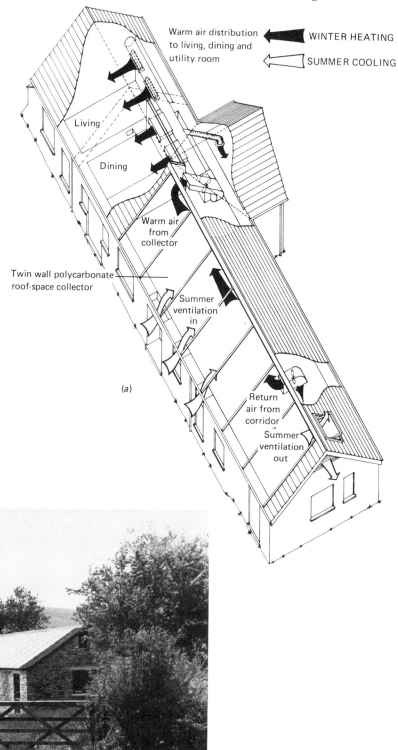

Fig. 12.13. Conversion of an agricultural building near Bristol for Glyn England. (Architects: Feilden Clegg Design. Energy Consultants: Energy Design Group Ltd with John Willoughby and Cedric Green.)

ping off, felting and replacing, and this was bound to result in rejecting a certain percentage of the original tiles. The savings on replacing felt tiles and battens could therefore realistically be taken into account and off-set against the cost of the roof-space glazing.

The collector itself consists of 16 mm twin-wall Makrolan polycarbonate sheeting supplied by Roehm Ltd, supported on battens at 1 m centres over the existing rafters. This form of glazing is inexpensive but of long life (see Chapter 4) and, unlike many plastics, does not have the drawback of high infra-red transmittance. The horizontal area immediately under the collector is insulated and lined with foil-faced building paper to reflect solar radiation at high angles of incidence in summer, all other surfaces are dark brown. Heat is distributed via a 400 mm diameter axial fan to a duct with registers in the dining room, living and utility rooms with a return-air fan at the end of the corridor leading to the bedrooms. The fans are controlled by differential thermostats, and when the house can absorb no more heat and the temperature rises above the upper limit of the differential controller (approximately 23 °C), the distribution fans shut down. When the collector temperature then rises above 40 °C, the roof space is ventilated to the outside air via roof vents controlled by heat-operated greenhouse-type vent controls.

There is a further thermostatically controlled fan to exhaust warm air from the collector should the heat-operated vents fail to operate or be insufficient in area to cope with the build up of heat. In summer, manual vents are open and the collection space is continuously ventilated to avoid the build up of heat.

The building was studied using a thermal simulation model SCRIBE devised by Cedric Green at the University of Sheffield, and various energy strategies were compared. The results are given in Table 12.1, together with the approximate costs of the various options.

The base case was simply to insulate the building to 1982 Building Regulation Standards, which resulted in an annual heat loss of approximately 94 GJ. Increasing the insulation standards to halve the U-values, and allowing for the installation of triple-glazed high-performance windows, reduces the seasonal heat loss by somewhat less than one-half (high insulation case). The incorporation of the roof-space collector was shown to further halve the seasonal heat requirements — a measure that was calculated to be cost-effective over a period of approximately eight years, based on a simple payback calculation assuming no increase in the real cost of energy.

The proposed system is completely untested, but every indication is that it illustrates an appropriate and straightforward use of passive solar energy. The concept throughout the design of the installation has been to use simple materials and off-the-shelf industrial components to avoid excessive costs resulting from purpose-made designs. One of the key elements was the design of the control system to allow for flexibility in setting various temperature limits and fan-speed settings.

The proposal illustrates a general approach to passive solar design in the UK, in that solar energy during the heating season (particularly the

Table 12.1. *Results of computer simulation study using SCRIBE[a]*

	Capital cost (£)	No. of glazings	Fabric walls	U-values roof	W/(m² K) floor
Base case	0	1	0.6	0.4	0.5
	840	2	0.6	0.4	0.5
	930	3	0.6	0.4	0.5
	2900	3N/1S[d]	0.3	0.2	0.3
High insulation	2600	3	0.3	0.2	0.3
High insulation, direct passive solar gain	4400	3N/1S[d]	0.3	0.2	0.3
High insulation + roof-space collector	4500	3	0.3	0.2	0.3
Collector contribution only	1900				

[a] Available from Cedric Green, University of Sheffield, UK.
[b] Night storage heaters.
[c] The cashflow of course would become more favourable year by year assuming that the rises in fuel costs outstrip those in mortgage costs.
[d] Triple glazed with night insulation.

shortened heating season resulting from high insulation measures) is available in such limited quantities that it is not economical to introduce complex and expensive storage systems. With masonry building there is often sufficient storage capacity within the structure of the building itself to absorb additional heat gains during the day. Glazing the roof, whilst obviously causing potential problems of overheating during summer, does allow for increased collection of diffuse solar energy in winter.*

12.6 Rehabilitated houses – a nineteenth-century terraced house

The rehabilitation of existing houses is of special importance to energy conservation because it is here that spatial, constructional and financial constraints are most clearly apparent.

Let us examine the recent renovation of the Cambridge home of one of the present authors. Fig. 12.14 shows the original ground- and first-floor plans, a fairly typical arrangement of 'two up, two down' with a kitchen and bathroom tacked on at the back.

The garden end of the house faces 68° east of south and so offers only a limited opportunity for the use of passive solar gain. Nonetheless, the glazed area on this facade was made as large as was practicable (see Fig. 12.15) so that the house 'opened on' to the garden and so that deep interior spaces were lighted naturally. Additionally, a roof light was installed in the back bedroom to provide natural lighting.

*Note added in print: This rehabilitation is working extremely well. Reports are available from the Energy Technology Support Unit of the Department of Energy.

Vent rate air changes (per h)	Specific heat loss (W/K)	Auxiliary heat (GJ/y)	Energy[b] saving (GJ/y)	Cost saving (£/y)	Annual saving[c] / Capital cost (£)
2	529	96	—	—	—
2	507	88	9	72	0.09
2	491	81	15	125	0.13
1	341	61	35	291	0.10
1	303	50	46	383	0.15
1	290	49	47	393	0.09
1	290	28	68	569	0.13
			22	186	0.10

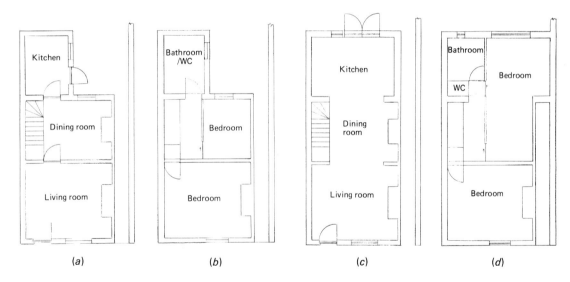

Fig. 12.14. Original and refurbished floor plans for a terraced house (a) Original ground floor. (b) Original first floor. (c) New ground floor. (d) New first floor.

Studies at Cambridge of the existing housing stock have shown that the scope for the use of passive solar gain is significant. Approximately 60% of all houses have either a facade and roof (55%) or a roof only (5%) which faces within 45° of due south.[3] For terraced houses in this group, analysis has shown that total glazing is 11 m² on average with 5 m² on the 'south' side (the terraced houses included a mixture of two- and three-storey buildings). (For the refurbished house of Fig. 12.14, the 'south' side window area is 7.8 m² and the 'north' side 3.9 m².) For all houses in the

Fig. 12.15. View of the garden-facing facade of a renovated terraced house in Cambridge.

Cambridge study it was found that, on average, the solar contribution was 6% of the space heating energy consumption. For the 60% of all houses facing within 45° of due south, the solar contribution is estimated to be 8%.

After years of living in what were beginning to feel like exceptionally small rooms, it was decided to enlarge the house on both floors and to make the ground floor open plan as shown in Fig. 12.14. Previously, the living and dining rooms were each served by a gas fire (electric resistance heaters were used upstairs) and effectively could be sealed off from each other so that energy could be saved by just using heating where needed. The open plan lacked this advantage but the desire for spaciousness prevailed. The original staircase, however, which permitted sealing the ground floor from the first floor, was retained.

A central-heating system, supplying both heating and domestic hot water (thus replacing an electric immersion heater), was installed with a gas-fired wall-hung boiler located in the kitchen, two radiators downstairs and four upstairs. All radiators are equipped with thermostatic radiator valves and a small standard programmer schedules the system for two periods of operation each day. For amenity purposes or perhaps simply because of a palaeolithic desire for an open fire, one of the fireplaces housing a gas fire was opened up and a well-fitting chimney damper was installed to reduce the ventilation heat loss when the fireplace was not in use.

Consideration was given to the use of internal insulation on the street side of the house to improve the thermal characteristics of the 225 mm brick wall but was ruled out when the builders' estimated costs were received. For the new addition, however, it proved relatively easy (some problems were encountered with items such as lintel widths) to incorporate high levels of insulation. The external garden-facing wall consists of 12 mm plaster lining, 100 mm lightweight concrete block, a 50 mm cavity lined with polystyrene slabs and 105 mm exterior brick, giving a U-value of 0.43 W/(m^2 K). The sloping ceiling–roof in the bedroom which looks onto the garden is of 12 mm plasterboard, a polythene vapour barrier, 150 mm glass fibre insulation between the joists, and slate tiles on roofing felt and battens, giving a U-value of 0.22 W/(m^2 K). The new solid floor in the kitchen consists of quarry tiles, a 50 mm screed, 50 mm expanded polystyrene board, a damp-proof polythene membrane and a 100 mm concrete slab over 100 mm hardcore; the estimated U-value is 0.35 W/(m^2 K). The insulation in the existing loft has been increased from 100 mm to 200 mm of glass fibre.

All windows and doors are weatherstripped. Double glazing was considered but delayed because of limited funds. When restoring a house it is only too easy, and entirely understandable, to use the mortgage money for the work that only the builder can do and to plan on the installation of, say, secondary double glazing later by either a specialist or oneself.

Other cost sacrifices included a small conservatory and active solar collectors mounted on the roof for domestic hot water. Needless to say, a solar heated swimming pool in the garden was reluctantly axed.

Water and waste services were all concentrated in one corner to reduce costs and length of runs. The hot-water cylinder and all hot-water piping are insulated. In the future it might be possible to install a small heat-recovery unit in the kitchen to take advantage of the waste heat from the bath above and perhaps the sinks.

The house as described has a design heat loss of 7 kW. The calculation used follows the method of Appendix 2 and assumes an hourly air-change rate of 1.2 and an internal to external temperature difference of 20 K. The heat loss could be reduced in the future by employing the insulation measures which were not possible because of cost during the initial renovation.

The overall result is a comfortable, pleasant home which is much more open to the garden than the original house was. One of the charms of nineteenth-century terraced housing is that it is not too difficult to convert from the mean, cold and damp conditions the first occupants endured to more spacious and enjoyable surroundings. The energy supply and conservation measures which should accompany that transition are in many ways modest but they are significant and of immediate effect.

References

[1] Air Infiltration Centre (1983). *Building Design for Minimum Air Infiltration.* Air Infiltration Centre, Bracknell, UK.
[2] Pressure testing carried out at the Polytechnic of Central London, 1982.
[3] Penz, F. (1982). Private communication based on work on passive solar gain at the Martin Centre, Department of Architecture, University of Cambridge, UK.

13

Non-domestic case studies

13.1 Introduction

This chapter briefly presents some non-domestic buildings designed with energy conservation in mind.

13.2 The swimming pool, Sheiling Schools

Given the opportunity to design a swimming pool for a school for handicapped children, it was automatically part of the brief to keep the running costs to a minimum, despite the fact that water temperatures had to be maintained at 27 °C for use throughout the year. The Energy Design Group made an initial study which investigated the potential of heat pumps and heat-reclaim systems, glazed and unglazed solar collectors, variable speed ventilation control and the use of a pool cover.

Initially it was felt that a heat-pump system would be preferable, using extract air as a primary source of heat. This, however, entailed expensive duct work to return the air to the boiler room, and also added considerably to the capital costs since an auxiliary gas-fired heater could not be dispensed with. Other heat-reclaim systems were dependent on the use of ozone or other very expensive purification systems which reduced their overall cost-effectiveness. (Chlorine from the pool is corrosive over the long term.)

The entire building was built to a very low budget, competitive with quotations from design-and-build contractors who offered cheap standard solutions. The overcost of energy-saving measures had thus to be kept to a minimum.

It was therefore decided to opt for high thermal insulation (100 mm of polystyrene on the roof and 80 mm of glass fibre to the walls with double glazing to all windows), a minimum volume for the building, and a high internal thermal mass. The greatest energy savings were effected by a four-step ventilation-control system operated by a timeclock and humidistat sensor. This, in conjunction with a simple pool cover was calculated to halve the actual energy consumption of the pool for an overcost of approximately £800 (1980 prices). When the pool is not in use, the pool cover is unrolled by the caretaker and the ventilation rate is automatically reduced, as is the throughput of warm air. At night the system shuts down to the extent that it is operating on a 150 mm diameter Ventaxia wall fan.

The swimming pool, Sheiling Schools

Fig. 13.1. The Sheiling Pool, Thornbury Schools (Feilden and Clegg Partnership).

With such drastic reduction in the waste heat from the building, other ways of reclaiming the heat become uneconomical. A roof light with a south-facing slope was incorporated for the time when solar collection becomes economical in the future. Detailed predictions showed that because of the high water temperature required and year-round use, pool insulation would only be marginally cost-effective. Further reductions in energy consumption were achieved by using a central thermostatic control on the water supply to the showers, and percussion valves on the showers themselves reduced water consumption.

The building is primarily of interest as an example of constructing a low-energy building to a very low budget and of devising an approach which first and foremost is concerned with reducing energy consumption by simple means rather than being attracted to more exotic and expensive methods of producing energy.

Fig. 13.1 illustrates the complete scheme.

Fig. 13.2. View from the south-east of the new BRS office building.[1] (Courtesy of the Property Services Administration.)

13.3 The new BRS office building

The new BRS office building at Garston, Watford (see Fig. 13.2) is one of a handful of non-domestic buildings for which energy conservation was a major consideration. In designing the engineering services the principal objective was to minimize energy consumption over the life of the building, while maintaining a good standard of environmental comfort.

The building is rectangular in shape with the long axis running east–west. By calculating the total primary energy consumption for lighting and plant and taking into account passive solar gain through the double-glazed windows which are used throughout, it was found that the optimum, that is, minimum energy size window was at a 50% window/wall ratio (giving an overall U-value of 1.8 W/(m^2 K)). For the north facade the comparable figure is 30%.

To guard against overheating, individually motorized external sun blinds are provided for each 3.6 m wide structural bay. When raised the blind fits neatly into a recess above the window and when lowered, the blind is at eye level from standing position in the room. In this position it prevents direct radiation entering the room when the sun's altitude is greater than 30°.

The blinds lower automatically when required to reduce the risk of uncomfortably high temperatures. The following conditions must be met for operation:
(1) Solar radiation must be high enough to justify protection from the sun.
(2) The room must be at or above 18 °C and space heating is not required.
(3) Wind conditions must be sufficiently calm for use of the blinds.

Agricultural buildings 329

The room occupants may override the automatic blind controls if they so wish.

Heating is from three atmospheric gas-fired boilers with flue-gas dilution, which supply variable temperature hot water to natural and forced-air convectors along perimeter walls. A recirculation mechanical ventilation system is included with the rate of fresh air intake variable for experimental purposes. Heat recovery on the extract air is effected by a heat wheel of 75% total efficiency for latent and sensible heat combined. For research purposes, a solar heating system providing water for wash basins is included. During the winter windows are locked shut but in the summer months the mechanical ventilation system is switched off and the windows are unlocked so that they may provide natural ventilation.

Automatic controls are provided for the lighting in order to conserve electrical energy and to provide BRE staff with a facility for future experiments. Occupants can only switch on their lights when sky illuminance does not afford adequate lighting. This is accomplished by allowing manual switching of a back 'bank' of lights when the daylight level falls below a level equivalent to 300 lux at a point 2 m into the room; at 200 lux, at the same point, the front 'bank' becomes available for switching.

13.4 Agricultural buildings

One of the most promising applications of solar energy is for crop drying, particularly grain crops, since the period of greatest solar energy is almost concurrent with the harvest season.

In many years in the UK a high proportion of the grain harvested must be dried before it can be stored safely in bulk. When necessary, about one-third of the grain is dried on farms using 'high'-temperature (40 °C upwards) 'continuous' and batch driers, although relatively little use is now made of the high-temperature batch driers. Most of the other two-thirds of the grain is dried more slowly in low-temperature bulk driers, often using only ambient air. Limited inputs of heat may be used at times to raise the temperature of the incoming air to produce a relative humidity low enough to enable drying of the grain to continue. Depending on the incoming moisture of the grain, the air flow, volume of grain and design of the plant, etc., drying to a safe moisture content may take from one to ten days or even longer. (Safe moisture content for long-term bulk storage of grain is 14%, and is lower for certain other seeds.) Unless heat is used for long periods with this system, most energy is used in blowing the air. With warm air available from solar collectors, the time taken for drying grain could be much reduced, if some way of recirculating the air could be found, and the need for additional heat much reduced or eliminated during daylight hours.

Fig. 13.3 shows a design (patent pending) developed by Mr Richard Wedgwood and one of the present authors for a solar roof for a typical storage barn. The system is relatively inexpensive and has the advantage

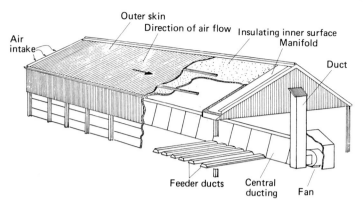

Fig. 13.3. Barn using solar heating for crop drying.[2] (Courtesy of R. Wedgwood.)

of being suitable for new and existing barns. No storage facilities are required and maintenance is an absolute minimum.

Air is drawn laterally along a double-skin roof consisting of an outer surface which absorbs solar radiation, and an inner insulating surface. Suitable materials for the outer surface include sheet aluminium painted black, black plastics, black-painted asbestos and corrugated sheet steel. The insulating surface can be insulating fibreboard, fibreglass, expanded polystyrene boards or a number of other materials.

The heat absorbed is transmitted through the outer surface to the air stream which it warms. The air then passes through a manifold and duct before distribution by a fan unit to central ducting and feeder ducts underneath the stored, harvested material.

Work elsewhere[3, 4] indicates that the efficiency of the double-skin collector is likely to be in the range of 10–40% depending on such factors as materials and air-flow rate.

The economic viability of the design depends principally on the costs of collector materials and the electrical energy to run the fan. Preliminary studies indicate that the proposal is economically attractive now and monitoring studies are underway to confirm this.

13.5 School buildings (contributed by Nick Baker)

The average delivered energy consumption per school child is around 4.9 GJ/y and most of this goes to provide space heating.[5] Taking account of the various prices of fuels used, this represents about £40/y. This significant cost has prompted education authorities to make strenuous efforts to reduce energy consumption in existing buildings, and to make energy conservation a major factor in new school design, and it is this area with which this section is concerned.

Before discussing the contemporary ideas in low-energy design, it is useful to look at the environmental problems of schools in a recent historical context.

The 1950s and 1960s saw a certain preoccupation with providing adequate daylight. There existed a design rule for a minimum daylight factor of 2% and, to achieve this, shallow-plan buildings with large areas of

School buildings

Fig. 13.4. Romsey School, Hampshire.

glazing were necessary. This, together with contemporary tastes in architectural style, resulted in school buildings such as Romsey School, Hampshire, which was completed in 1959 (see Fig. 13.4).

Then, system building was seen as a way to meet the growing demand, by reducing building time and design costs. Many authorities formed consortia in order to develop their own building systems which, although they did not have to be, tended to be of lightweight construction type.

At no time during this period was insulation given much consideration. Energy was cheap and an understanding of thermal comfort was not applied to the school environment. The result of this combination of a high level of glazing, large surface area and a lightweight and poorly insulated structure was a very poor-quality environment with the discomfort of winter underheating only being exceeded by that of summer overheating. Moreover, because of the inherent lack of thermal stability in such buildings and the customary use of rudimentary control systems with a single-boiler thermostat, winter underheating was sometimes combined with winter overheating – in sunny conditions it was not unusual to have occupants of south-facing rooms sweltering and opening windows while those in north-facing rooms shivered.

It seemed to take some time to realize that an environment of this kind might not be the best in which to learn and to teach, and various studies[6, 7] were carried out at the BRE in the late 1960s and early 1970s. Parallel work was also being carried out on the ideal office environment where similar problems had been experienced with highly glazed office blocks.[8, 9]

The notion of creating an 'artificial' or 'controlled environment' began to be applied to school building.[10, 11] A mild reaction set in which was strengthened by the conviction that a 'controlled environment' meant an economical one. By 1974 the oil crisis – oil being a fuel on which the majority of schools relied – had begun to make energy economy a serious factor for the first time since the war.

A number of prestigious office blocks had been completed, boasting 'integrated environmental design' with deep plans, small windows demanding high levels of artificial lighting, and full mechanical ventilation. Control

Fig. 13.5. Elmstead Primary School, near Colchester, Essex. (*a*) Exterior. (*b*) Interior. (Courtesy of Essex County Council.)

was taken away from the occupants who were informed that they were now enjoying 'an optimum environment'.

The principles embodied in these office designs were at this time adapted to a number of schools. Undoubtedly it was a reaction; a belief that openable windows and daylight lead inevitably to energy wastage, but it would be unfair to suggest that the early examples of controlled-environment schools were as barren and sterile as their bigger commercial office counterparts. Scale and educational requirements resulted in many of these designs being very pleasant spaces and the improved environment – thermal, lighting and acoustic – was appreciated by their occupants.

A good example is Elmstead Primary School, near Colchester, Essex (see Fig. 13.5). Here, a novel constructional system of load-bearing concrete panels, with a heavyweight roof in conjunction with a small area of glazing and deep plan, provided a stable thermal environment. Energy

School buildings

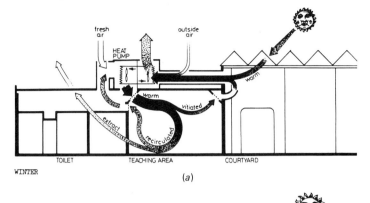

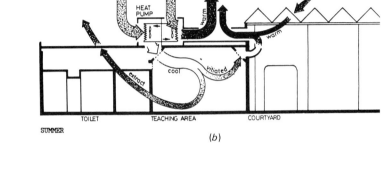

Fig. 13.6. Heat-pump operation at Roach Vale Primary School. (*a*) Winter operation. (*b*) Summer operation. (Courtesy of Essex County Council.)

conservation was beginning to dominate considerations of comfort alone and, in subsequent buildings of this type, more emphasis was placed on energy conservation. Essex County Council, in particular, began to experiment with heat pumps and heat recovery, as exemplified by the Roach Vale Primary School at Colchester[11] (see Fig. 13.6).

To summarize then, the school environment had deteriorated with the development of highly glazed buildings and, subsequently, lightweight construction, during a period when low energy costs did not draw attention to the need to provide a more efficient envelope. Awareness of discomfort problems, closely followed by the rising energy costs of the early 1970s brought about a reaction which tended to reject the need for daylighting and natural ventilation and adopt an 'engineered' or 'controlled-environment' approach. This brings us to the present day.

There is currently yet another swing in emphasis. The two main factors, comfort and energy conservation, are as important as ever, particularly the latter, but our understanding of them has developed. Firstly, we are now beginning to understand that energy conservation is not a purely technical problem in the sense that we simply have to increase insulation, reduce surface area, seal up the windows and mechanically ventilate. Indeed, studies[12] have shown that some early controlled-environment schools consumed more energy than their earlier counterparts, and certainly this has been the case with many office examples. We now realize that the occupants have to be taken into account. Field studies made by Haigh[13]

Fig. 13.7. Yateley Newlands School, Hampshire. (a) Exterior. (b) Plan. (c) Conservatory. (Courtesy of R. Bryant and *The Architects Journal*.)

have focussed attention on the need for better-designed controls if the occupants are expected to interact intelligently with the building and its systems.

Thus it might be said that the problem of designing a building and its systems, however automated and controlled, and ensuring that it is economical, is much more difficult than building scientists thought a decade ago. Related to this, we now realize that our notions of 'the optimum environment', one which is held within a narrow band of environmental parameters, was oversimplistic. A view that is gaining acceptance is that, in fact, most occupants are happier in an environment which shows considerable variation in both time and space.[14]

This has led to the current position in school design where a much less closely controlled environment is provided. A further factor is that a 'selective' approach is being made to the ambient environment as a source of light, solar energy, and air movement, rather than an 'exclusive' approach, where the external environment is seen as something to exclude at all costs.[15]

A good example of this more relaxed approach is the primary school at

School buildings

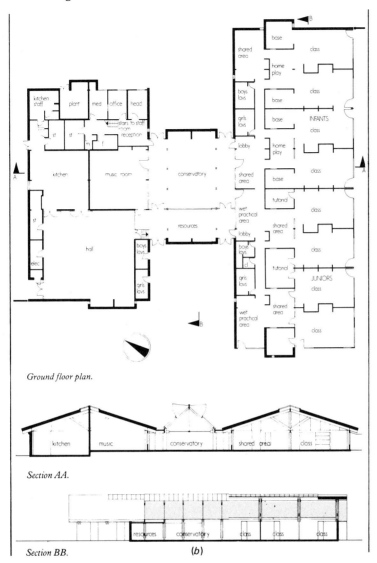

Ground floor plan.

Section AA.

Section BB. (b)

Yateley Newlands, Hampshire,[16] shown in Fig. 13.7. Although it was not designed with energy conservation as the major criterion, it is well insulated, and has a number of characteristics which result in low energy use. One of these is that, although it is a daylit school, it is quite different from the overglazed 2% daylight factor schools of the 1960s and 1970s. This is a result of understanding that daylighting quality, as well as quantity, is of major influence on the degree to which artificial lighting has to be used. Rooms which have a large range of brightness due to the provision of large glazed areas on one side only, without diffusing surfaces, will require artificial lighting in the dimmer areas even when the average lighting level is quite high. At Yateley, this problem has been solved by providing high-

(c)

level roof lights which illuminate a light-coloured wall, as well as side windows.

Yateley, due in part to its need for daylight, has moved away from the deep plan and has adopted an elongated form, compensating for increased surface area by adopting high levels of insulation. This plan form immediately offers the possibility of much greater use of solar gains, provided of course that the correct controls are installed, and that overheating discomfort is prevented by shading the direct sun from the occupants and providing sufficient ventilation.

Another interesting feature of Yateley is the atrium. This space, which is unheated and is formed simply by glazing over an otherwise open courtyard, has proved to be a valuable asset to the school and functions both as part of the general circulation area and a useable space in its own right,

School buildings

Fig. 13.8. St John's School, Clacton on Sea, Essex. (*a*) Exterior. (*b*) Plan. (*c*) Schematic of energy systems. (Courtesy of Essex County Council.)

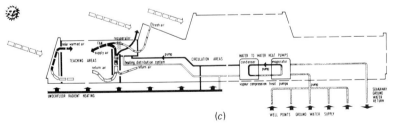

although its environmental standards are quite different from the fully protected interior.

A recently completed project, which demonstrates a more explicit move towards ambient energy sources, is St Johns School, at Clacton on Sea, Essex,[17] shown in Fig. 13.8.

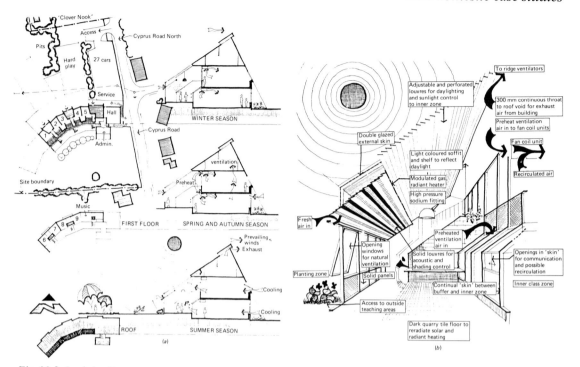

Fig. 13.9. Locksheath School, Hampshire. (a) Seasonal operation. (b) Interior of conservatory.

The Essex County Architects' Department, which was responsible for the earlier controlled environment approaches typified by Elmstead, to which reference has already been made, have adopted a complex range of technical options — heat pumps using groundwater, thermosiphon air-heating panels and conservatories. Whilst still showing a highly 'engineered approach', it has moved some way from its earlier 'exclusive' predecessors.

Another Hampshire project at Locksheath (see Fig. 13.9), demonstrates that even a simple concept can become complex in its realization.[18] Having determined that in a well-insulated building it is ventilation which accounts for the major loss (because of the statutory high fresh-air requirement), the designers use a conservatory running along the south side of the elongated building, as both circulation space, and as a solar collector to preheat ventilation air drawn into the class spaces via a warm-air heating system. This has the advantage of always making use of the gains, even when the conservatory temperature is below that of the heated spaces, while not suffering from the problems of direct-gain spaces, where seated occupants in direct sunlight experience overheating even at modest air temperatures. The conservatory also provides a shared craft area, and thermal analysis shows that temperatures will for the most part be acceptable for non-sedentary activities, with no auxiliary heating.

What has been the result in terms of energy performance, of these developments in school design? In a survey[5] carried out in the mid 1970s of schools in Essex, most of which had not yet adopted any significant energy-conservation measures, the average delivered energy was about 1 GJ/(m² y), or 1.3 GJ/(m² y), primary energy. However, there was wide

School buildings

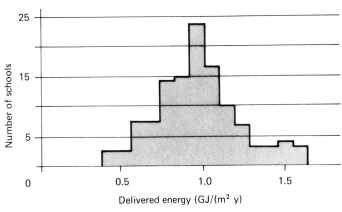

Fig. 13.10. Energy consumption in schools.[5]

divergence of individual schools' energy consumption, as shown in Fig. 13.10.

This variation was not attributable to a systematic variation of building type, suggesting that occupant behaviour and management was the major influence. Detailed surveys of schools often indicate this — for example, in one school the caretaker will carefully vary switch-on time throughout the heating season to prevent wasteful preheating, whereas in an otherwise similar school, the time clock may simply be set for 6.00 a.m. for the whole heating season. This, together with a few other extravagances such as leaving windows open, can easily lead to a doubling of energy consumption.

Thus it is difficult to establish a baseline for comparing the actual effect of building design. Rather, we have to take the view that we are making improvements in building design, system design, management and occupant behaviour, all together. The degree to which these are interdependent is an interesting issue.

There is still a shortage of monitored energy data from schools, and subsequent comparison of figures relies upon calculated values. In 1980, Page[17] showed a comparison between primary energy consumption for controlled-environment schools in Essex, and a target value given in *Department of Education and Science Design Note 17* (1979). The comparison showed that whilst earlier controlled environment designs were above the target, they were well below the standard SEAC (system-built) school of that time. Furthermore, later improved designs were well below the DES target, as shown in Fig. 13.11.

Primary energy consumption is sensitive to the use of electricity due to the poor conversion efficiency at the power station. Thus, although heating energy is small, the lighting and fan power required in controlled-environment buildings constitutes a relatively high primary energy demand. The 'selective' approach, typified by the approach of the Hampshire schools, may result in higher heating demand, but the overall primary energy performance is likely to be as good or better than the 'exclusive' controlled-environment buildings. For example, the overall

Fig. 13.11. Comparison of annual primary energy in schools.[17]

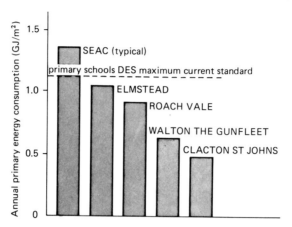

primary energy consumption of Locksheath school is calculated to be as low as 0.4 GJ/(m² y).

We must not delude ourselves that calculated and actual performance are one and the same. As we have seen already, the performance of a building is very much dependent upon the behaviour of the occupants. The 'selective' buildings will require even more cooperation from the occupants than the controlled-environment buildings. Nevertheless, a target has been identified and the principles of how this target can be met are known.

To summarize, school building design is currently directing its attention to the 'selective approach', producing buildings which in plan form and use of glazing are not so different from the buildings of two decades earlier. However, with a much greater understanding of thermal comfort requirements, and of the need to integrate the ambient gains with the auxiliary systems, these contemporary buildings will perform much better. Furthermore, the educational values of these buildings, which clearly relate to the climate, is being recognized. There is considerable enthusiasm and support in the development of these designs, and it is likely that we shall see more progress and innovation in selective (or passive) design in new schools than in most other non-domestic sectors.

References

[1] Price, P. (1977). 'Energy conservation in new offices at BRS'. *Construction*, 23, 25–8. This section is based entirely on this article; we are greatly indebted to the author.

[2] British Patent Filing No. 7 941 163. November 1979. Improvements in and Relating to Crop Drying Facilities. R.B. Wedgwood & R.B. Thomas.

[3] Ekstrom, N. & Gustafsson, G. (1979). *The Application of Solar Collectors for Drying of Grain and Hay*. Uppsala, Sweden: Swedish Institute of Agricultural Engineering.

[4] Ferguson, W.E. & Bailey, P.H. (1979). 'Solar air heater for near-ambient crop drying: description, testing methods and 1977

results'. *Departmental Note No. SIN/275*. Penicuik: Scottish Institute of Agricultural Engineering.

[5] Yannas, S. & Wilkenfield, G. (1978). 'Energy strategies for secondary schools in Essex'. *Research Paper 1/78*. London: Architectural Association.

[6] Langdon, F.J. & Loudon, A.G. (1970). 'Discomfort in schools from overheating in summer'. *Journal of the Institute of Heating and Ventilation Engineers*, 38, 181–9.

[7] Humphreys, M.A. (1974). *Classroom Temperature, Clothing and Thermal Comfort – A Study of Secondary School Children in Summertime*. BRE CP 22/74. Garston, Watford: BRE.

[8] Humphreys, M.A. & Nicol, J.F. (1970). 'An investigation into the comfort of office workers'. *Journal of the Institute of Heating and Ventilation Engineers*, 37, 265–79.

[9] Hardy, A.C. (1971). 'Architecture and building science'. *RIBA Journal*, November.

[10] (1971). *Integrated Environmental Design: A Feasibility Study for School Buildings*. County Architects Department, Gloucester County Council, December.

[11] Crowe, R.V. & Page, P.A. (1976). 'Integrated environmental design of schools'. *Proceedings of the Symposium on Energy Conservation in the Built Environment*. London: Construction Press–CIBS.

[12] O'Sullivan, P. & Austin, M.J. (1976). 'Energy targets (schools)'. *Proceedings of the Symposium on Energy Conservation in the Built Environment*. London: Construction Press–CIBS.

[13] Haigh, D. (1981). 'User response in environmental control'. In: *The Architecture of Energy*. London: Construction Press.

[14] Cooper, I. (1981). 'Comfort theory and practice'. *Applied Energy*, 2 (4), 267–9.

[15] Hawkes, D. (1981). 'Building shape and energy use'. In: *The Architecture of Energy*. London: Construction Press.

[16] Smith, C.S. & Hawkes, D.)1981). 'Yateley Newlands Primary School'. *Architects Journal*, 173 (25), 1199–214.

[17] Page, P.A. (1981). 'School buildings – Essex County Council Architects Department'. In: *The Architecture of Energy*. London: Construction Press.

[18] Baker, N.V. (1982). 'The influence of thermal comfort and user control on the design of a passive solar school building'. *Energy and Building* (in press).

Appendix 1: Weather data

A 1.1 Introduction

UK weather data should be covered in a fairly comprehensive way, in a book currently under production, by Page & Lebens.[1] In the meanwhile, very restricted data is given here. Maps of global solar radiation are available from the Meteorological Office for the UK at Bracknell.

French data is presented in the *Atlas Solaire Francais*,[2] for 33 sites in terms of solar energy penetrating glazing at various orientations. Details are also given for overshadowing and degree days.

US data is well presented in the book by Mazria.[3]

Data for Germany can be found in Bossel's book[4] which contains ambient temperature and global solar radiation, amongst other data, for each month.

A 1.2 Solar spectrum

The spectral distribution of solar radiation varies with local conditions and time of day, but Fig. A 1.1 provides representative values. The diagram also indicates how glass affects the distribution of energy with wavelength.

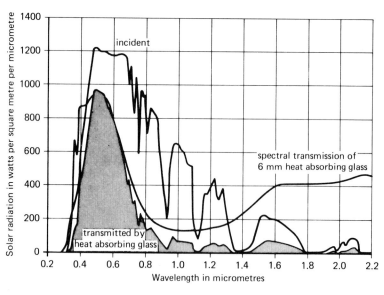

Fig. A1.1. The spectral distribution of transmitted energy.[5]

Weather data

Fig. A 1.2 gives the transmittance of float glass for comparison. It may be seen that most of the ultra-violet, and all of the infra-red characteristic of bodies at room temperature, is filtered out.

Fig. A 1.3 illustrates values on inclined surfaces at Kew, calculated by Page.

A 1.3 Global solar radiation data for the UK — monthly means in MJ/(m² d). Kew and Bracknell latitude, about 51.5° (tilt 0 = horizontal; 90 = vertical)

Month	Kew[6] azimuth tilt 0	Bracknell[7] south 90[a]	Bracknell[7] east 90[a]	Bracknell[7] west 90[a]	Bracknell[7] north 90[a]
January	1.9	3.0	1.2	1.1	0.8
February	3.6	5.4	2.8	2.5	1.2
March	7.4	7.8	4.8	3.8	1.9
April	11.5	7.7	6.6	6.0	3.0
May	15.9	7.7	8.3	7.7	4.3
June	17.4	7.7	9.8	9.0	5.0
July	15.8	7.7	8.0	8.4	4.9
August	13.3	7.7	6.8	7.0	3.4
September	10.2	7.8	5.2	5.3	2.3
October	5.7	6.9	3.4	3.6	1.6
November	2.5	4.9	2.0	1.8	0.8
December	1.5	3.0	1.1	1.1	0.6

[a] Excluding ground-reflected radiation.

A 1.4 Global solar radiation data on the horizontal for three UK stations in MJ/(m² d)

	Kew	Eskdalemuir	Lerwick
January	1.9	1.5	0.8
February	3.6	3.9	2.7
March	7.4	6.3	5.6
April	11.5	10.8	11.8
May	15.9	14.0	14.5
June	17.4	16.3	16.7
July	15.8	13.8	14.0
August	13.3	11.3	11.0
September	10.2	8.1	7.3
October	5.7	4.3	3.2
November	2.5	2.0	1.1
December	1.5	1.2	0.6
Latitude	51.5°	55°	61°

A 1.5 Direct and diffuse solar radiation for Kew, UK (Fig. A 1.4)

It is important for readers to note the very high fraction of diffuse radiation from the overcast skies of the UK.

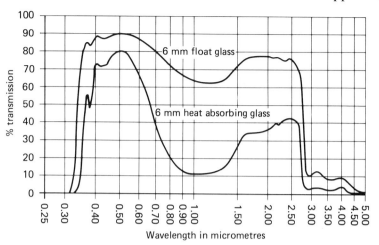

Fig. A1.2. The spectral transmission characteristics of glasses.[5]

A 1.6 Temperature data for the UK (monthly means °C)

Month	Kew	Eskdalemuir	Aberporth
January	4.0	3.0	5.8
February	4.9	3.1	5.6
March	6.8	4.0	6.3
April	9.4	5.5	9.0
May	12.5	8.4	11.2
June	15.9	10.1	13.5
July	16.9	12.1	15.3
August	16.5	12.0	15.3
September	14.7	10.3	14.0
October	11.8	8.1	11.8
November	7.5	5.5	8.5
December	4.9	4.1	6.7
Latitude	51.5°	55°	52°

A 1.7 Wind data

Wind data is presented in Chapter 8.

References

[1] J. Page and R. Lebens are preparing comprehensive tables of data for a publication in conjunction with the UK Department of Energy. The book, entitled *Data Handbook*, will be published early 1984.

[2] Claux, P., Pesso, J. & Raoust, M. (1982). *Atlas Solaire Francais – Energie Transmise et Calcules de Laisser de Masque.*

[3] Mazria, E. (1979). *Passive Solar Energy Book*, professional edn. Emmaus, Pa: Rodale Press.

References

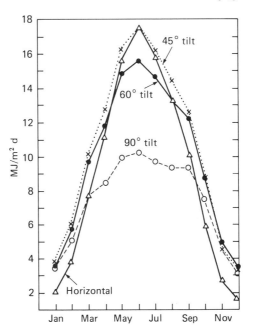

Fig. A1.3. Mean monthly daily values of total irradiation on slopes of 40, 60 and 90° at Kew, compared with that on a horizontal surface. Estimated from horizontal surface data, 1957–71.[7]

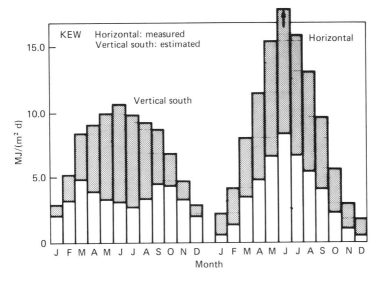

Fig. A1.4. Monthly mean daily irradiation on a vertical south-facing surface (estimated) and a horizontal surface at Kew, measured 1955–70. Direct part clear, diffuse stippled.[7]

[4] Bossel, U. (1980) 'Klimadaten Mitteleuropas 3'. Adelebsen, W. Germany: Solentec GmbH.

[5] Turner, D.P. (ed.) (1969). *Windows and Environment*. Ormskirk: Pilkington Brothers UK.

[6] Collingbourne, R.H. (1975). 'UK solar radiation network on the availability of solar radation data from the Meteorological Office for solar energy application'. *International Solar Energy Society, UK Section, Conference Proceedings*, February 1975. London.

[7] 'Solar Energy UK Assessment'. *International Solar Energy Society, UK Section*, 1976. London.

Appendix 2: Thermal performance

An estimate of the thermal performance of a building is vital for sizing the heating plant (and the ventilation and refrigeration plant if there is one), for thermal comfort, for the avoidance of condensation and for energy conservation. The simplest way of estimating thermal performance is the U-value technique based on the steady-state heat-loss equation

$$\bar{q} = (\Sigma \, AU + C_v)(t_i - t_o),$$

where

$\bar{q}$ = heat requirement (W);
A = area of a structural element (m^2);
U = thermal transmittance of a structural element (W/(m^2 K));
C_v = ventilation loss (W/K);
t_i = inside temperature (°C);
t_o = outside temperature (°C).

The equation is valid in situations where temperatures do not change significantly with time. It is more accurate when applied to buildings with little fenestration in, say, winter conditions of low external temperature and little or no solar gain. It is less suited to buildings designed for passive solar gain on an autumn day when the sun appears intermittently through the clouds. There are also important reservations about the value of a technique based on steady-state, one-dimensional heat flow to deal with the problems of, for example, corners in buildings, thermal energy storage in the building fabric and changes in the U-value due to, say, moisture penetration.

However, this method has the great virtue of simplicity and is readily understood by those familiar with only the simplest mathematics. The U-value, or thermal transmittance, of a structural element is given by

$$U\text{-value (W/(m}^2 \text{ K))} = 1/R,$$

where R(m^2 K/W) is the sum of the resistances of surfaces and layers of material. More explicitly,

$$R = R_{si} + R_1 + R_2 + \ldots + R_n + R_{so},$$

where

R_{si} is the resistance of the inner surface (dependent on both radiation determined by the emittance of the surface and convection determined by the layer of air immediately adjacent to the

Thermal performance

Table A 2.1. *Inside surface resistance*, R_{si}[1]

Building element	Heat flow	Surface resistance (m^2 K/W)	
		High emissivity factor	Low emissivity factor
Walls	Horizontal	0.12	0.30
Ceilings or roofs, flat or pitched, floors	Upward	0.10	0.22
Ceilings and floors	Downward	0.14	0.55

Notes:
(1) High emissivity factor assumes $\epsilon_1 = \epsilon_2 = 0.9$.
    Low emissivity factor assumes $\epsilon_1 = 0.9$, $\epsilon_2 = 0.05$.
(2) Surface temperature is assumed to be 20 °C.
(3) Air speed at the surface is assumed to be not greater than 0.1 m/s.
(Reproduced from Section A3 of *The CIBS Guide*, by permission of the Chartered Institution of Building Services.)

Table A 2.2. *Outside surface resistance* (R_{so}) *for stated exposure*[1]

Building element	Emissivity of surface	Surface resistance for stated exposure (m^2 K/W)		
		Sheltered	Normal	Severe
Wall	High	0.08	0.06	0.03
	Low	0.11	0.07	0.03
Roof	High	0.07	0.04	0.02
	Low	0.09	0.05	0.02

Note: Form (shape) factor for radiative heat transfer is taken to be unity.
(Reproduced from Section A3 of *The CIBS Guide*, by permission of the Chartered Institution of Building Services.)

surface);
$R_1, \ldots, R_n$ are the resistances of the layers of material (or voids); and
R_{so} is the resistance of the outer surface.

Internal and external surface resistances are given in Tables A 2.1 and A 2.2 respectively.

The thermal resistance of a layer of material is given by

$$R_i \text{ (m}^2 \text{ K/W)} = \text{thickness of layer (m)}/k \text{ (W/(m K))},$$

where k is the thermal conductivity. Typical thermal conductivities and other properties of common building materials are shown in Table A 2.3.

Table A 2.4 shows the resistance of a variety of air spaces.

Table A 2.5 shows a U-value calculation for the wall of Fig. A 2.1.

U-values have been calculated and compiled for a wide variety of constructions. Table A 2.6 gives both U-values and admittance (Y-) values (dis-

Table A 2.3. *Properties of common building materials*[a][1]

Material	Density (kg/m^3)	Thermal conductivity (W/(m K))	Specific heat capacity (J/(kg K))
Walls (external and internal)			
Brickwork (outer leaf)	1700	0.84	800
Brickwork (inner leaf)	1700	0.62	800
Concrete block (lightweight)	600	0.19	1000
Plasterboard	950	0.16	840
Surface finishes			
Plaster (dense)	1300	0.50	1000
Plaster (lightweight)	600	0.16	1000
Roofs			
Asphalt	1700	0.50	1000
Felt — bitumen layers	1700	0.50	1000
Floors			
Cast concrete	2000	1.13	1000
Timber flooring	650	0.14	1200
Insulation			
Expanded polystyrene (EPS) slab	25	0.035	1400
Glass fibre quilt	12	0.040	840
Polyurethane board	30	0.025	1400
Urea formaldehyde (UF) foam	10	0.040	1400

[a] Values used in the calculation of U- and Y-values by the CIBS. (Reproduced from Section A3 of *The CIBS Guide*, by permission of the Chartered Institution of Building Services.)

cussed below) for several examples. Tables A 2.7 and A 2.8 give U-values for glazing and windows (for floors see Chapter 3).

Now let us consider the heat loss from the simple structure, shown in Fig. A 2.2, which incorporates the wall in Fig. A 2.1. Table A 2.9 shows a calculation for the fabric heat loss for the building, assuming an external temperature of $-1\,°C$ and an internal temperature of $19\,°C$ as the design conditions. These could be considered as the basis for the 'maximum' heat loss for which the heating system should be sized.

To the fabric heat loss must be added the ventilation heat loss, C_v, which is given by the formula

$$C_v\,(W) = 0.36\,NV\,(t_i - t_o),$$

where

N = the number of air changes per hour;
V = volume of the space (m^3);
t_i = inside temperature (°C);
t_o = outside temperature (°C).

Thermal performance

Table A 2.4. *Standard thermal resistances for unventilated airspaces*[1]

Type of air space		Thermal resistance / (m² K/W) for heat flow in stated direction		
Thickness	Surface emissivity	Horizontal	Upward	Downward
5 mm	High	0.10	0.10	0.10
	Low	0.18	0.18	0.18
25 mm	High	0.18	0.17	0.22
or more	Low	0.35	0.35	1.06
High emissivity plane and corrugated sheets in contact		0.09	0.09	0.11
Low emissivity multiple foil insulation with airspace on one side		0.62	0.62	1.76

(Reproduced from Section A3 of *The CIBS Guide*, by permission of the Chartered Institution of Building Services.)

Table A 2.5. *Determination of the thermal transmittance (U-value) of a wall*

Element	Thickness (m)	Thermal conductivity (W/(m K))	Thermal resistance (m² K/W)
R_{si} = inner surface resistance	—	—	0.12
R_1 = plaster	0.013	0.16	0.08
R_2 = concrete blocks	0.100	0.19	0.53
R_3 = polystyrene	0.050	0.035	1.43
R_4 = brickwork	0.105	0.84	0.13
R_{so} = outer surface resistance	—	—	0.06

R = Total resistance = 2.35 m² K/W
$U = 1/R = 0.43$ W/(m² K)

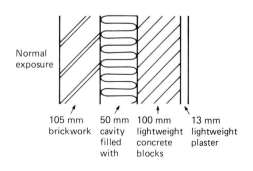

Fig. A2.1. Typical wall construction.

Table A 2.6. *U-values and admittance (Y-) values for construction elements*[1]

Construction (outside to inside)	U-value (W/(m² K))	Admittance Y-value (W/(m² K))
(1) Wall: 105 mm brickwork, 50 mm mineral fibre, 100 mm heavyweight concrete block, 13 mm lightweight plaster	0.53	4.3
(2) Wall: 105 mm brickwork, 75 mm mineral fibre, 105 mm brickwork, 13 mm lightweight plaster	0.37	3.7
(3) Roof: 10 mm tile, loft space, 100 mm glass-fibre quilt, 10 mm plasterboard	0.34	0.7

(Reproduced from Section A3 of *The CIBS Guide*, by permission of the Chartered Institution of Building Services.)

Table A 2.7. *U-values for glazing (without frames)*[1]

Construction	U-value[a] for stated exposure (W/(m² K))		
	Sheltered	Normal (standard)	Severe
Single window glazing	5.0	5.6	6.7
Double window glazing with airspace			
25 mm or more	2.8	2.9	3.2
12 mm	2.8	3.0	3.3
6 mm	3.2	3.4	3.8
3 mm	3.6	4.0	4.4
Triple window glazing with airspace			
25 mm or more	1.9	2.0	2.1
12 mm	2.0	2.1	2.2
6 mm	2.3	2.5	2.6
3 mm	2.8	3.0	3.3
Roof glazing skylight	5.7	6.6	7.9
Horizontal laylight with skylight or lantern light over			
Ventilated	3.5	3.8	4.2
Unventilated	2.8	3.0	3.3

[a] In calculating these values, the thermal resistance of the glass has been ignored.
(Reproduced from Section A3 of *The CIBS Guide*, by permission of the Chartered Institution of Building Services.)

Thermal performance

Table A 2.8. *U-values for typical windows*[1]

Window type	Fraction of area occupied by frame[a] (%)	U-value for stated exposure (W/(m² K))		
		Sheltered	Normal	Severe
Single glazing				
Wood frame	10	4.7	5.3	6.3
	20	4.5	5.0	5.9
	30	4.2	4.7	5.5
Aluminium frame (no thermal break)	10	5.3	6.0	7.1
	20	5.6	6.4	7.5
	30	5.9	6.7	7.9
Aluminium frame (with thermal break)	10	5.1	5.7	6.7
	20	5.2	5.8	6.8
	30	5.2	5.8	6.8
Double glazing				
Wood frame	10	2.8	3.0	3.2
	20	2.7	2.9	3.2
	30	2.7	2.9	3.1
Aluminium frame (no thermal break)	10	3.3	3.6	4.1
	20	3.9	4.3	4.8
	30	4.4	4.9	5.6
Aluminium frame (with thermal break)	10	3.1	3.3	3.7
	20	3.4	3.7	4.0
	30	3.7	4.0	4.4

[a] Where the proportion of the frame differs appreciably from the above values, particularly with wood or plastic, the *U*-value should be calculated (metal members have a *U*-value similar to that of glass).
(Reproduced from Section A3 of *The CIBS Guide*, by permission of the Chartered Institution of Building Services.)

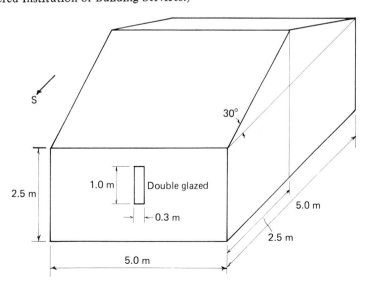

Fig. A2.2. Structure for heat-loss calculations.

Table A 2.9. *Determination of the fabric heat loss for a simple structure*

Element	U-value (W/(m² K))	Area (m²)	Temperature difference[a] (K)	Fabric heat loss (W)
South wall[b]	0.43	11.9	20	102
South window	2.9	0.6	20	35
East wall	0.43	12.5	20	108
North wall	0.43	12.5	20	108
West wall	0.43	12.5	20	108
Roof[c]	0.34	25.0	20	170
Floor[d]	0.84	25.0	20	420
Sum of fabric heat loss				1051 W

[a] Internal temperature 19 °C; for simplicity the external temperature of both the air and ground has been taken as −1 °C.
[b] Wall construction: 105 mm brickwork, 50 mm polystyrene, 100 mm concrete block, 13 mm plaster; unheated loft space therefore use 12.5 m² area for east and west wall.
[c] Roof construction: 10 mm tile, loft space, 100 mm glass fibre quilt, 10 mm plaster board; use plan area for calculation.
[d] Floor construction: suspended timber floor with carpet.

(The factor of 0.36 is the volumetric specific heat of air, 1300 J/m³ K, divided by 3600, the number of seconds in an hour.) Alternatively, for situations of high occupancy, the ventilation heat loss can be calculated from

$$C_v \text{ (W)} = \text{Number of people} \times \text{Air-change rate per person (m}^3\text{/s)}$$
$$\times \text{ Volumetric specific heat of air} \times (t_i - t_o).$$

In the case of the simple structure of Fig. A 2.2, if we assume one-and-a-half changes per hour the ventilation heat loss is 675 W, giving a total design heat loss of 1.73 kW. (The ventilation heat loss can be estimated more precisely using the crack estimation method.[2]) Note that this calculation makes no allowance for 'free-heat' gains, which is reasonable since, in sizing the heating system, it is best to be cautious and not rely on such a variable factor.

Of course, most buildings are considerably more complex, with several floors, partitions and variable ventilation rates in, for example, kitchen and living rooms. Burberry[3,4] provides more-detailed computation procedures. Additional information on U-values may be found in a BRE publication[5] and for the whole subject of energy demand and thermal performance, the *CIBS Guide*[6] and its US counterpart, the *ASHRAE Handbook*[7] are invaluable.

To calculate the seasonal heat loss, the simple method devised by the BRE for small houses shown in Table A 2.10 may be used. For the purposes of this exercise we will assume one occupant but use figures for the heat gains from electricity, cooking and water heating, which are in accord with the greater values given in Chapter 3 for groups of several people.

Thermal performance 353

Table A 2.10. *Seasonal heat requirements (after Ref.* [8] *)*

Heat losses

(1) Fabric heat losses

Element	×	Area (m^2)	×	Seasonal transmittance coefficient (W/(m^2 K))	×	Average temperature difference[a] (K)	×	Rate of heat loss (W)
Roof		25.0		0.34		10		85
Window		0.6		2.9		10		17
External walls		49.4		0.43		10		212
Ground floor		25.0		0.84		10		210
Total								524

(2) Ventilation heat loss
$$C_v = 0.36 \times N V (t_i - t_o)$$
$$= 0.36 \times 1.5 \times 62.5 \text{ m}^3 \times 10 \text{ K}$$
$$= 338 \text{ W}$$

(3) Total heat loss = 862 W × 0.02[b]
= 17.2 GJ

Heat gains

(1) Windows South facing 0.6 m^2 × 0.68[c] GJ/m^2	0.4 GJ
(2) Body heat (1 GJ/p)	1.0 GJ
(3) Electricity (based on estimated consumption)	3.3 GJ
(4) Cooking (gas)	6.0 GJ
(5) Water heating	2.0 GJ
Total seasonal gains	12.7 GJ

Net heat requirement   17.2 − 12.7 = 4.5 GJ

Gross heat requirement = Net heat requirement × 100 / (% efficiency of heating system)

---

[a] An average temperature difference of 10 K was found to adequately represent the actual variation in the difference between internal and external temperature during the heating season. (Other sources sometimes use varying assumptions. For example, in Fig. 3.1 a 7 K difference was used.)
[b] 0.02 is the length of the heating season, 20 × 10^6 s, divided by 10^9 to convert the heat loss to GJ.
[c] Heat gains from unobstructed east- and west-facing windows are given as 0.41 GJ/m^2 and north-facing windows as 0.25 MJ/m^2.

Obviously, the method can easily incorporate more-sophisticated refinements. For example, in more-recent housing the average air-change rate during the heating season is not as high as the figure of two originally suggested by the BRE[8] or even the one-and-a-half used above (see Chapter 3) but a better estimate is easily substituted. Similarly, if temperature data is available, the calculation can be made monthly and summed over the heating season and if solar radiation data is available a more precise figure for solar gains can be used.

Seasonal heating requirements can also be estimated using the degree-day method.[4, 9, 10] A degree day, according to the 'generally accepted' definition[9] is the product of one day (24 h) and 1 K difference of temperature between the base temperature and the average outside air temperature, $(t_{max} + t_{min})/2$. The base temperature is 15.5 °C which allows for a 2.8 °C lift due to incidental gains from such sources as people, lights, cooking appliances and motors to bring the internal temperature up to 18.3 °C which is considered comfortable for normal domestic purposes.

Maps of degree-day areas and 20-year mean degree-day figures are available.[9] For example, for the period 1957–77 in the Thames Valley area (which includes London) the average number of degree days from September to May was 2030. Returning to the simple structure of Fig. A 2.2 we can calculate that the total heat loss is 87 W/K. (Note that we haven't considered heat gains since we will accept the 2.8 °C figure as standard. In practice, however, this must be carefully checked because in highly insulated buildings casual gains may keep the building sufficiently warm down to outside temperatures of, say, 12–13 °C rather than the 15.5 °C base temperature.) The seasonal heat loss, assuming that the building is kept at 18.3 °C throughout the heating season, is thus approximately

$$87 \frac{W}{K} \times 24 \frac{h}{day} \times 2030 \, °C - day = 4238 \text{ kWh} = 15.3 \text{ GJ}.$$

The difference between this figure and the net heat requirement in Table 2.10 is due principally to the use of the 2.8 °C lift figure. In such a small well-insulated structure the inclusion of large incidental gains would in fact considerably increase the lift and consequently lower the effective number of degree days.

Refinements in the method can easily be introduced – Page is developing a program which incorporates the concept of day and night degree days so that the effects of curtaining windows at night can be properly evaluated.[11]

The degree-day method is perhaps of greatest use in evaluating the efficiency of energy-conservation measures. For this application actual fuel consumption before and after, for example, the installation of insulation can be compared and by dividing by the observed number of degree days for an equal period before and after, the effect of weather can be separated from that of the insulation. Steady-state heat loss calculations can be carried out quite simply on advanced calculators and minicomputers. The Royal Institute of British Architects (RIBA) has a number of such programs ranging from the design day heat loss for a building design to a calculation which derives the heat loss and solar gain from monthly weather data and then balances losses against solar and internal gains to assess the need for additional space heating.[12]

More-sophisticated approaches to thermal performance exist. In the UK

Thermal performance 355

Table A 2.11. *Admittance (Y-values) for internal walls*[1]

Construction	Admittance (Y-value) (W/(m² K))
(1) 105 mm brickwork	4.1
(2) 105 mm brickwork with 13 mm lightweight plaster on each side	3.6
(3) 105 mm brickwork, 25 mm air gap, 105 mm brickwork	4.9
(4) 105 mm brickwork, 25 mm air gap, 105 mm brickwork with 13 mm lightweight plaster on each side	3.8
(5) 100 mm lightweight concrete block	2.0
(6) 100 mm lightweight concrete block with 13 mm lightweight plaster on each side	2.3

(Reproduced from Section A3 of *The CIBS Guide*, by permission of the Chartered Institution of Building Services.)

the increasing importance of summertime design for the prediction of air-conditioning loads and internal temperatures led to the development of the admittance procedure.[13, 14] This technique has also proved useful for investigating the performance of intermittently heated buildings and, on a smaller scale, the comparison of different internal surfaces. This last aspect, which is the only one that space allows to be even briefly treated here, depends on a knowledge of the admittance of the surface. This is the rate of heat flow between the internal surface of the construction and the space temperature for each degree of swing in space temperature about its mean value. The admittance is a function of the thickness, thermal conductivity, density and specific heat capacity of the materials used in the construction and of the frequency of the thermal input (usually a 24 h cycle is assumed). For thin structures the admittance is equal to the U-value, thus, for example, the admittance of single glazing is 5.6 W/(m² K).[15] With multilayer constructions the admittance is determined primarily by the characteristics of the materials in the layers next to the internal surface. Table A 2.6 has shown the admittance of several external wall constructions. Table A 2.11 gives figures for some internal walls.

If one is designing a lecture room which is used intermittently and heated with, say, electric fires, if the internal walls are of 105 mm brick, each 1 K rise will require 4.1 W per square metre of wall. Change to 100 mm lightweight concrete block and the figure drops to 2.0 — other things being equal, the audience will be more comfortable more quickly in the second case.

Another approach is the response factor method whose basis has been described[16] as follows:

The heat flux through the surfaces of a wall can be expressed in terms of the surface temperature and thermal characteristics of the wall using time series techniques. Thus, all unknown heat fluxes at the surface can be related to the surface temperatures and, if the temperature history of the surface is known, the heat balance equations can be solved to find the current value of surface temperature. When the heat transfer coefficient depends on the surface temperature it is necessary to solve the equations by an iterative procedure. Once the surface temperature has been found it becomes part of the temperature history to be used for future calculations.

While the mathematics of this approach are considerably more difficult and often beyond the range of all but specialists in heat transfer, the availability of computer programs[17] that can incorporate it should lead to increased acceptance. As Kusuda said in the discussion following Mitalas' paper,[16] such advanced techniques are needed to deal with the 'hour by hour evaluation of instantaneous heat gain and heat loss, responding to the changing climatic conditions and energy use schedules'.

Electrical analogue techniques which model thermal response also exist but, in general, advances in digital computers have favoured methods developed for them. Complex thermal prediction models are now common in consulting engineers' firms and the research centres of the energy supply industries. Further computer advances should soon make them available to many architects' offices.

Among the more complicated models are those of Nottingham University (the BUILD program), the University of Strathclyde (the ESP program) and Ecotope (Seattle, US) (the SUNCODE program). These could well be used in the final design stages of a house. Even more elaborate programs are available, in particular for commercial buildings but they tend to be difficult to use and too expensive to run repeatedly for design purposes.

References

[1] (1980). 'Thermal properties of building structures'. *CIBS Guide. Section A3.* London: Chartered Institution of Building Services.
[2] (1976). 'Air infiltration'. *CIBS Guide. Section A4.* London: Chartered Institution of Building Services.
[3] Burberry, P. (1975). *Environment and Services.* London: Batsford.
[4] Burberry, P. (1979). 'Predictions of thermal performance'. *Architects Journal*, 172 (43, 44, 45, 46), 893–905, 951–62, 993–4, 1055–62.
[5] Anon. (1975). *Standard U-values.* BRE Digest 108. Garston, Watford: BRE.
[6] *CIBS Guide.* London: Chartered Institution of Building Services. The guide is continually updated.
[7] *ASHRAE Handbook, 1977, Fundamentals.* New York: American Society of Heating Refrigerating and Air Conditioning Engineers.
[8] Anon. (1956). *Domestic Heating – Estimation of Seasonal Heat Requirement and Fuel Consumption in Houses.* BRE Digest 94. Garston, Watford: BRE.
[9] Anon. (undated). *Degree Days.* Department of Energy Fuel Efficiency Booklet 7. London: HMSO.

References

[10] Peach, J. (1977). 'Degree days'. *Building Services Engineer*, 10 (44), 213–17.

[11] Page, J.K. (1978). 'The development of systematic climatological design procedures for solar buildings'. *Sun at Work in Britain*, 6, 48–50.

[12] Royal Institute of British Architects Publications Ltd, London.

[13] Loudon, A.G. (1968). *Summertime Temperatures in Buildings*. BRS Current Paper 47/68. BRE: Garston, Watford.

[14] Millbank, N.O. & Harrington-Lynn, J. (1974). *Thermal Response and the Admittance Procedure*. BRS Current Paper 61/74. BRE: Garston, Watford.

[15] (1976). *IHVE Guide Book A*. London: Institute of Heating and Ventilating Engineers.

[16] Mitalas, G.P. (1968). *Calculation of Transient Heat Flow Through Walls and Roof*. New York: ASHRAE.

[17] Kusuda, T. National Bureau of Standards. Washington, DC.

Appendix 3: Interstitial condensation

Just as a temperature gradient exists within a structural element, a dew-point gradient depending on the water vapour diffusion properties of the element exists too. If at any point in the structure the actual temperature is below the dew point then condensation will occur at that point.

Table A 3.1 gives some typical values of vapour resistance and thermal and vapour resistivities (thermal resistivity is the reciprocal of thermal conductivity – see Table A 2.3).

With more and more insulation being used, the designer must remember to consider both the thermal and vapour properties he or she is specifying. This is particularly true since some very good thermal insulants, for example glass fibre, are also very permeable to water vapour.

Let us now return in Fig. A 3.1 to the wall construction of Fig. A 2.1 and using the standard BRE procedure[1] assess whether there is a risk of interstitial condensation.

Procedure

(A) Assumed interior/exterior air temperature difference, ΔT, is $20° - 0° = 20\,°C$.

(B) Thermal resistance, r_t (thermal resistance equals the thermal resistivity times the thickness of the material).

(a) Inside air to point 1 (from Table A 2.1) $= 0.12\ m^2\ K/W$;

(b) Point 1–2 13 mm lightweight plaster
$$\frac{1}{0.16} \times 0.013 = 0.08;$$

(c) Point 2–3 100 mm lightweight concrete block
$$\frac{1}{0.19} \times 0.100 = 0.53;$$

(d) Point 3–4 50 mm expanded polystyrene
$$\frac{1}{0.035} \times 0.050 = 1.43;$$

(e) Point 4–5 105 mm brickwork
$$\frac{1}{0.84} \times 0.105 = 0.13;$$

(f) Point 5 to outside air (from Table A 2.2) $= 0.06$

$$R_t = \Sigma r_t = 2.35$$

Interstitial condensation

Table A 3.1. *Selected thermal and vapour properties of materials*[1]

		Vapour resistance (r_v) (MN s/g m)
Membranes		
Average gloss paint film		7.5–40
Polythene sheet (0.06 mm)		110–120
Aluminium foil		4000

	Thermal resistivity (m K/W)	Vapour resistivity[a] (MN s/g m)
Materials		
Brickwork	0.7–1.4	25–100
Concrete[b]	0.7	30–100
Rendering	0.8	100
Plaster	2	60
Timber	7	45–75
Plywood	7	1500–6000
Fibre building board	15–19	15–60
Hardboard	7	450–750
Plasterboard	6	45–60
Compressed strawboard	10–12	45–75
Wood–wool slab	9	15–40
Expanded polystyrene	30	100–600
Foamed urea-formaldehyde	26	20–30
Foamed polyurethane (open or closed cell)	40–50	30–1000
Expanded ebonite	34	11 000–60 000
Glass fibre	29	6[c]

[a] Resistivity = 1/diffusivity.
[b] An approximate figure (manufacturers' data depend on material density, etc.) for the vapour resistivity of lightweight concrete block is 50 MN s/g.
[c] Manufacturers' data. For glass fibre the vapour resistivity is typically taken as that of fresh air.

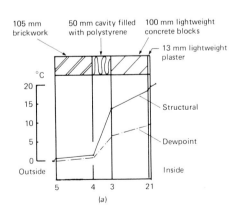
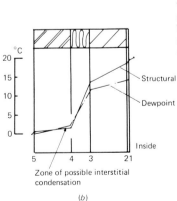

Fig. A3.1. Wall construction with temperature plots. (*a*) Internal moisture contribution 3.4 g/kg. (*b*) Internal moisture contribution 6.4 g/kg.

(C) Temperature drops between points Δt (°C) =

$$\Delta t = \frac{\Delta T}{R_t} \times r_t$$

$$= \frac{20}{2.35} \times r_t = 8.5 \times r_t$$

	Initial temperature (°C)	Temperature drop (°C)	Resulting temperature (°C)
(a) Inside air			20.0
(b) Inside air to point 1 = 8.5 × 0.12 ⇒	20.0	1.0	19.0
(c) Point 1–2 = 8.5 × 0.08 ⇒	19.0	0.7	18.3
(d) Point 2–3 = 8.5 × 0.53 ⇒	18.3	4.5	13.8
(e) Point 3–4 = 8.5 × 1.43 ⇒	13.8	12.2	1.6
(f) Point 4–5 = 8.5 × 0.13 ⇒	1.6	1.1	0.5
(g) Point 5 to outside air = 8.5 × 0.06 ⇒	0.5	0.5	0.0
(h) Outside air	0.0		

(D) Plot the temperature profile as shown in Fig. A 3.1.

(E) Assume that the outside air is at 0 °C and is saturated at a mixing ratio of 3.8 g/kg (see Fig. A 3.2). A mixing ratio of 3.8 g/kg corresponds to an outdoor vapour pressure of 6 mb (read on the extreme right-hand scale of Fig. A 3.2). The BRE states that at normal ventilation rates a person not engaged in physical exertion adds roughly 45 g/h of moisture to the air and that this results in an excess moisture content over outdoor air of some 1.7 g of water vapour per kilogramme of dry air. For dwellings they suggest a safer design value of 3.4 g/kg to allow for the moisture gains from cooking and bathing (see Chapter 3 for moisture emission rates) and the likelihood of restricted ventilation in cold weather. We thus have a total moisture content of 7.2 g/kg which corresponds to a vapour pressure of 11.4 mb. The indoor–outdoor vapour pressure difference, ΔP, is therefore $11.4 - 6.0 = 5.4$ mb.

(F) Vapour resistances, r_v, equal the vapour resistivities (from Table A 3.1) times the thickness of the component.

(a) Point 1–2	60 × 0.013	= 0.78 MN s/g
(b) Point 2–3	50 × 0.100	= 5.00 MN s/g
(c) Point 3–4	350 × 0.050	= 17.50 MN s/g
(d) Point 4–5	25 × 0.105	= 2.63 MN s/g
	$R_v = \Sigma r_v$	= 25.91 MN s/g

(G) Pressure drops between points, Δp (mb):

$$\Delta p = \frac{\Delta P}{R_v} \times r_v$$

$$= \frac{5.4}{25.9} \times r_v = 0.21 \times r_v.$$

Interstitial condensation 361

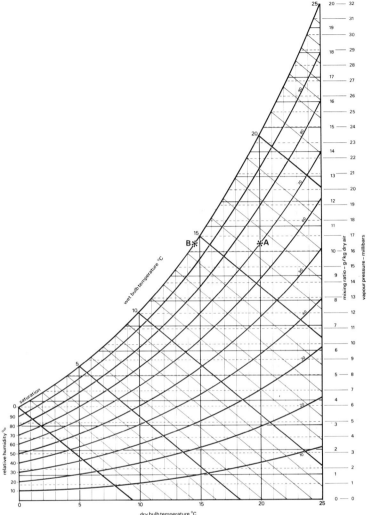

Fig. A3.2. Chart relating moisture contents and temperatures[1] (see Chapter 3 for an explanation of the lettered points). (BRE, Crown Copyright, HMSO.)

		Initial vapour pressure (mb)	Vapour pressure drop (mb)	Resultant vapour pressure (mb)	Corresponding dew-point temperature from Fig. A 3.2 (°C)
(a) Inside air		11.4			9.4
(b) Point 1–2	0.21 × 0.78 ⇒	11.4	0.2	11.2	8.8
(c) Point 2–3	0.21 × 5.0 ⇒	11.2	1.1	10.1	7.2
(d) Point 3–4	0.21 × 17.5 ⇒	10.1	3.7	6.4	0.9
(e) Point 4–5	0.21 × 2.63 ⇒	6.4	0.6	5.8[a]	0.0[a]

[a] The difference between 6.0 and 5.8 is due to approximation errors. The dewpoint temperature corresponding to 6.0 mb has been used.

(H) Plot the dew-point temperature profile as shown in Fig. A 3.1(a).
(I) Estimation of condensation risk: condensation can occur in the conditions assumed at any point where the structural temperature is lower than the dew-point temperature. Since this does not occur anywhere for the case studied there is no computed risk of condensation.

If, however, we were to approximately double the moisture contribution from internal activities to 6.4 g/kg and repeat the procedure we would have the graph of Fig. A 3.1(b). In this case there is a risk of interstitial condensation in the zone indicated. One possible way of reducing this risk would be to use plasterboard with an aluminium-foil backing as the internal surface, taking care to seal the joints of the plasterboard sheets.

Similar calculations can be made using a number of computer routines including the RIBA programs[2] mentioned in Appendix 2.

In all cases, it must be remembered that the initial assumptions are both quite important and subject to considerable uncertainty, particularly with regard to internal moisture conditions.

References

[1] Anon. (1969). *Condensation.* BRE Digest 110. Garston: Building Research Station.
[2] Royal Institute of British Architects Publications Ltd, London.

Index

AC/DC supply, 282
active solar heating, 6–7, 159, 203–4, 232, 234
active solar heating – collection angle, 89–91
admittance, 44, 47, 347, 350, 355
adobe brick, 127, 134
agricultural buildings, 329–30
air flow in Trombe walls, 139
air movement in rooms, 52
antifreeze, 176
appliance energy, 304
atria, 37–8, 336

batteries – electric, 281
Beadwall, 142
berms, 35, 65, 127
BLAST, 151
blinds, 109–12, 126
boilers, 194, 198–201, 203–6, 221, 230–1, 234, 329
Buildings
    Ambient Energy Design House, 70, 71, 203, 205
    Ark Project, 255
    Autarkic House, 6, 72, 82, 128, 132, 241, 247, 278, 289–90, 310, 312
    Autarkic House Turbine, 277
    Baer House, 142
    Balcomb House, 134 *et seq.*
    Basildon Northlands 1 House, 20, 29, 59, 63–4, 78, 234
    Bebington Houses, 143
    BRE Solar Energy Houses, 216, 232–3
    BRS Office Building, 328–9
    Conservation House, Wales Turbine, 284
    Curtis House, 124
    Danish Zero Energy House, 115, 223–4, 248
    Denver Solar House, 162, 250
    Dover Massachusetts House, 254
    Ecology Houses, 35–6
    Elmstead Primary School, 332
    Energy Design Group Solar Skin, 310, 312
    Feilden and Clegg Farmhouse, 318
    First Village House, 134 *et seq.*
    Frusher Polyurethane Home, 83
    Glyn England House, 319
    Hay House, 144
    HDD Better Insulated Houses, 78
    Lewisham flats, 46, 68, 234–5
    Lioux House, 250–1
    Locksheath School, 388
    Milton Keynes Houses, 175, 180, 191, 221, 248
    Minimum Energy Dwelling, California, 305
    MIT Solar 5 House, 116, 120, 125 *et seq.*
    National Coal Board House, 202–3
    Newnham House, 222, 315–16
    Odeillo Houses, 139, 140
    Ourobouros House, 65–6
    Pennyland, 18–19
    Peterborough Houses, 167, 250, 310
    Roach Vale Primary School, 333
    Romsey School, 331
    Salford Strawberry Hill Houses, 216–17, 223, 233
    Sheiling School, 330
    Solar One House, 245
    St. Johns School, Clacton on Sea, 337
    Termoroc House, 114
    Terry House, 117, 127
    Tyrell House, 142
    Vale House, 200, 201
    Wallasey School, 98, 124
    Yately Newlands School, 334
Building Regulations, 42, 72, 84
buildings – heavyweight, 43, 46–7, 217
buildings – lightweight, 43, 46–7, 331
buildings – thermal response, 42–7, 77, 194

clothing, 4, 51
coatings on glass, 107, 126
coefficient of performance of turbines, 229–30, 232, 234
cold bridges, 72, 73, 75, 81
collectors – area solar hot water, 175
collectors – evacuated tubes, 177
collectors – panels, 173
collectors – plastic, 177
collectors – refrigerant charged loop, 177
collectors – surface temperatures, 312
collectors – swim pool, 326

collectors – thermal expansion, 312
comfort, 47, 48, 86, 331, 333
computer models, *see*: ESP, BLAST, DEROB, FCHART, SCRIBE, SERIRES, SUNCODE
condensation, 53, 56–60, 76–7, 81, 88, 308, 358–62
condensation, *see also* moisture emission
conservatories, 129 *et seq.*, 200, 238
conservatory at Peterborough, 315
conservatory retrofit, 318
contaminants in air, 53, 54, 74
control of air solar systems, 165
control of solar hot water systems, 177
controls-heating, 193, 198, 201, 204, 208, 212–15, 220, 354
convective loops, 147
convectors, 207, 210–11, 216, 221–2, 329
cooking, 307
cooling, 145

dampers, 163, 314
daylight, 330–1, 335
degree days, 127, 354
DEROB, 150
design methods – passive solar, 148 *et seq.*
design methods – solar hot water systems, 179 *et seq.*
diffuse solar data, 345 *et seq.*
domestic appliances, 304
domestic hot water supply, 144, 211, 233, 244
drain back systems, 171, 176
drain down systems, 172, 176
ducts, 147, 162

economics of retrofit, 319 *et seq.*
efficiencies of collectors, 174
electrical generators, 278
energy – consumption, 2, 4, 330
energy – delivered, 191–2, 330, 338
energy – primary, 2, 191, 222, 230, 328, 338–40
energy – useful, 191–2
energy collected in solar systems, 160
energy in the wind, 268
energy saving in appliances, 304
ESP, 150
external shutters, 114, 328

F-chart 4, 182
fan coil unit, 204, 223–4
fans for air collectors, 162
form, 60–5, 315, 328
free heat gains, 43–5, 70, 352–3
furnishings, thermal mass, 123

glare, 99, 102, 127
glazing, 100, 331, 333
glazing – coatings, 107
glazing – multiple, 84–7, 103, 317, 328, 350
glazing – effect of dust, 100
glazing – effect of wind chill, 106

glazing – heat mirrors, 107, 126
glazing – infiltration, 108
glazing – Kalwall plastic glazing, 103
glazing – plastic, 102–4
glazing – properties of glass, 101
glazing – sealants, 102
glazing – U values, 103 *et seq.*
ground coupling, *see also* heat storage, 126
ground reflection, 117

heat centre, 212–13
heat exchangers, 163, 175
heat mirrors, 107, 126
heat pumps, 91, 126, 216, 220, 228–34, 241, 258, 326, 333, 338
heat recovery, 59, 220–1, 317, 329, 333
heat storage, *see* thermal storage
heat transfer – air movement, 133
heat transfer – air to water, 314
heat transfer – via doors, 318
heat transfer – wall openings, 133, 139
heating systems – ceiling, 49, 227
heating systems – efficiency, 192–3
heating systems – electrical, 225–8
heating systems – forced air, 217–24, 311
heating systems – solid fuel, 70, 196–205, 324
heating systems – temperature response, 195–6, 218
heating systems – underfloor, 211, 216, 227, 236
heating systems – water distribution, 205–17, 324
hot water system installed solar example, 179
hot water usage, 305
housing density, 3, 18, 21

infiltration at windows, 108
infiltration, *see also* ventilation, 24, 26, 53, 55–6, 68, 75, 86
insulation, 7, 58, 74–91, 203, 226, 315, 324, 331

Kalwall plastic glazing, *see* glazing

layout, 67–8
lighting, 69, 329, 331
lights – heat gain, 123

mass, *see* heat storage
mass walls, 137 *et seq.*
meteorological data, 16–24, 130, 242, 342–4
moisture emission, 58–9, 360

occupancy patterns, 4, 45–6, 77, 80, 189, 237
open loop system, 172, 176
orientation, 19, 64, 311
overshadowing, 20, 68, 116

Index

passive solar gain, 18, 46, 63–6, 70, 94 *et seq.*, 123 *et seq.*, 195, 216, 237, 315, 323
pebble beds, 120 *et seq.*, 133 *et seq.*, 163 *et seq.*
plastic glazing, 102–4
pool covers, 326
pump, 189, 206, 211–12, 235

radiators, 207–8, 210–11, 234
reflectors, 117, 141
refrigerators and freezers, 307
rehabilitation of buildings, 317, 319
response factor method, 355–6
rock bed storage, 120 *et seq.*, 133 *et seq.*, 163 *et seq.*
roof ponds, 144
roof space hybrid collector, 146, 319
roofs, 88–91
room air movement, 52

Savonius Rotor, 272, 276
school buildings, *see* buildings
SCRIBE, 321
SERI-RES, 151
shutters, external, 114
shutters, internal, 109, 112
simulation models, *see* ESP. BLAST, DEROB, FCHART, SCRIBE, SERIRES, SUNCODE
site planning, 14–39
site planning – Beaufort scale, 30
site planning – glasshouses, 28
site planning – soil, 33–6
site planning – underground structures, 35–6
site planning – water table, 33
site planning – wind, 24–33
site planning – wind turbine, 31–2
site planning – windbreaks, 25–30
skirting heating, 211
solar air heating, 160 *et seq.*
solar hot water – anti-freeze, 176
solar hot water – collector area, 175
solar hot water – control, 177
solar hot water – design method F-chart, 182
solar hot water – economics, 161
solar hot water – hard water effects, 176
solar hot water – heating, 159, 161, 171 *et seq.*
solar hot water – storage, 175, 182
solar hot water – stratification, 176
solar hot water – system size, 175
solar hot water – transfer fluids, 176
stack effect, 25, 54, 69
storage, *see* thermal storage
storage fan heaters, 225
storage radiators, 225–6
stratification, 176
structural thermal storage, 122
SUNCODE, 151
sunspaces, 128 *et seq.*

temperature, dry resultant, 48–50
temperature, mean radiant, 48–9
thermal performance, 346–56
thermal storage, 33, 35, 99, 118 *et seq.*, 241–58
thermal storage – adsorbent beds, 257
thermal storage – annual cycle energy system, 255–6
thermal storage – bricks and adobe, 127, 134, 223
thermal storage – earth, 255
thermal storage – effect of furnishings, 123
thermal storage – for wind turbines, 283
thermal storage – ground coupling, 126
thermal storage – Peterborough houses, 311
thermal storage – phase change, 119, 127, 252–5
thermal storage – photovoltaic, 242
thermal storage – rock and pebble beds, 120 *et seq.*, 133 *et seq.*, 163 *et seq.*, 204, 249–51
thermal storage – solar ponds, 255, 257
thermal storage – structural, 122
thermal storage – water, 128, 137, 141
thermal storage for solar hot water, 175, 182
thermal storage walls, 137 *et seq.*
thermosiphoning, 147, 160, 168
transfer fluids, 176
transport energy use, 308
Trombe walls, 137 *et seq.*

urethane foam, 127

vapour check, 76, 77, 80, 84, 88
vegetation and plants, 21–3, 26–7, 30, 32, 36, 91, 129
venetian blinds, 116
ventilation, 53–7, 77
ventilation air preheat, 314
ventilation control, 326
ventilation, mechanical, 54–5, 69, 217–18, 329, 331

wall insulation – retrofit, 317
waste disposal systems, 289, 293–301
waste disposal systems – aerobic, 295
waste disposal systems, methane digester, 293, 295–6, 300–1
waste disposal systems, septic tanks, 295
water heating energy, *see also* hot water, 305
water supply systems, 289–92
wave power, 262
wind – gust ratio, 266
wind – hourly speeds in the UK, 266
wind – power in the, 242, 264
wind – variation with height, 267
wind – variation with location, 264
wind energy, 262 *et seq.*
wind systems, real outputs, 279
wind turbines – Autarkic House, 277

wind turbines – Conservation House, Wales, 284
wind turbines – cut in speed, 268
wind turbines – Darrieus, 277
wind turbines – Wind Energy Group, 3 MW, 263
wind turbines – gears, 280
wind turbines – Musgrove, 277
wind turbines – NASA 100 kW, 274
wind turbines – power coefficient, 268, 270
wind turbines – power extracted by, 268
wind turbines – rated speed, 268
wind turbines – types of rotor, 271
window insulation – Beadwall, 142
window insulation – blinds, 109–12, 126
window insulation – external shutters, 114
window insulation – internal shutters, 109, 112
window insulation – movable, 109 *et seq.*
window insulation – venetian blinds, 116
windows, 63, 351
wood stoves, 128